Cyclodextrin Materials Photochemistry, Photophysics and Photobiology

Chemical, Physical and Biological Aspects of Confined Systems

Cyclodextrin Materials Photochemistry, Photophysics and Photobiology

Edited by

Abderrazzak Douhal

Departamento de Química Física, Sección de Químicas,
Facultad de Ciencias del Medio Ambiente,
Campus Tecnologico de Toledo,
Universidad de Castilla-La Mancha, Toledo, Spain

ELSEVIER

Amsterdam – Boston – Heidelberg – London – New York – Oxford – Paris
San Diego – San Francisco – Singapore – Sydney – Tokyo

Elsevier
Radarweg 29, PO Box 211, 1000 AE Amsterdam, The Netherlands
The Boulevard, Langford Lane, Kidlington, Oxford OX5 1GB, UK

First edition 2006

Notice
No responsibility is assumed by the publisher for any injury and/or damage to persons
or property as a matter of products liability, negligence or otherwise, or from any use
or operation of any methods, products, instructions or ideas contained in the material
herein. Because of rapid advances in the medical sciences, in particular, independent
verification of diagnoses and drug dosages should be made

Library of Congress Cataloging-in-Publication Data
A catalog record for this book is available from the Library of Congress

British Library Cataloguing in Publication Data
A catalogue record for this book is available from the British Library

ISBN-13:   978-0-444-52780-6
ISBN-10:   0-444-52780-X
ISSN: 1872-1400

For information on all Elsevier publications
visit our website at books.elsevier.com

Printed and bound in Italy

06 07 08 09 10 10 9 8 7 6 5 4 3 2 1

Working together to grow
libraries in developing countries

www.elsevier.com | www.bookaid.org | www.sabre.org

ELSEVIER   BOOK AID International   Sabre Foundation

Contents

Preface

This book reflects the height of the latest knowledge of the effects of cyclo-dextrins (CD) confinement on the photochemical, photophysical and photobi-ological events in caged molecules. The research work using light to interrogate the behavior of CD inclusion nanostructures developed around the world dur-ing the first five years of the 21st century is described here in 14 chapters. These were reviewed by specialists working in the field and revised when necessary, adding a collective responsibility for the quality of the book.

It is difficult to estimate the number of published contributions in the fields involving CD, but it certainly reaches thousands using these sweet nanocavities to host different kinds of guests. This great interest makes today's CD science and technology deeply involved in a large number of important areas of re-search covering both basic and applied science in chemistry, biology and phys-ics. Their ability to modify the behavior of trapped guests is being used for example, in chemistry, biology, physics, food, cosmetics, pharmacy/medicine and agriculture. The goal of this volume is twofold: (1) to provide to the sci-entific community the state-of-the-art on photochemistry, photophysics and photobiology of CD complexes in one book, (2) to trigger further research in applied science connected to these small nanocapsules.

The book contains 14 theoretical and experimental chapters describing and examining the main findings of a large number of published works, based on the use of an arsenal of techniques to characterize the CD confinement effect on the selected molecular systems. These guests range from small molecules like water to large ones like porphyrins. Steady-state (Uv–visible absorption and emission, and NMR) spectroscopy and time-resolved techniques exploring events from the femtosecond regime to the millisecond one have been used to understand the studied reactions and related events. Therefore, the chapters examine the effect of confining guests in CD cage at the ground state, triplet state, and the elec-tronically singlet-excited state using light as a tool to get information on the behavior of the formed nanostructures. The reader is invited to discover the exciting material and details in each chapter, and to get a personal impression on the development of the field and its future. The ordering of the chapters was done from "simple" to more complex systems and techniques.

The idea to make this book, originates while editing a special issue on this topic published last year in the Journal of Photochemistry and Photobiology A: Chemistry, Vol. 173(3) (2005). During the preparation and edition of this issue I realized that there is an urgent need for the scientific community working in the field, to have a book describing our recent understanding of the effect of light on the behavior of inclusion complexes of CD. The book will then be of interest to a multidisciplinary audience working in confined nanostructures. Truely, several articles and few reviews and book chapters have been published in the field, but a book dedicated to the science studying photo-induced processes in CD materials alongside the concepts of confined and assembled nanostructures is presented for the first time here.

This volume was made possible thanks to the authors devoting valuable time from their schedules to write exciting chapters, to the efforts of the referees in shaping and evaluating the works published here, and to the team at Elsevier who contributed toward publishing the book. I sincerely wish to thank all of them. My special thanks go to Michiel Thijssen and Joan Anuels at Elsevier. When I told Michiel in March 2005 about the idea of the book and of the series in which this book is the first volume, he encouraged me through e-mails, thereby exciting me more about the idea. Three months later, the invited contributors accepted the challenge and started working on their chapters. By the end of January 2006, the revision, shaping and editing of the contributions were done.

My wish to edit this book is that the published material here will be of value not only to readers working in this field, but also to researchers working on related areas like those of nano- and biotechnology, for example, and to undergraduate students, academics and engineers when studying, examining or using the effect of a nanocavity confinement in chemical, physical and biological processes. Finally, I hope that the book will arouse the interest of scientists and engineers who wish to diversify their research fields.

Abderrazzak Douhal

UCLM, Toledo, Spain, February 2006

List of Contributors

Pietro Bortolus, *Istituto per la Sintesi Organica e la Fotoreattività, C.N.R. Area della Ricerca, I-40129 Bologna, Italy*

Guo-Song Chen, *Department of Chemistry, State Key Laboratory of Elemento-Organic Chemistry, Nankai University, Tianjin 300071, P. R. China*

Yong Chen, *Department of Chemistry, State Key Laboratory of Elemento-Organic Chemistry, Nankai University, Tianjin 300071, P. R. China*

Abderrazzak Douhal, *Departamento de Química Física, Sección de Qumicas, Facultad de Ciencias del Medio Ambiente, Universidad de Castilla-La Mancha, Avda. Carlos III, S.N., 45071 Toledo, Spain*

Hiroshi Ikeda, *Department of Bioengineering, Graduate School of Bioscience and Biotechnology, Tokyo Institute of Technology, 4259-B-44 Nagatsuta-cho, Midori-ku, Yokohama 226-8501, Japan*

Yoshihisa Inoue, *ICORP Entropy Control Project, JST, 4-6-3 Kamishinden, Toyonaka 560-0085, Japan, Department of Applied Chemistry, Osaka University, Suita 565-0871, Japan*

Wei Jun Jin, *College of Chemistry, Beijing Normal University, Beijing 100875, P. R. China, School of Chemistry and Chemical Engineering, Shanxi University, Taiyuan 030006, P. R. China*

Maged El-Kemary, *Departamento de Química Física, Sección de Qumicas, Facultad de Ciencias del Medio Ambiente, Universidad de Castilla-La Mancha, Avda. Carlos III, S.N., 45071 Toledo, Spain, Chemistry Department, Faculty of Science, Tanta University, 33516 Kafr ElSheikh, Egypt*

Irene W. Kimaru, *Department of Chemistry and Biochemistry, Southern Illinois University, Carbondale, IL 62901, USA*

Yu Liu, *Department of Chemistry, State Key Laboratory of Elemento-Organic Chemistry, Nankai University, Tianjin 300071, P. R. China*

Antonino Mazzaglia, *Istituto per lo Studio dei Materiali Nanostrutturati, ISMN-CNR, Salita Sperone 31, 98166, Messina, Italy, Dipartimento di Chimica Inorganica, Chimica Analitica e Chimica Fisica, Salita Sperone 31, 98166, Messina, Italy*

Matthew E. McCarroll, *Department of Chemistry and Biochemistry, Southern Illinois University, Carbondale, IL 62901, USA*

Norberto Micali, *Istituto per il Processi Chimico Fisici, IPCF-CNR, Via la Farina 237, 98123 Messina, Italy*

Sandra Monti, *Istituto per la Sintesi Organica e la Fotoreattività, C.N.R. Area della Ricerca, I-40129 Bologna, Italy*

Miquel Moreno, *Departament de Química,. Universitat Autònoma de Barcelona, 08193 Bellaterra (Barcelona), Spain*

Tooru Ooya, *School of Materials Science, Japan Advanced Institute of Science and Technology, 1-1 Asahidai, Nomi, Ishikawa 923-1292, Japan, Department of Intelligent System Design Engineering, Toyama Prefectural University, 5180 Kurokawa Imizu, Toyama 939-0398, Japan*

Joon Woo Park, *Department of Chemistry, Ewha Womans University, Seoul 120–750, Republic of Korea*

Luigi Monsù Scolaro, *Dipartimento di Chimica Inorganica, Chimica Analitica e Chimica Fisica, Salita Sperone 31, 98166, Messina, Italy*

He Tian, *Laboratory for Advanced Materials and Institute of Fine Chemicals, East China University of Science & Technology, Meilong Road 130, Shanghai 20037, P.R. China*

Akihiko Ueno, *Department of Bioengineering, Graduate School of Bioscience and Biotechnology, Tokyo Institute of Technology, 4259-B-44 Nagatsuta-cho, Midori-ku, Yokohama 226-8501, Japan*

Brian D. Wagner, *Department of Chemistry, University of Prince Edward Island, Charlottetown, PE, Canada C1A 4P3*

Qiao-Chun Wang, *Laboratory for Advanced Materials and Institute of Fine Chemicals, East China University of Science & Technology, Meilong Road 130, Shanghai 20037, P.R. China*

Yafei Xu, *Department of Chemistry and Biochemistry, Southern Illinois University, Carbondale, IL 62901, USA*

Cheng Yang, *ICORP Entropy Control Project, JST, 4-6-3 Kamishinden, Toyonaka 560-0085, Japan*

Cyclodextrin Materials Photochemistry, Photophysics and Photobiology
Abderrazzak Douhal (Editor)
DOI 10.1016/S1872-1400(06)01001-6

Chapter 1

Fluorescence Methods for Studies of Cyclodextrin Inclusion Complexation and Excitation Transfer in Cyclodextrin Complexes

Joon Woo Park

Department of Chemistry, Ewha Womans University, Seoul 120–750, Republic of Korea

1. Introduction

Cyclodextrins (CD) are cyclic oligosaccharides consisting of D-(+)-glucopyranose units. They are produced by enzymatic hydrolysis of starch [1]. The three major CD produced industrially are α-CD (six glucose units), β-CD (seven glucose units), and γ-CD (eight glucose units). The larger homologues were also isolated as minor products. CD are rigid molecules with hollow, torus-shape cavity. The height of CD torus is 7.9 Å. The primary hydroxyl groups are on the narrow side of the torus, and the secondary hydroxyl groups are on the wider side. The cavity diameters of the narrow and wide sides are 4.7 and 5.3 Å for α-CD, 6.0 and 6.5 Å for β-CD, and 7.5 and 8.3 Å for γ-CD, respectively. CD can be modified either at the narrow primary side or the wider secondary side [2]. Possessing the hydrophobic cavity, CD and their derivatives form inclusion complexes with a variety of hydrophobic or amphiphilic species of appropriate size by admitting them, at least partially, into the cavity in aqueous media. Because of these characteristics, CD and their derivatives have been widely used as biomimetic micro-reactors, novel media for photochemical and photophysical studies, and building blocks and functional units for supramolecular structures as well as in various fields of industries [1].

The restricted space and relatively reduced polarity of the CD cavity can influence the molecular properties of guest molecules included in the cavity. Particularly, the photophysical and photochemical properties of the guest fluorophore are remarkably changed upon complexation with CD [3]. This has been utilized in the studies of inclusion complexation equilibria between CD and fluorophore, and the dynamics of the fluorophore in the CD cavity. CD derivatives with a suitable pendant fluorescent group can form self-inclusion complex either intramolecularly or intermolecularly. The self-included group is excluded from CD cavity by added guest molecules. This results in a large change in the fluorescence property of the pendant fluorophore. Also, efficient excitation energy transfer or photoinduced electron transfer can occur between the pendant group and CD-encased guest molecules. Therefore, the fluorophore-modified CD have drawn much interest as fluorescent chemosensors and photoactive host molecules [3,4].

In this chapter we will see the general features of CD complexation, how to use fluorescent methods for studies of the complexation equilibria, the structures of the complexes, and excitation energy and electron transfers in the CD complexes.

2. Fluorescence methods for studies of cyclodextrin inclusion complexation

Inclusion complexation of guest molecules or pendant group of modified-CD into CD cavity is the most basic and important topic in CD chemistry. Thermodynamic stability, stoichiometry, and even the structural features of the CD complexes can be obtained using fluorescence methods.

2.1. Thermodynamic, kinetic, and structural aspects

2.1.1. Thermodynamics of complexation

Complexation equilibrium of a guest G with CD or CD derivative (denoted as C) is generally represented by

$$nG + mC \rightleftarrows G_nC_m \tag{1}$$

The equilibrium constant of the reaction, which is also called as binding or association constant of G with C or the formation constant of the complex, is given as

$$K = \frac{[G_nC_m]}{[G]^n[C]^m}; \quad K_{th} = \frac{\gamma_{G_nC_m}[G_nC_m]}{\gamma_G^n\gamma_C^m[G]^n[C]^m} \tag{2}$$

where γ_i is the activity coefficient of the species i. The concentration-based equilibrium constant (K) differs from the thermodynamic equilibrium constant (K_{th}). However, as the γ_i values are generally unknown, only the concentration-based K values are usually reported.

The standard thermodynamic parameters of the complexation reaction, such as ΔG°, ΔH°, and ΔS°, are usually calculated from the K values determined at various temperatures in the standard state (1 atm) using the following relationships:

$$\Delta G^\circ = -RT \ln K; \quad \{\partial \ln K / \partial (1/T)\}_p = -\Delta H^\circ / R; \quad \Delta S^\circ = (\Delta H^\circ - \Delta G^\circ)/T$$

$$(3)$$

Rekharsky and Inoue [5] tabulated a large number of K values and thermo-dynamic parameters for 1:1 complexation equilibria reported until October 1997. With exception of a few cases, ΔH° and ΔS° are negative and the enthalpy–entropy compensation is often observed.

The major driving forces for the cyclodextrin inclusion complexation are believed to be van der Waals and hydrophobic interactions [5]. It is also believed that hydrogen bonding and release of strain energy of CD upon complexation have certain roles to the stability of the complexes. In the complexation of ionic guests with ionic group-modified CD, the electrostatic interaction also affects the stability of the complexes

2.1.2. Kinetics of complexation and decomplexation

The elementary reactions of CD complexation and the decomplexation of the complex are represented as

$$G + C \underset{k_-}{\overset{k_+}{\rightleftarrows}} GC \tag{4}$$

For a guest molecule, which is small enough to fit inside a CD cavity with little potential energy barrier, the rate constants (k_+) for bimolecular complex-ation reactions are in the range of 10^8–10^9 $M^{-1} s^{-1}$ [6,7]. If one considers the fraction of the surface area of inner torus of the CD to the total surface area of CD, the k_+ values are almost the same as the typical diffusion-controlled rate constants (k_D) [6]. The pseudo-first-order rate constant for the complexation reaction is given by $k_+[C]$. In the usual concentration range of CD used for complexation studies ($\leqslant 10^{-2}$ M), it is generally smaller than $10^7 s^{-1}$. The decomplexation rate constants (k_-) is related to the stability of the complexes by the relationship $k_- = k_+/K$. Even for complexes of weak stability, the k_- values are in the range 10^6–$10^7 s^{-1}$.

The entry and exit rates of a guest into and from CD cavity are much slower than the rate constant of fluorescence decay of a typical fluorophore, the order of magnitude of $10^9 s^{-1}$. Therefore, there is little chance for a guest to enter into or exit from the CD cavity in the lifetime of the excited state, and it can be assumed that the change in the steady-state fluorescence of a fluorophore upon complexation with a CD reflects ground-state association of the fluorophore with CD. This makes it possible to study the CD complexation by steady-state fluorescence.

If both the free fluorophore and fluorophore–CD complexes are fluorescent, the analysis of the fluorescence decay curve of a system composed of these species yield a set of decay rate constants and pre-exponential factors. Ideally, the decay rate constants should be invariant with CD concentration. However, the pre-exponential factors vary with CD concentration, as the pre-exponential factor of a species is proportional to the fluorescence quantum yield, molar absorptivity at the exciting wavelength, and the fraction of the species among all the fluorescent species. The above statements are equally applicable to the complexation of a guest with fluorophore-modified CDs.

For a common CD complexation, the exchange rate between the free guest and CD-complexed guest is faster than the NMR timescale (ca. ms), and thus the free and complexed species are not resolved in NMR spectra. However, if complexation and decomplexation occur via passing a CD through bulky and/ or a highly hydrophilic group to form pseudorotaxanes, the rates of these processes can be slower than the NMR timescale. In this case, the free guest and CD-complexed guest appear as isolated peaks and the ratio can be determined directly from NMR spectra [8,9].

2.1.3. Structures and stoichiometries of complexes

The 1:1 (guest:CD) stoichiometry is the most common type of CD complexes. Complexes of other stoichiometries such as 1:2, 2:1, 2:2, 1:1:1, and 1:1:2 are also frequently observed. Even for a given guest–CD pair, complexes of different stoichiometries can be formed. Examples of this are naphthalene and pyrene complexes of CD, which were studied extensively by fluorescence techniques [10–12]. Schematic representation of various types of CD complexes is shown in Fig. 1.

For a given guest, the stoichiometry, stability, and structure of the complex can depend on the type of CD. A major factor deciding these is the cavity sizes of CD. The linear alkyl chains fit snugly into α-CD, but rattles around inside β-CD cavity, resulting in the higher binding constant of an *n*-alkyl compound with α-CD than with β-CD [5]. Aromatic rings are common fluorescent groups of guest molecules. Phenyl group fits best into β-CD cavity, and inclusion of the group into α-CD or γ-CD cavities is too tight or loose, respectively. Thus the binding constant of a phenyl compound with β-CD is larger than that with α-CD or γ-CD: the reported log K values of 4-nitrophenol complexes at pH 4.3 are 2.25 for α-CD complex, 2.48 for β-CD complex, and 1.79 for γ-CD complex [13]. For 2-substituted naphthalenes, the naphthalene ring is included deeply into β-CD cavity (structure **I**), but shallowly into α-CD cavity (structure **II**). This is revealed as opposite sign of the induced circular dichroism of the complexes [14]. The cavity size of γ-CD is large enough to accommodate two linear

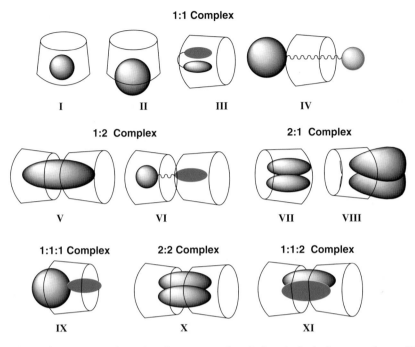

Fig. 1. Schematic representation of various types of cyclodextrin inclusion complexes. CD molecules can take the other orientation depending on the guest.

hydrocarbon chains, naphthalene or anthracene rings inside the cavity giving 2:1 complexes of structure **VII**. An evidence for this is the efficient photodimerization of 2-substituted naphthalenes and anthracenes in the presence of γ-CD (see Chapter 11 of this monograph). The length of the long axis of naphthalene ring is larger than the width of the cavity of common CD. Thus, naphthalene ring can be included deeply into β- or γ-CD cavities only with the long axis parallel to CD axis. By the same structural reason, pyrene and 1-substituted naphthalenes are not included deeply into CD cavities to form the 1:1 complex of structure **I**, but they form the 1:1 complexes having the structure similar to **II**. The 1:1 pyrene/CD complex binds with another CD molecule resulting in the formation of the 1:2 complex of structure **V** or dimerizes to 2:2 complex of structure **X** [12].

Multiple binding of CD molecules with the long hydrophobic chains with appropriate size is possible. The maximum number of CD molecules that can be complexed with the chain depends on the chain length. The height of CD is about 0.8 nm, which is slightly shorter than the length of the fully extended heptamethylene units. Thus *n*-alkyl compounds having a hydrophilic end group (n-$C_nH_{2n+1}X$) can form 1:2 complex in addition to 1:1 complex when $n > 8$ [15]. A hydrocarbon chain-linked bi-functional compound having bulky and/or hydrophilic end groups forms pseudorotaxane-type complex (structure **IV**) [8,9]. For the compounds having polymethylene linkage (–(CH$_2$)$_n$–), no complexation is expected when $n < 7$, and 1:2 complex can be formed when $n > 15$. The molecules having two guest moieties can form a 2:2 complex of structure **VI**.

2.2. Fluorescence change upon complexation with CD and CD derivatives

The most common feature of a fluorescent molecule resulting from complexation with CD or non-fluorescent CD derivatives in aqueous media is intensity enhancement and spectral blue shift of its fluorescence spectrum. This is mainly due to the hydrophobic nature of CD cavity: the feature is similar to that observed when the solvent medium is changed from water to less polar solvent. The reduced micropolarity of the fluorescent molecule inside a CD cavity also results in variations of relative intensities of the vibronic fine structures of the fluorescence spectra. Though numerous aromatic hydrocarbons exhibit this effect, the most intensively studied system is pyrene. The peak ratio (III/I) and fluorescence lifetime of pyrene increase when pyrene molecule moves from water to CD cavity [12].

Another environmental effect is the protection of a fluorophore against collisional quenching by species in bulk aqueous phase via encapsulation of the fluorophore within CD cavity. Oxygen is a ground-state triplet. Collision of oxygen with the fluorophore in an excited singlet state leads to an enhanced rate of singlet–triplet interconversion and thus quenches the fluorescence. Though the mechanism is different, the heavy atoms such as iodide or cesium ions have similar effects [16].

Substituted arylamines (e.g. 1-anilino-8-naphthalenesulfonate (1,8-ANS: **1**), [4-(dimethylamino)styryl]-1-methylpyridiniums (2-ASP: **2**; 4-ASP: **3**) and *p-N, N*-dimethylaminobenzonitrile (DMABN: **4**) exhibit very large change in their fluorescence behaviors upon complexation with CD or solvent polarity change [15,17,18]. These molecules, on excitation to the singlet excited state (S$_1$), can undergo intramolecular charge transfer to S$_{1,ICT}$ state: negative charge is transferred from amine or nitrogen atom to aromatic ring. The high polar environment promotes the charge-transfer transition S$_1 \rightarrow$ S$_{1,ICT}$ and rate of deactivation of S$_{1,ICT}$ and decreases the frequency of the S$_{1,ICT} \rightarrow$ S$_o$ fluorescence. The fluorescence lifetime and quantum yield (thus the emission intensity) of these compounds usually increase upon complexation with CD [17]. However, certain molecules having a group with lone-pair electron connected to the aromatic ring by a single bond show medium sensitive dual fluorescence. The firstly recognized and extensively discussed example of this class of compounds is DMABN (**4**). Rotation of the amino group of the planar ICT excited-state molecule leads to a twisted intramolecular charge transfer (TICT) excited-state molecule. The dual fluorescence is from ICT and TICT excited-state molecules. The emission from TICT state appears at longer wavelength. The emission from TICT state increases and the emission form normal ICT state decreases upon complexation of the molecule with CD. The structural and fluorescence changes accompanying the TICT process are recently reviewed [19].

1: 1,8-ANS   2: 2-ASP   3: 4-ASP   4: DMABN

Another class of compounds that exhibit large fluorescence change upon complexation with CD is donor(D)–acceptor(A) dyad molecules linked by long flexible hydrophobic linkage (the end groups of structure **IV** are D and A). Examples of this type of molecules are aliphatic chain-linked aromatic (D)-viologen(A) compounds [20]. The molecules form ground-state charge-transfer (CT) complex between D and A. In the absence of CD, the D and A moieties of the dyad molecules are in close proximity due to the CT complexation and/or flexibility of the linkage, making the dyad molecules to adapt folded conformations and much less fluorescent due to efficient electron transfer from excited D to A. Formation of CD complexes (structure **IV**) separates the D and A pair further and the electron transfer quenching is retarded leading to a large increase in fluorescence emission.

Excimer fluorescence is often observed from 2:1 complexes (structure **VII** or **VIII**) [10] and 2:2 complexes (structure **X**) [11,12] of pyrene and naphthalene derivatives. Also, intramolecular exciplex emission from 1:1 complex of the structure **III** [21] and room temperature phosphorescence or exciplex emission from 1:1:1 (structure **IX**) and 1:1:2 (structure **XI**) ternary complexes have been reported [22]. Systems exhibiting the room temperature emission were recently reviewed [22].

The binding of guest molecules to the CD cavity of fluorophore-modified CD also results in large fluorescence change. Decrease of emission intensity is frequently observed [4]. This is due to the transfer of the self-included pendant fluorophore from inside of CD cavity to outside, or efficient intracomplex quenching of the fluorescence from the pendant fluorophore by the guest molecules. Sometimes, the fluorescence intensity of fluorophore-modified CDs can be increased upon the binding with guest molecules [23]. This is often observed with fluorophore-modified CD that exhibit shallow self-inclusion or no inclusion of the pendant fluorophore. Guest binding with such molecules can provide less polar environment to the fluorophore by fluorophore–guest interaction.

2.3. Fluorescence methods for determination of complexation constants and stoichiometry of the complexes

Any physical and chemical methods, which can follow the change of the property of a system accompanying the composition change, can be used for the determination of formation constants and stoichiometry of CD complexes. These include solubility, microcalorimetry, absorption, circular dichroism, fluorescence, NMR, ESR, chromatographies, conductometry, potentiometry, kinetics, etc. [5,24]. The choice of methods depends on the facility available and characteristics of the system. A convenient method is that the value of the observed property (λ) is a weighted average of contributions from existing species as

$$\lambda = \sum \lambda_i f_i \tag{5}$$

where λ_i is the expected value when all of the guest molecules are present as the species i, and f_i is the fraction of the species i present in the mixture. When the total concentration of a guest G ($[G]_o$) is kept constant and the concentration of a native CD or CD derivative (which do not have pendant chromophore for spectroscopic methods, or ionic group for conductivity method) is varied, steady-state fluorescence, electric conductivity, absorbance, or ellipticity of circular dichroism measurements meet the condition. The chemical shifts of NMR resonance peaks of the guest molecule follow Eq. (5), even when $[G]_o$ is varied, if the exchange rate between the free G and complexed G is much faster than the NMR timescale (see Section 2.1.2).

The fluorescence method for studies of CD complexation has several advantages, compared to other methods. Because of high sensitivity, the method is applicable to complexation studies with large K values. The fluorescence method can be used to determine K values as high as $10^7 \, M^{-1}$, whereas ^1H NMR is used for $K < 10^4 \, M^{-1}$ and UV-visible and circular dichroism spectroscopies are used for $10^2 < K < 10^4 \, M^{-1}$, in general. Moreover, upon complexation with a CD, the fluorescent properties usually change more sensitively than the other spectroscopic properties. As the concentration of a fluorophore is kept low, the fraction of CD complexed with the fluorophore is usually negligibly small. This enables one to assume that the equilibrium concentration of CD is the same as the total CD concentration and thus the analysis of experimental data to evaluate K values and stoichiometries of the complexes becomes simple.

The change in fluorescence properties also provides rich information about the fluorophore, microenvironment, and interaction in ground and excited states, and excitation transfer in the complexes.

Despite the advantages, the fluorescence method is not free of shortcomings and complications. One is that the fluorescence intensity is proportional to the fluorophore concentration only when the absorbance (A) of a solution is low, typically $A < 0.05$. The absorption of exciting and emitting light by the solution reduces the observed fluorescence intensity of the fluorophore. The attenuation of fluorescence intensity by the inner filter effects can be partially corrected by $I_{corrected} = I_{obs} \times$ antilog$\{(A_{ex} + A_{em})/2\}$: A_{ex} and A_{em} are absorbances at excitation and emission wavelengths, respectively. Even with the correction, it is sometimes difficult to study the phenomenon such as dimerization, which requires high fluorophore concentration, by fluorescence method. Another problem is the sensitivity of the fluorescence intensity on temperature due to the change of quenching by dissolved quenchers such as oxygen and solvent molecules. Reasonably good temperature control is usually necessary. One more problem is that the Raman scattering peak sometimes overlaps with the fluorescence emission band. This complicates the accurate fluorescence intensity measurement of weakly fluorescent solutions. The scattering peak can be identified by the shift of a peak by changing excitation wavelength.

In this section, the mathematical relationships which can be used for studies on the complexation equilibria between fluorescent guests and CD are described. The relationships can also be applied to the complexation of fluorescent guests with non-fluorescent CD-derivatives, and complexation between fluorophore-modified CD and non-fluorescent guests. The relationships can be applied to other methods, if the observed property follows Eq. (5), by substituting the fluorescent intensity (I) with the corresponding property (λ).

2.3.1. Intermolecular complexation of a fluorophore with CD or non-fluorescent CD derivatives

For convenience of presentation, the fluorescent guest and CD are denoted as G and C, respectively. Unless otherwise mentioned, the relationships are derived for variation of fluorescence intensity of a solution of constant total concentration of G, $[G]_o$, as functions of equilibrium concentration of free CD, $[C]$. Under the condition of $[G]_o \ll [C]_o$, the total concentration of C, $[C]_o$, can be used for $[C]$.

1:1 Complexes (structures I–IV): The 1:1 complex formation equilibrium (Eq. (6)) is the most common case in CD complexation and its analysis is straightforward:

$$G + C \overset{K_{11}}{\rightleftharpoons} GC \tag{6}$$

The fractions of G present as free form (f_o) and as 1:1 complex (f_{11}) are represented as

$$f_o = \frac{[G]}{[G]_o} = \frac{1}{1 + K_{11}[C]}; \quad f_{11} = \frac{[GC]}{[G]_o} = \frac{K_{11}[C]}{1 + K_{11}[C]} \tag{7}$$

The observed fluorescence intensity (I_{obs}) taken at constant $[G]_o$ is given as

$$I_{obs} = \frac{I_o + I_{11}K_{11}[C]}{1 + K_{11}[C]} \tag{8}$$

where I_o is the fluorescence intensity in the absence of C and I_{11} is the expected fluorescence intensity when all of G forms the complex.

The K_{11} and I_{11} values can be obtained from the non-linear-least-squares fitting of I_{obs} versus [C] data to Eq. (8). However, the fluorescence titration data are usually analyzed by either of following two linear equations obtained by the rearrangement of Eq. (8):

$$\frac{\Delta I}{[C]} = \Delta I_{11}K_{11} - \Delta I K_{11} \tag{9}$$

$$\frac{1}{\Delta I} = \frac{1}{\Delta I_{11}K_{11}[C]} + \frac{1}{\Delta I_{11}} \tag{10}$$

where ΔI and ΔI_{11} are ($I_{obs} - I_o$) and ($I_{11} - I_o$), respectively.

The K_{11} value and the enhancement of the fluorescence intensity at the emission wavelength upon complexation, $I_{11}/I_o = \Delta I_{11}/I_o + 1$, are obtained from plots of ΔI versus [C] data using either Eq. (9) or Eq. (10). The choice of the equation usually depends on data set and uniformly spaced data points give more reliable results. Generally, the accurate K_{11} value is obtained when f_{11} is in the range of 0.1–0.9, and the data outside of this range may give less reliable K_{11} value. This requires pre-estimation of the K_{11} value and selection of the CD concentrations to give a nearly same increment of experimental data points.

When G does not exhibit intermolecular self-association, deviation from the linearity expected from Eq. (9) or Eq. (10) can be taken as an evidence of the formation of complexes of higher stoichiometries. In case of a system where the fluorescence intensity is enhanced upon complexation, the convex curvature from Eq. (9) or concave curvature from Eq. (10) is usually indicative of the formation of 1:2 complex. Opposite trends are observed for systems forming 2:2 or 2:1 complex.

For complexation studies with a guest that exhibits intermolecular self-association, the association should also be considered in the analysis. An example for this is the complexation of phenothiazine dyes with CDs. The dyes form non-fluorescent dimers. The monomers form 1:1 complexes with β-CD, while the dimers form 1:1 complexes with γ-CD. Evaluation of equilibrium constants involved in the multiple equilibria is highly complicated. The method can be found in Ref. [25].

In most cases, the condition of $[G]_o \ll [C]_o$ is easily met and $[C]_o$ can be used for $[C]$ in the data analysis. However, for the system of high K_{11} value, typically $K_{11} \geqslant 10^4 \, M^{-1}$, it is sometimes difficult to maintain the condition throughout the experiment. In this case, ΔI_{11} and K_{11} values can be obtained from non-linear fitting of the experimental data to

$$\Delta I = \frac{\Delta I_{11} \left\{ ([G]_o + [C]_o + 1/K_{11}) - \sqrt{([G]_o + [C]_o + 1/K_{11})^2 - 4[G]_o[C]_o} \right\}}{2[G]_o} \tag{11}$$

An alternative method is estimation of ΔI_{11} from the ΔI value obtained at high concentration of C. The f_{11} values are calculated by $\Delta I / \Delta I_{11}$ and then the equilibrium concentrations of C are calculated by $[C] = [C]_o - [G]_o f_{11}$ for each data point. The K_{11} value is obtained from averaging the K values calculated from each data using a relationship $K = f_{11}/\{(1 - f_{11})[C]\}$. Any regular trend of the K values with $[C]_o$ can be taken as an evidence for improper ΔI_{11} value or for the formation of a complex of higher stoichiometry.

1:2 Complexes (structures V and VI): The 1:2 complex can be formed with a guest molecule having the hydrophobic portion longer than the cavity depth of CD or with a guest molecule having two inclusable groups or parts. In this case, in addition to the 1:1 complex formation, the complexation equilibrium between the 1:1 complex with another CD molecule needs to be considered:

$$GC + C \overset{K_{12}}{\rightleftharpoons} GC_2 \tag{12}$$

The fractions of the 1:1 complex (f_{11}) and the 1:2 complex (f_{12}) are expressed as Eq. (13) and the I_{obs} is given by Eq. (14):

$$f_{11} = \frac{K_{11}[C]}{1 + K_{11}[C] + K_{11}K_{12}[C]^2}; \quad f_{12} = \frac{K_{11}K_{12}[C]^2}{1 + K_{11}[C] + K_{11}K_{12}[C]^2} \tag{13}$$

$$I_{obs} = \frac{I_o + I_{11}K_{11}[C] + I_{12}K_{11}K_{12}[C]^2}{1 + K_{11}[C] + K_{11}K_{12}[C]^2} \tag{14}$$

Unlike the 1:1 complexation, Eq. (14) cannot be rearranged to a linear form corresponding to Eq. (9) or Eq. (10). The complex formation constants, K_{11} and K_{12}, and relative emission intensities of the 1:1 and 1:2 complexes, I_{11} and I_{12}, are obtained from the non-linear least-squares fitting of I_{obs} versus $[C]$ data to Eq. (14).

If the binding of CD to a guest is highly cooperative, i.e. $K_{12} \gg K_{11}$, the presence of the 1:1 complex is negligible and the complexation equilibrium can be considered as a one-step 1:2 complexation, (Eq. (15)). In this case, the complexation equilibrium can be studied in analogous way with the 1:1 complexation by substituting [C] terms in Eqs. (8)–(10) with $[C]^2$: the equation corresponding to Eq. (10) becomes Eq. (16). The K_2 and ΔI_{12} values are obtained from either $\Delta I/[C]^2$ versus ΔI or $1/\Delta I$ versus $1/[C]^2$ plot:

$$G + 2C \overset{K_2}{\rightleftarrows} GC_2 \tag{15}$$

$$\frac{1}{\Delta I} = \frac{1}{\Delta I_{12} K_2 [C]^2} + \frac{1}{\Delta I_{12}} \tag{16}$$

1:1:1 Complex (structure IX): When the hydrophobic part of a guest molecule is too large to be included deeply in a CD cavity (as in pyrene/β-CD or pyrene/γ-CD) [26]) or fits loosely in the cavity (as in naphthalene derivatives/γ-CD [27]), the binding of the guest is enhanced by the addition of short-chain alcohol or amines. This was ascribed to the formation of 1:1:1 complex. The overall 1:1:1 complex formation equilibrium is represented as Eq. (17): the 1:1 complexes between G and C, and between S and C can also be formed:

$$G + C + S \overset{K_{111}}{\rightleftarrows} GCS \tag{17}$$

The fractions of G present as GS (1:1 complex) and GCS (1:1:1 complex) are given as

$$f_{11} = \frac{K_{11}[C]}{1 + K_{11}[C] + K_{111}[C][S]}; \qquad f_{111} = \frac{K_{111}[C][S]}{1 + K_{11}[C] + K_{111}[C][S]} \tag{18}$$

If one adds a non-fluorescent S to a solution of G and C, keeping $[G]_o$ and $[C]_o$ constant, the observed emission intensity is a sum of the contributions from G, GS, and GCS and is expressed as

$$I_{obs} = \frac{I_o + I_{11}K_{11}[C] + I_{111}K_{111}[C][S]}{1 + K_{11}[C] + K_{111}[C][S]} \tag{19}$$

Eq. (19) is rearranged to Eq. (20), which is equivalent to Eq. (8) given for 1:1 complex formation: I_o^0 is the fluorescence intensity of the solution of G and C in the absence of S. Linear equations equivalent to Eqs. (9) and (10) can be derived from this

$$I_{obs} = \frac{I_o^0 + I_{111}K'[S]}{1 + K'[S]}, \qquad \text{where } K' = \frac{K_{111}[C]}{1 + K_{11}[C]} \tag{20}$$

When $[G]_o \ll [C]_o$ and $[S]_o \ll [C]_0$ or S exhibits little binding affinity to C in the absence of G, one can use $[C]_o$ for [C]. The little binding affinity of S to C makes $[S]_o \approx [S]$, when $[G]_o \ll [S]_o$. However, if S has appreciable binding affinity to C, it is necessary to determine the binding constant (K_S) of S with C and calculate [S]. The I_{111} and K' values are obtained from the plot by either $\Delta I/[S]$ versus ΔI or $1/\Delta I$ versus $1/[S]$ as described for 1:1 complex formation. To calculate K_{111} from K', it is necessary to obtain the 1:1 binding constant (K_{11}) of G with C, independently.

The equilibrium of the 1:1:1 complex formation can be considered as a sum of two binding equilibria. A pathway for this is formation of the 1:1 GC complex (binding constant, K_{11}) and then binding of S to the GC complex (binding constant, $K_{11/S}$). Alternative pathway is formation of CS complex (binding constant K_S) and then binding of G to the CS complex (binding constant, $K_{CS/G}$). From a cyclic rule of thermodynamics, K_{111} is related to the microscopic equilibrium constants by $K_{111} = K_{11} \times K_{11/S} = K_S \times K_{CS/G}$. Once K_{111} and K_{11} or K_S are obtained, the $K_{11/S}$ or $K_{CS/G}$ value can also be calculated from the relationships.

The ternary complex often exhibits room temperature phosphorescence or exciplex emission. As the phosphorescence and exciplex emissions occur at longer wavelength than the normal fluorescence, one can follow the complexation from the intensity of phosphorescence or exciplex emission. In these case, I_o^o in Eq. (20) becomes zero and I_p or $I_{exciplex}$ is used instead of ΔI in the equations equivalent to Eqs. (9) and (10) [26].

Complexes of other higher stoichiometries: The formation of the 2:1 (G_2C), 2:2 (G_2C_2), or 1:1:2 (GSC_2) complexes can be recognized from the emission behavior of G in the presence of C and/or S. The quenching of fluorescence or observation of excimer emission from G upon the addition of CD can be regarded as the formation of G_2C or G_2C_2 complex. The guest molecules forming such complexes usually also form a dimer in the absence of CD. The exciplex emission or static-like quenching in which the emission intensity versus [S] profile deviates from the 1:1:1 complex can be taken as an evidence of the formation of GSC_2-type complex. The formation of GS-type complex is also expected even in the absence of CD.

In principle, an equation relating the emission intensity (of fluorescence, excimer or exciplex) to equilibrium constants and concentrations of G and C (and S for GSC_2 complex formation) can be derived from definitions of the equilibrium constants and mass balances. The equilibrium constants can be obtained by numerical fitting of the experimental data to the equation [12]. However, as the upper limit of fluorophore concentration for fluorescence measurement is low, the evaluation of the equilibrium constants involved in the formation of these complexes by using only fluorescence technique is often difficult. In most cases, the concentration dependences of other spectroscopic properties (e.g. absorption, circular dichroism, NMR) of the guest in the presence and in the absence of CD are used as complementary techniques. Detailed description of the methods of analysis can be found in Refs. [25–29].

2.3.2. Fluorescence-probed method for cyclodextrin inclusion complexation

The fluorescence method can also be used for studies on complexation between non-fluorescent guest and non-fluorescent host using a fluorescence probe (P). This is primarily based on the competition of P and G for the cavity of CD [15]. This method is particularly useful for binding study of G with C when the other methods are not readily applicable or the binding constant is too high to be determined accurately from other methods. The choice of P is such that P and G do not interact appreciably in the experimental condition; P, G, and C do not form a ternary or quaternary complex; P and C form 1:1 complex accompanying a large fluorescence change.

The 1:1 complexation equilibrium between C and P (Eq. (21)) can be studied by the usual fluorescence method described in the previous section. The equilibrium constant (K_P) for the PC formation reaction and the fluorescence intensity of the PC complex (I_{PC}) are determined from the analysis of the dependence of the observed fluorescence intensity of P on [C]:

$$P + C \overset{K_p}{\rightleftarrows} PC \tag{21}$$

$$I_{obc} = \frac{I_o + I_{PC}K_P[C]}{1 + K_P[C]} \tag{22}$$

The addition of G to a given solution of P and C reduces the concentration of free C by the binding of G to C. This results in the shift of the P/C binding equilibrium to left and the fluorescence intensity change. As K_P and I_{PC} are pre-determined, the equilibrium concentration of C, [C], is calculated from the observed fluorescence intensity by using Eq. (22). The difference between the total and equilibrium concentrations of C, $[C]_o - [C]$, becomes the concentration of C complexed with G.

In case of 1:1 complex formation between G and C, the ($[C]_o - [C]$) value is the same as the concentration of the GC complex. The equilibrium concentration of free G becomes $[G]_o - [C]_o + [C]$. The 1:1 binding constant (K_{11}) of G with C is calculated from the definition of K_{11} as shown in Eq. (23). Consistency of the K_{11} values calculated at different $[G]_o$ suggests the validity of the 1:1 complexation:

$$K_{11} = \frac{[GC]}{[G][C]} = \frac{([C]_o - [C])}{([G]_o - [C]_o + [C])[C]} \tag{23}$$

When a guest forms 1:1 and 1:2 complexes with C, the definitions of the equilibrium constants and mass balance of G give Eq. (24). The [C] values are calculated at various initial concentrations of G ($[G]_o$) from fluorescence intensities and pre-determined K_P and I_{PC} values. Nonlinear least-squares fitting of a set of $[G]_o$ versus $[C]$ data to Eq. (24) gives the K_{11} and K_{12} values [15].

$$[G]_o = \frac{1 + K_{11}[C] + K_{11}K_{12}[C]^2}{K_{11}[C] + 2K_{11}K_{12}[C]^2}([C]_o - [C]) \tag{24}$$

The fluorescence-probed method was used to determine the 1:1 and 1:2 binding constants of n-alkyl sulfates ($C_nH_{2n+1}OSO_3^-$) and n-alkyl sulfonates ($C_nH_{2n+1}SO_3^-$) with β-CD by using 1,8-ANS (**1**) as a fluorescent probe [15a]. The result showed the formation of 1:2 complex, in addition to 1:1 complex, when $n \geqslant 8$ for n-alkyl sulfates and $n \geqslant 10$ for n-alkyl sulfonates. The formation of the 1:2 complexes was not noticed from previous studies using conductometric method. Also, the inclusion complexations of n-alkylammoniums, n-alkyltrimethylammoniums, n-alkylamines, and n-alkyl alcohols with β-CD were studied using 2-ASP (**2**) as a probe [15b]. The results also showed the formation of 1:2 complex with the compounds having long aliphatic chain. These works indicated that the stability of complexes of β-CD with guest molecules having the same hydrocarbon tail depends on the nature of the polar head group. The order of stability was found to be $-NH_3^+ < -SO_3^- < -OH \approx -NH_2 < -OSO_3^-$ [15b].

2.3.3. Self-inclusion of fluorophore-modified cyclodextrins and binding of guest molecules with fluorophore-modified cyclodextrins

The CD derivatives with a suitable pendant group can form self-inclusion complexes intramolecularly or intermolecularly as shown in Fig. 2 [30–33].

Binding of an external guest molecule to the CD cavity of the self-included complexes transfers the included pendant group to the outside of the CD cavity. If the pendant group is a fluorophore, the transfer results in a change in the fluorescent properties of the pendant group [4]. The change would be essentially the opposite of that observed from intermolecular association of the fluorophore with CD. The fluorescence behavior of a fluorophore-modified CD can also be changed without involving self-inclusion and exclusion processes. One

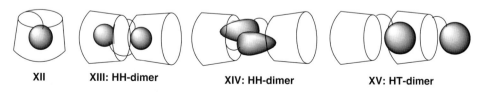

Fig. 2. Schematic representation of various types of cyclodextrin self-inclusion complexes.

is through interaction of the CD-bound guest molecule with the appended fluorophore, which is similar to formation of intermolecular ternary complex. The other is via intracomplex excitation transfer from the excited fluorophore to the guest (see Section 3.3). Because of these properties, the fluorophore-appended CD have been extensively investigated in view of using them as fluorescent chemo-sensors and light harvesting host molecules. As several chapters of this book deal with this topic, this section focuses on the subject relevant to the determination of binding constants of guest molecules to the fluorophore-modified CD.

Guest binding to intramolecularly self-including or not self-including fluorophore-modified CD: The methods used for this are parallel to those used for studies of binding of fluorescent guests to non-fluorescent CD (see Section 2.3.1). The only difference is that the concentration of the modified CD is usually maintained constant and the concentration of G is varied, while keeping the condition of $[CD]_o \ll [G]_o$. The equations in Section 2.3.1 can be applied here by substituting each [C] term with [G]. However, caution should be exercised in the interpretation of the determined binding constants. If the modified CD forms intramolecular self-inclusion complex (structure **XII**), the microscopic binding equilibrium is established between the guest and the non-included form of the modified CD. The observed apparent binding constant (K_{app}) is related to the microscopic binding constant (K) and intramolecular self-association constant (K_{in}) as $K_{app} = K/(1 + K_{in})$: K_{in} is defined as $K_{in} = $ [self-included complex]/[non-included form].

Intermolecular self-inclusion and guest binding of fluorophore-modified CD: When the linkage between the pendant group and CD-skeleton is short or not flexible, intermolecular association of the modified CD often occurs. One of the intermolecular associations is a head-to-head (HH) type dimer formation. The structure **XIII** is usually obtained when the pendant group is included deeply into CD cavity. Examples are the naphthalene group-appended β-CD (**5**) [30] and azobenzene-modified α-CD [31]. When the pendant group is shallowly included into CD cavity and/or there is favorable interaction between the pendant groups, the dimer of structure **XIV** is often formed. Another form of the intermolecular association is the formation of the head-to-tail (HT) type dimer (structure **XV**). However, the HH dimer (**XIII** or **XIV**) is thermodynamically more favorable as the dimer is stabilized by mutual inclusion of the pendant groups into CD cavity of counter molecules and stacking interaction between the pendant groups [30,32,33]. Similar to this, the two different CD derivatives in which the pendant group of one CD derivative fits better to the CD cavity of the other derivative would form a heterocomplex rather than homodimers, by mutual inclusion of the pendant groups into the CD cavities of counter molecules. This was demonstrated with β-CD-pyrene (**6**) and γ-CD-naphthalene [34].

5a: X = SO_3^-
5b: X = H

6

7: methylalkylviologen

The dimer formation constant (K_D) of a fluorophore-modified CD can be obtained by fitting the concentration dependence of the fluorescence intensity to Eq. (25) [30,35]:

$$I_{obs} = \frac{(I_{dimer} - I_{mon})\left(1 - \sqrt{1 + 8K_D C_{tot}}\right) + 4I_{dimer}K_D C_{tot}}{4K_D C_{tot}} \qquad (25)$$

where C_{tot} is the total concentration of the modified CD, and I_{mon} and I_{dimer} are the relative emission intensities of the monomeric and dimeric species, respectively. From the dimer of structure **XIV**, excimer emission, which does not overlap with the monomer emission, can be observed. In this case, Eq. (25) is simplified by making either I_{mon} (when one follows the excimer) or I_{dimer} (when one follows monomer) to 0. Similarly, one can use $I_{dimer} = 0$, when the dimer is not fluorescent.

For a molecule having a small K_D value (e.g. $< 10^3 \, M^{-1}$), reliable K_D value is not obtained by the fluorescence method due to the concentration limit. In this case, NMR and circular dichroism titration methods are usually used [33,34]. The necessary procedure for this is substitution of I's in Eq. (25) with chemical shift or molar ellipticity.

To analyze the guest binding of the fluorophore-modified CD quantitatively and determine the binding constants, both the monomer–dimer equilibrium of the modified CD and the monomer–guest binding equilibrium should be considered. The procedure can be found in recent reports on the binding of alkyl viologens (**7**) with **5** [35], and the binding of **3** with **5a** or **6** [36].

2.3.4. Studies on the directional binding constants of alkyl viologens (7) with β-CD

Guest molecules can be included from both sides of CD cavity. The face selectivity of a guest binding is believed to be very important in molecular recognition and chemical reactions mediated by CD or CD derivatives. NMR [37] and induced circular dichroism [14,38] have been widely utilized to study the orientation of a guest molecule in CD complexes. However, there is a paucity of experimental evidences on the face selectivity in guest binding with CDs. Recently, the directional binding constants of **7** with β-CD were obtained from results of binding studies of **7** with β-CD or **5** using induced circular dichroism and fluorescence techniques [35,39].

The binding of **7** to β-CD or **5** can give two types of complexes that differ in the location of the hydrophilic 4,4'-bipyridinium moiety on CD faces. The types of the complexes with native CD are difficult to distinguish, unless one has the prior detailed knowledge of the dependence of physicochemical characteristics of the complexes on structures. However, as viologens are good quenchers for naphthalene fluorescence and form ground-state charge-transfer complex with naphthalene [20], the two isomeric complexes with **5** (Scheme 1) can be distinguished by their absorption and fluorescence properties.

An equation relating the fluorescence intensity of a given solution of **5** to the concentration of **7** was derived [35]. The dependence of the naphthalene fluorescence of **5** on the concentration of **7** was fitted to the equation to obtain the apparent binding constant K_C and the ratio (γ) of fluorescence intensities of the complex and monomeric **5**. The K_C corresponds to ($K_I + K_{II}$). The K_I/K_{II} ratios, and then K_I and K_{II} values, were obtained from lifetimes of monomeric **5** and the type II complex, K_C and γ values: for this, the fact that the type I complex is non-fluorescent, due to direct contact of naphthyl group with 4,4'-bipyridinium moiety, was utilized. The K_I/K_{II} ratios were obtained as 3.2 for **5a** and 1.0 for **5b**, independent of the length of alkyl chain of **7** [35,39].

The type I complex is stabilized by the inclusion of alkyl chain of **7** into CD cavity from the primary face and the charge-transfer interaction between the 4,4'bipyridinium dication and naphthyl group. On the other hand, the type II

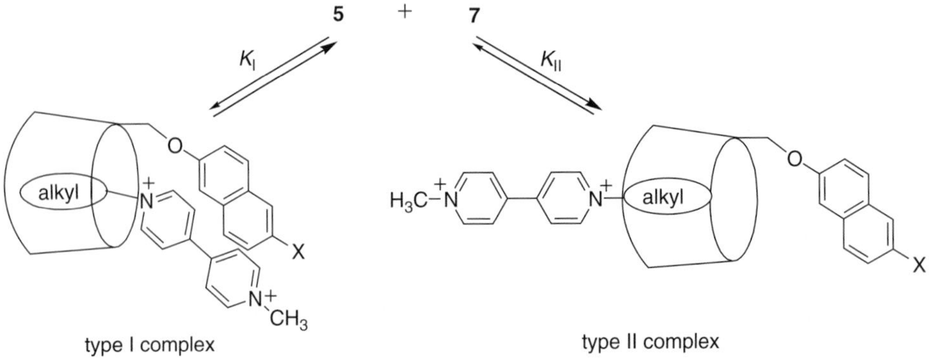

Scheme 1. Inclusion complexation of alkyl viologens **7** with naphthalene group-tethered β-CD **5**. The native CD/**7** complex can also have two different orientations of the guest with respect to CD cavity.

complex is stabilized only by the inclusion of alkyl chain of **7** into CD cavity from the secondary face. The contribution of the charge-transfer interaction to K_I values was evaluated from the charge-transfer absorption. Then, the binding affinities of the alkyl chain of **7** into CD cavity from the primary and secondary face were obtained. The results indicated that, without the charge-transfer interaction, the bipyridinium moiety prefers the secondary face about 14 times more favorably to the primary face of β-CD [35,39].

The K_{II} values of the alkyl viologen complexes with **5** are similar to those of corresponding complexes with native β-CD. This implies that the alkyl chains of **7** are included from the secondary side of native β-CD, placing the 4,4'-bipyridinium moiety above the secondary side of CD cavity in the complex. This was supported by the comparison of induced circular dichroism of **7**/β-CD complexes with those of 1-adamantylammonium viologen/β-CD complex and a β-CD-viologen compound in which the 4,4'-bipyridinium unit is directly connected to the primary face of β-CD [35]. The preference of the secondary side of CD cavity for 4,4'-bipyridinium moiety was also demonstrated in the [2]pseudorotaxane composed of α-CD and an aliphatic chain-linked carbazole-viologen compound [8], and in the self-inclusion complexation of β-CD-viologen compounds in which the viologen moiety is connected by an octamethylene chain to either the primary or secondary side of β-CD [33].

3. Excitation transfer in cyclodextrin complexes

This section starts with a brief description on the general aspect of the excitation transfer reactions relevant to those in CD complexes. More details on excitation transfer can be found in Refs. [40,41]. Two examples of the excitation transfer in CD complexes investigated by fluorescence methods will be followed.

3.1. General aspects of excitation transfer reactions

Photoinduced electron transfer and excitation energy-transfer reactions play a central role in photobiological processes and in various light-driven physical and chemical processes, and have attracted much interests in view of designing photomolecular devices [40]. The excitation transfer processes can be either energy transfer or electron transfer.

The excitation transfer quenches the fluorescence emission and shortens the fluorescence lifetime of the excited fluorophore *F. The bimolecular quenching reactions are usually analyzed by the Stern–Volmer equation

$$I_o/I = \tau_o/\tau = 1 + K_{SV}[Q] \tag{26}$$

where I_o and τ_o are the emission intensity and lifetime of $^*F$, respectively, in the absence of Q: I and τ are the corresponding values in the presence of the quencher Q. The Stern–Volmer constant K_{sv} is given as $k_q \cdot \tau_o$, where k_q is the bimolecular quenching rate constant. The linearity in the plots of I_o/I versus [Q] and τ_o/τ versus [Q] with the same slope is indicative of the dynamic quenching.

If F and Q form non-fluorescent ground-state complex (with complexation constant K_C), the complexation results in static quenching. When both static and dynamic quenching occur for the same fluorophore, the Stern–Volmer equation is modified to

$$I_o/I = (1 + K_{SV}[Q])(1 + K_C[Q]) \tag{27}$$

The modified Stern–Volmer equation is second order in [Q]. The upward curvature is observed in the plot of I_o/I against [Q]. However, as the complex is not fluorescent, the τ_o/τ versus [Q] plot still yields a straight line with a slope of K_{SV}. The Eq. (27) is rearranged to a linear form, which accounts for the linearity in the $(I_o/I-I)/[Q]$ versus [Q] plot. The intercept is $(K_C + K_{SV})$ and the slope is $K_C \cdot K_{SV}$.

CD affect the bimolecular quenching reaction by binding either with F or Q. If Q, but not F, forms 1:1 complex with CD (binding constant K), the K_{SV} value depends on the concentration of CD, [C], as Eq. (28): K_{SV}^o and K_{SV}^{CD} are the Stern–Volmer constants for quenching by the free Q and the Q/CD complex, respectively. The Eq. (28) is analogous to the effect of CD on the rate constant of bimolecular reaction [43]:

$$\left(1 - \frac{K_{SV}}{K_{SV}^0}\right)^{-1} = \left(1 - \frac{K_{SV}^{CD}}{K_{SV}^0}\right)^{-1} + \left\{\left(1 - \frac{K_{SV}^{CD}}{K_{SV}^0}\right)K\right\}^{-1}[C]^{-1} \tag{28}$$

The K_{SV}^{CD} and K values are obtained from the intercept and slope of the plot of $(1-K_{SV}/K_{SV}^o)^{-1}$ against $[C]^{-1}$.

When F, but not Q, forms 1:1 complex with CD, the observed fluorescence intensity is a sum of the contributions from F and F/CD complex, which are quenched by Q, and becomes

$$I_{obs} = \frac{I_o}{(1 + K_{SV}^o[Q])(1 + K_{11}[C])} + \frac{I_{11}K_{11}[C]}{(1 + K_{SV}^{CD}[Q])(1 + K_{11}[C])} \tag{29}$$

Here, the K_{SV}^o and K_{SV}^{CD} are the Stern–Volmer constants for the quenching of the free F and the F/CD complex, respectively, by Q. In principle, the parameters in Eq. (29) can be obtained from the analysis of the quenching data taken at various concentrations of CD. A simpler method for determining these values would be the determination of K_{SV} values in the absence of CD (for K_{SV}^o) and in the high concentration of CD, i.e. under the condition of $K_{11}[C] \gg 1$ (for K_{SV}^{CD}):

I_{11} and K_{11} are obtained from fluorescence titration of F with CD in the absence of Q. Co-inclusion of F and Q in the same CD cavity forming 1:1:1 complex would result in static-like quenching.

A theory developed by Förster for singlet–singlet energy transfer from *F to energy acceptor (A) gives the rate constant for the energy transfer (k_T) as Eq. (30) and the energy transfer efficiency (E) as Eq. (31) [41]:

$$k_T = (1/\tau_o)(R_o/R)^6 \tag{30}$$

$$E = \frac{R_o^6}{R_o^6 + R^6} \tag{31}$$

where R is the distance between the two chromophores, R_o the characteristic transfer distance and proportional to $J^{1/6}$. J is a measure of the overlap between the emission band of F and absorption band of A. For commonly used pairs of chromophores, R_o varies from $10\,\text{Å}$ to more than $50\,\text{Å}$.

The rate constant (k_{et}) of the electron transfer reaction from an electron donor (D) and an electron acceptor (A) is related to the electron coupling matrix element between D and A, ($|V|$), standard Gibbs free energy change of the electron transfer step (ΔG^o), and the sum of molecular and solvent reorganization energies (λ) by [40]

$$k_{et} = \left(2|V|^2/h\right)\left(\pi^2/\lambda RT\right)^{1/2} \exp\left[-(\Delta G^o + \lambda)^2/4\lambda RT\right] \tag{32}$$

The $|V|^2$ term depends strongly on the distance R between D and A, decreasing exponentially with R. The electronic coupling and the electron transfer can occur through space, solvent, and bond. Solvent and linkage mediate the coupling by superexchange interaction, enabling the long-range electron transfer [40,42]. The distance dependence of k_{et} is given as $k_{et} = A\exp(-\beta R)$: β is the attenuation factor and depends on the nature of linkage or solvent medium.

The rate constant of the excitation transfer reaction is usually obtained from Eq. (33): τ is the fluorescence lifetime of F in F/Q pair or in the presence of Q:

$$k_{et} = 1/\tau - 1/\tau_o \tag{33}$$

For a bimolecular quenching reaction, the k_{et} corresponds to $k_q[Q]$. However, for a covalently linked F/Q dyad or F/Q assembly, the k_{et} is the intramolecular or intra-assembly first-order excitation transfer rate constant.

The efficiency of excitation energy or electron transfer between isolated molecules is small, unless the concentration is high enough to bring average intermolecular distance within about $10\,\text{Å}$. Therefore, for efficient excitation transfer, it is necessary to bring the appropriate F and Q (or energy accptor) pair within short distance by covalent bonding [40,42,44] or assembling the pair into supramolecular structures. CD cavity has been widely used for the latter purpose [37,45].

3.2. Studies on linkage length dependence of electron transfer in flexible chain-linked donor-acceptor dyads using cyclodextrin

Among various factors affecting the excitation transfer rate in a given D/A pair, the distance is generally considered to be most important. To study the distance dependence of the transfer rate, it is necessary to bring the D/A pair into a controlled distance. One approach for this is covalent bonding of the pair with rigid linkage. Using a series of compounds having different linkage length, β values for electron-transfer reactions were obtained [40,46].

The D–A systems with flexible linkage are easier to be synthesized than those with rigid linkages, but they do not exhibit the usual exponential decrease of the electron transfer rate with the linkage length [20,47]. This is due to the flexibility of the linkage that brings the D and A pair in closer proximity than that in fully extended conformation. The through-space and through-solvent electron transfer, rather than the through-bond transfer, dominate in the folded conformation. Similarly, high efficiency of intramolecular energy transfer in the flexible bichromophoric compounds was also observed and explained in terms of facilitation of the through-space interaction due to the conformational flexibility [48].

Encapsulation of the flexible linkage connecting D and A into CD cavity would stretch the molecule changing its conformation similar to D–A system with rigid linkage. This may allow one to determine the β value for the flexible linkage. This was demonstrated with polymethylene chain-linked aromatic-viologen compounds **8** [20].

8a: Aromatic = 1-naphthoxyl; n = 3, 6, 8,10

Aromatic —(CH$_2$)$_n$—$\overset{+}{N}$⟨⟩⟨⟩$\overset{+}{N}$—CH$_3$ **8b**: Aromatic = 2-naphthoxyl; n = 3–10,12

8c: Aromatic = 2-dibenzofuranoxyl; n = 3, 6, 8,10

From the fluorescence lifetimes of the dyads and the model compounds of the isolated aromatic groups, the intramolecular k_{et} values were calculated by using Eq. (33). Though k_{et} becomes smaller as the number of methylene unit (n) in the linkage is larger, it does not decrease exponentially with n. For molecules of $n \geqslant 8$, the dependence of k_{et} on n was much less pronounced than that of $n < 8$.

Fully extended structures of **8** were induced by complexation of the compounds with methylated β-CD. The fluorescence intensities and lifetimes of the extended conformations were determined. From the lifetimes and those of model compounds of the isolated aromatic groups, the k_{et} values of the extended conformations were calculated. The k_{et} decreases exponentially with n and the β value was 0.89 Å^{-1} (1.09 per methylene unit), regardless of the nature of aromatic moiety. The β value is similar to that observed with the rigid norbornyl bridges [40]. Comparison of the effects of Me-β-CD on the steady-state fluorescence intensity and on the excited-state lifetime suggested that the through-space and through-solvent electron transfer are the predominant quenching pathways for the molecules having the linkage shorter than hepta-methylene group.

3.3. Excitation transfer in donor–acceptor assemblies derived from modified cyclodextrins

Cyclodextrin cavities have been extensively used to assemble the electron or energy donor (D) and acceptor (A) pairs. This usually starts with the synthesis of D- (or A-) appended CD and is followed by assembling the modified CD with appropriate A (or D). Intracomplex excitation energy or electron transfer has been investigated with a variety of CD-mediated assemblies [35,36,45]. An example for this is the type II complexes between β-CD-naphthalenes (**5**) and alkyl viologens (**7**) shown in Scheme 1. The electron transfer rate constant was found to be about $2 \times 10^8 \, \text{s}^{-1}$, independent on the alkyl chains [35,39]. The rate constant is almost the same as that of naphthalene-viologen dyad linked by nine methylene unit, **8b** ($n = 9$), extended by complexation with Me-β-CD [20]. The electron transfer in the **5/7** complexes occurs through-space or through-solvent, whereas that in extended **8b** occurs through-aliphatic bond. This implies that the electron transfer rate constant by through-space pathway in aqueous media is close to that of through-aliphatic bond transfer in similar distance. A similar result was also reported with *p*-nitrobenzoyl/porphyrin pair [45].

The characteristic energy transfer distance R_o of commonly used energy donor and acceptor pair is longer than the height of CD torus. Thus, efficient energy transfer is expected in the energy donor–acceptor pair assembled with CD. For an ideal host–guest system of efficient energy transfer, large absorptivity and emission intensity of the energy donor, a good overlap between the emission spectrum of the donor and absorption spectrum of the acceptor, and high binding affinity of the guest to CD cavity are required. For the sensitized emission from the acceptor, high emission quantum yield of the acceptor is also required. In a recent paper, it was demonstrated that the aromatic group-tethered β-CD (**5** or **6**) and 4-ASP (**3**) pairs meet much of the requirement [36].

The binding constant of **3** with **5** is $3.6 \times 10^4 \, \text{M}^{-1}$, and that with **6** is $4.7 \times 10^4 \, \text{M}^{-1}$. These are 50–60 times greater than that with native β-CD. The emission intensity of 4-ASP is enhanced by 14- and 56-fold upon binding with **5** and **6**, respectively, whereas the binding with β-CD enhances the intensity by 6.4 times. These were explained in terms of the interaction between the CD-encased **3** and the appended aromatic groups. The higher binding constants of guest molecules to aromatic group-modified β-CDs than those to native β-CD were also observed with other guests [45,49]. The efficiency of intracomplex excitation transfer from the CD-appended aromatic groups to **3** is near 100% and emission from **3** is observed upon excitation of the aromatic groups. Because of high absorptivity of pyrenyl group and little tendency of dimerization, the pyrene-modified β-CD **6** appears to be better light-harvesting (excitation donor) host than the extensively studied naphthalene-modified β-CDs [36].

4. Concluding remarks

CDs are believed to be the most important hosts. They appeal to investigators in both pure research and applied technologies. This is primarily due to their ability to form inclusion complexes with a variety of guest molecules in water. It would be hard to find an article on cyclodextrin chemistry, which is not related to the guest binding into CD cavity, directly or indirectly. Studies on the inclusion complex-ation that provide information on the stability, stoichiometry, and structure of CD complexes seems to be the most important subject of cyclodextrin chemistry. Enormous number of reports dealing with this subject using a variety of techniques have been published. However, in some cases, the binding data of a report are not reproduced from studies on the same system but by using other techniques and the conclusions are doubtable. This suggests that strict experimental condition and rigorous justification of the mathematical model used for the analysis of exper-imental data are necessary for the studies. Much of this chapter was devoted to this.

Because of restricted space and relatively non-polar environment of CD cavity, the photochemical and photophysical behaviors of fluorescent molecules are greatly changed upon inclusion complexation of the molecules with CD. Many of utili-zation of CD or CD derivatives in industrial and emerging technologies are based on this. The studies related to this were usually carried out by following the flu-orescence. In fact, among 9466 articles topically indexed on cyclodextrin in the ISI Web of Science® in the period from 1998 to mid-October of 2005, 1079 articles (11.4%) are also indexed on fluorescence. This reveals the importance of fluores-cence techniques in the studies of cyclodextrin chemistry. This chapter describes the logical basis of using fluorescence for studies of CD complexation, and how to use the fluorescence methods for determination of the stability, stoichiometry and structure of CD complexes, and the excitation transfer in CD complexes

In the future, we can expect to have more guest-selective and sensitive host molecules derived from CD, and the CD-mediated assemblies of well-defined structures and functions. The studies using fluorescence techniques would pro-vide valuable guides for designing the CD-derivatives and construction of the assemblies. Promises for this can be found in the several parts of this chapter.

References

[1] For Compilation of Recent Research on Cyclodextrins, see (a) J. Szejtli, T. Osa (Eds), Comprehensive Supramolecular Chemistry, Pergamon Press, Oxford, 1996, Vol. 3; (b) V.T. D'Souza, K.B. Lipkowitz (Eds), Chem. Rev. 98 (1998) 1977–2076 (a special issue on CD).

[2] (a) A.R. Khan, P. Forgo, K.J. Stine, V.T. D'Souza, Chem. Rev. 98 (1998) 1977;
(b) E. Engeldinger, D. Armspach, D. Matt, Chem. Rev. 103 (2003) 4147.

[3] Besides this monograph, a special issue dealing with photochemistry and phophysics of cyclodextrin inclusion complexes was published. (J. Photochem. Photobiol. A: Chem. 173(3) (2005) 230–389).

[4] T. Hayashita, A. Yamauchi, A.J. Tong, J.C. Lee, B.D. Smith, N. Teramae, J. Incl. Phenom. Macro. Chem. 50 (2004) 89.

[5] M.V. Rekharsky, Y. Inoue, Chem. Rev. 98 (1998) 1875.

[6] T. Fukahori, T. Ugawa, S. Nishikawa, J. Phys. Chem. A 106 (2002) 9442.

[7] H. Bakirci, W.M. Nau, J. Photochem. Photobiol. A: Chem. 173 (2005) 340.

[8] J.W. Park, H.J. Song, Org. Lett. 6 (2004) 4869.

[9] (a) T. Oshikiri, Y. Takashima, H. Yamaguchi, A. Harada, J. Am. Chem. Soc. 127 (2005) 12186; and references cited therein
(b) A.J. Baer, D.H. Macartney, Org. Biomol. Chem. 3 (2005) 1448
(c) M. Horn, J. Ihringer, P.T. Glink, J.F. Stoddart, Chem. Eur. J. 9 (2003) 4046.

[10] (a) A. Ueno, K. Takahashi, T. Osa, J. Chem. Soc. Chem. Commun. (1980) 921
(b) H. Ikeda, Y. Iidaka, A. Ueno, Org. Lett. 5 (2003) 1625.

[11] (a) R.S. Murphy, T.C. Barros, B. Mayer, G. Marconi, C. Bohne, Langmuir 16 (2000) 8780
(b) T.C. Barros, K. Stefaniak, J.F. Holzwarth, C. Bohne, J. Phys. Chem. A 102 (1998) 5639
(c) J.H. LaRose, T.C. Werner, Appl. Spec. 54 (2000) 284
(d) G. Pistolis, A. Malliaris, J. Phys. Chem. B 108 (2004) 2846.

[12] A.S.M. Dyck, U. Kisiel, C. Bohne, J. Phys. Chem. B 107 (2003) 11652.

[13] T.K. Korpela, J.-P. Himanen, J. Chromatogr. 290 (1984) 351.

[14] M. Kodaka, J. Phys. Chem. A 102 (1998) 8101.

[15] (a) J.W. Park, H.J. Song, J. Phys. Chem. 93 (1989) 6454;
(b) J.W. Park, K.H. Park, J. Incl. Phenom. Mol. Recogn. Chem. 17 (1994) 277.

[16] M. N. Berberan-Santos, PhysChemComm. (2000) Art. No. 5.

[17] K. Kalyanasundaram, Photochemistry in Microheterogeneous System, Academic Press, New York, 1987, pp. 299–334.

[18] C. Reichardt, Solvents and Solvent Effects in Organic Chemistry, 3rd Ed., Wiley–VCH, Weinheim, 2002, pp. 329–358.

[19] Z.R. Grabowski, K. Rotkielwicz, Chem. Rev. 103 (2003) 3899.

[20] J.W. Park, B.A. Lee, S.Y. Lee, J. Phys. Chem. B 102 (1998) 8209.

[21] G.S. Cox, N.J. Turro, N.C. Yang, M.-J. Chen, J. Am. Chem. Soc. 106 (1984) 422.

[22] Y.L. Peng, Y.T. Wang, Y. Wang, W.J. Jin, J. Photochem. Photobiol. A: Chem. 173 (2005) 301.

[23] H. Ikeda, T. Murayama, A. Ueno, Org. Biomol. Chem. 3 (2005) 4262.

[24] (a) H. Tsukube, H. Furuta, A. Odani, Y. Takeda, Y. Kudo, Y. Inoue, Y. Liu, H. Sakamoto, K. Kimura, Determination of Stability Constants, in: J.E.D. Davies, J.A. Ripmeester (Eds), Supramolecular Chemistry, Pergamon Press, Oxford, 1996, Vol. 8, pp. 425–482
(b) K.E. Connors, Measurement of Cyclodextrin Complex Stability Constants, 1996, pp. 205–241.

[25] C. Lee, Y.W. Sung, J.W. Park, J. Phys. Chem. B 103 (1999) 893.

[26] (a) S. Hamai, J. Phys. Chem. 92 (1988) 6140;
(b) S. Hamai, J. Phys. Chem. 93 (1989) 2074;
(c) S. Hamai, J. Am. Chem. Soc. 111 (1989) 3954.

[27] A. Ueno, K. Takahashi, Y. Hino, T. Osa, Chem. Commun. (1981) 194.

[28] (a) S. Hamai, Bull. Chem. Soc. Jpn. 55 (1982) 2721;
(b) S. Hamai, J. Phys. Chem. B 101 (1997) 1707.

[29] (a) W.G. Herkstroeter, P.A. Martic, T.R. Evans, S. Farid, J. Am. Chem. Soc. 108 (1986) 3275;
     (b) S. Hamai, T. Ikeda, A. Nakamura, H. Ikeda, A. Ueno, F. Toda, J. Am. Chem. Soc. 114 (1992) 6012;
     (c) A. Nakamura, S. Sato, K. Hamasaki, A. Ueno, F. Toda, J. Phys. Chem. 99 (1995) 10952.
[30] J.W. Park, H.E. Song, S.Y. Lee, J. Phys. Chem. B 106 (2002) 5177.
[31] T. Fujimoto, A. Nakamura, Y. Inoue, Y. Sakata, T. Kaneda, Tetrahedron Lett. 42 (2001) 7987.
[32] J.W. Park, N.H. Choi, J.H. Kim, J. Phys. Chem. 100 (1996) 769.
[33] J.W. Park, S.Y. Lee, H.J. Song, K.K. Park, J. Org. Chem. 70 (2005) 9505 and references cited therein.
[34] J.W. Park, H.E. Song, S.Y. Lee, J. Org. Chem. 68 (2003) 7071.
[35] J.W. Park, H.E. Song, S.Y. Lee, J. Phys. Chem. B 106 (2002) 7186.
[36] J.W. Park, S.Y. Lee, S.M. Kim, J. Photochem. Photobiol. A: Chem. 173 (2005) 271.
[37] (a) H.-J. Schneider, F. Hacket, V. Rüdiger, Chem. Rev. 98 (1998) 1755;
     (b) T. Ishizu, K. Kintsu, H. Yamamoto, J. Phys. Chem. B 103 (1999) 8992.
[38] (a) M. Kodaka, T. Fukaya, Bull Chem. Soc. Jpn. 62 (1989) 1154;
     (b) M. Kodaka, J. Am. Chem. Soc. 115 (1993) 3702.
[39] J.W. Park, S.Y. Lee, J. Incl. Phenom. Macro. Chem. 47 (2003) 143.
[40] V. Balzani, F. Scandola, Supramolecular Chemistry, Ellis Horwood, New York, 1991.
[41] J.R. Lakowicz, Principles of Fluorescence Spectroscopy, 2nd Ed., Kluwer, New York, 1999.
[42] M.N. Paddon-Row, Adv. Phys. Org. Chem. 38 (2003) 1.
[43] J.W. Park, S.Y. Cha, K.K. Park, J. Chem. Soc. Perkin Trans. 2 (1991) 1613.
[44] E.A. Weiss, L.E. Sinks, A.S. Lukas, E.T. Chernick, M.A. Ratner, M.R. Wasielewski, J. Phys. Chem. B 108 (2004) 10309 and references cited therein.
[45] (a) Y.H. Wang, M.Z. Zhu, X.Y. Ding, J.P. Ye, L. Liu, Q.X. Guo, J. Phys. Chem. B. 107 (2003) 14087;
     (b) Q.H. Wu, M.Z. Zhu, S.J. Wei, K.S. Song, L. Liu, Q.X. Guo, J. Incl. Phenom. Macro. Chem. 52 (2005) 93.
[46] S.A. Vail, P.J. Krawczuk, D.M. Guldi, A. Palkar, L. Echegoyen, J.P.C. Tome, M.A. Fazio, D.I. Schuster, Chem. Eur. J. 11 (2005) 3375.
[47] T.J. Chow, Y.T. Pan, Y.S. Yeh, Y.S. Wen, K.Y. Chen, P.T. Chou, Tetrahedron 61 (2005) 6967.
[48] (a) P.J. Wagner, P. Klan, J. Am. Chem. Soc. 121 (1999) 9626;
     (b) L. Vrbka, P. Klan, Z. Kriz, J. Koca, P.J. Wagner, J. Phys. Chem. A 107 (2003) 3404.
[49] (a) L. Jullien, J. Canceill, B. Valeur, E. Bardez, J.-M. Lehn, Angew. Chem. Int. Ed. Engl. 33 (1994) 2438;
     (b) L. Jullien, J. Canceill, B. Valeur, E. Bardez, J.-P. Léfevre, J.-M. Lehn, V. Marchi-Artzner, R. Pansu, J. Am. Chem. Soc. 118 (1996) 5432.

Cyclodextrin Materials Photochemistry, Photophysics and Photobiology
Abderrazzak Douhal (Editor)
DOI 10.1016/S1872-1400(06)01002-8

Chapter 2

The Effects of Cyclodextrins on Guest Fluorescence

Brian D. Wagner

Department of Chemistry, University of Prince Edward Island, Charlottetown, PE, Canada C1A 4P3

1. Introduction

Cyclodextrins are by far the most popular family of molecular hosts for the inclusion of small organic guest molecules. They have a long history dating back to the work of Villiers in 1891 [1] and Schardinger in 1903 [2]. In fact, their use as hosts for complexing guest species dates back to the early 1900s, to Schardinger's work on the complexation of iodine. This early history is well described in an excellent review article by Szjetli [3], the lead review of a special issue of *Chemical Reviews* (July/August 1998) on cyclodextrins. There have been a very large number of useful comprehensive reviews of cyclodextrin chemistry published over the past 25 years [3–22]. This list included 3 monographs [5,9,17] and various general overview articles [3,4,7] on cyclodextrins. It also includes more focused review articles, covering such aspects of cyclodextrin chemistry as chemically modified cyclodextrins [6,14,20], photochemistry and photophysics in cyclodextrin cavities [8,13,21], analytical chemistry [10,16], cyclodextrins as molecular building blocks for supramolecular architectures [11,12] and molecular machines [19], stability of cyclodextrin complexes [15], and dynamics of guest inclusion into cyclodextrins [18,22].

The inclusion of guests into cyclodextrins has many useful applications; this is the main reason for the widespread interest, study, and use of cyclodextrins as molecular hosts. Such applications include as agents for chromatographic separations [10], drug delivery [23], aqueous solubilization of persistent environmental pollutants [24] and other compounds [25], stabilization of food products and additives and other industrial applications [4,26], fluorescent sensors [27], and mediation of organic reactions [28].

Cyclodextrin inclusion complexes can be studied by a number of experimental techniques, including UV/visible spectroscopy [29], NMR spectroscopy [30], calorimetry [31], circular dichroism [32], X-ray crystallography [33], and fluorescence spectroscopy [34]. In addition to these experimental methods, computational studies have also been widely applied to cyclodextrin inclusion [35]. In each of the spectroscopic methods, inclusion of a guest molecule into a cyclodextrin cavity results in a change in a measurable property of the guest and/or cyclodextrin. In the case of UV/visible spectroscopy, a change in absorbance and/or absorption wavelength of the guest is observed. In the case of ^{1}H NMR, a change in the chemical shift of one or more protons on the guest, or those on the cyclodextrin, particularly the hydroxyl protons, is observed. In the case of fluorescence, a change in the emission intensity and/or emission wavelength maximum of the guest is observed. Of these, fluorescence is the most sensitive, as it requires the lowest concentration of sample, and furthermore for highly polarity-sensitive guests, exhibits by far the largest effects upon inclusion. It is thus an excellent technique for studying cyclodextrin inclusion phenomena.

Although fluorescence spectroscopy is the most sensitive method for studying the formation of cyclodextrin inclusion complexes, it does have two significant limitations, which must be taken into account when determining the best experimental technique for studying a given CD complex:

(1) The guest must exhibit polarity-sensitive fluorescence. Since not all guests of interest are fluorescent, this technique cannot always be used.

(2) While observation of changes in guest fluorescence provides evidence that an inclusion complex has formed, it provides only limited structural information on the geometry and mode of inclusion. Direct evidence must be obtained from NMR studies, which can show specific interactions between specific parts of the guest and host, and thus direct evidence of particular modes of inclusion, or of course from X-ray crystallography. The latter obviously can be used only in cases in which the complex can be crystallized.

The aim of this chapter is to describe in detail the effects that inclusion into cyclodextrins can have on guest fluorescence, and how those effects can be exploited to obtain information about the inclusion process. The focus will be on fluorescent guest molecules, which display a high sensitivity to polarity. This chapter is meant to provide a representative survey of the effects cyclodextrins can have on guest fluorescence, and by no means meant to serve as a comprehensive review of the literature in this area, which is massive. Therefore the focus is on the particular polarity-sensitive fluorescent guests which have been studied in my research group; for these guests, references to my work as well as that of other researchers will be included. This chapter will also be focused on the effect of CD on guest fluorescence spectra, i.e. using steady-state fluorescence spectroscopy, which has been the primary method used in my research. Although cyclodextrin complexes have been studied using time-resolved fluorescence spectroscopy [34], this has not found the same widespread application as has steady-state spectroscopy, and will not be covered in this chapter. (However, it is described in detail in several other chapters of this book.) Furthermore, only intermolecular inclusion complexes will be considered, i.e. cyclodextrins with tethered guests will not be discussed.

1.1. Supramolecular host–guest inclusion

Supramolecular chemistry, or "chemistry beyond the molecule" [36], refers to structures consisting of two or more molecules held together by intermolecular forces. This is in contrast to molecular chemistry, which involves the formation of new covalent bonds between components to form new single molecules. Supramolecular chemistry can result in a wide array of interesting and useful chemical structures and architectures. Despite the lack of covalent bonding between the supramolecular components, these structures can be surprisingly robust. A key feature of supramolecular chemistry is the concept of *self-assembly*, whereby the components are simply mixed together in the appropriate proportions, and structure formation occurs spontaneously, with no control or guidance required. Furthermore, and most importantly, the physical and chemical properties of the components can be significantly modified upon formation of supramolecular structures; this can lead to a wide range of applications. In the case of solid supramolecular structures, knowledge of the effects of supramolecular interactions can lead to the rational design of supramolecular structures with specific desired properties; this is the focus of the emerging field of *crystal engineering*.

The simplest type of supramolecular structure (and that which is the sole focus of this chapter) is a host–guest inclusion complex, in which a small guest molecule becomes included within the hollow internal cavity of a larger, cage-like host molecule in solution. This is illustrated in Fig. 1 for naphthalene as a guest becoming included in the cavity of a cyclodextrin (represented as a "molecular bucket", as will be described in Section 1.2).

As clearly indicated in Fig. 1, host–guest inclusion in solution is an equilibrium process, with individual guest molecules able to enter and exit a host cavity in a dynamic process. In order for a significant concentration of the host–guest complex to be obtained, the rate of entrance into the cavity must be significantly larger than the rate of exit or egress. As mentioned previously, this dynamical approach to host–guest inclusion into cyclodextrins has been recently reviewed by Bohne [18] and Douhal [22].

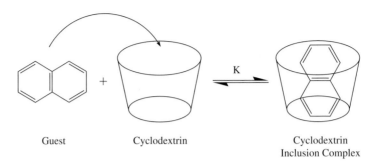

Guest   Cyclodextrin   Cyclodextrin
                                                  Inclusion Complex

Fig. 1. A cartoon representation of the inclusion of a fluorescent probe inside a cyclodextrin cavity, forming a host–guest inclusion complex.

This chapter will focus on the overall equilibrium approach to the study of cyclodextrin inclusion. In this approach, the primary measurement of interest is the binding constant K (alternatively referred to as the association constant), which is simply the equilibrium constant for the process illustrated in Fig. 1. In the simplest case of 1:1 host:guest inclusion as illustrated in Fig. 1, K is given by Eq. 1:

$$K = \frac{[CD:G]}{[CD][G]} \tag{1}$$

The value of K is of course related to the stability of the complex, and the Gibbs energy ΔG for the inclusion process is related to K by Eq. 2:

$$\Delta G = -RT \ln K \tag{2}$$

This value of ΔG is an overall measurement of the driving forces (both enthalpic and entropic) for the inclusion process, and can include contributions from van der Waals attractions, electrostatic attractions, and hydrogen bonding between the host and guest, as well as other factors including release of solvent molecules from the host cavity, conformational strain release, and the hydrophobic effect. The relative importance of these various driving forces for the specific case of cyclodextrins as hosts will be discussed in Section 1.2.

Host–guest inclusion complexes involving more than one host and/or guest can also be obtained; these are referred to in general as higher-order complexes. Although there is unfortunate variation in the literature, the consensus for describing the stoichiometry is the notation **x:y**, where **x** is the number of host molecules and **y** the number of guest molecules. For example, if the guest is significantly longer than the host cavity (although by necessity more narrow, allowing for inclusion to occur), then a significant portion can stick outside of the cavity, allowing for inclusion of the guest by a second host, yielding a 2:1 host:guest complex. If on the other hand the host cavity is very large with respect to the size of the guest, 1:2 host:guest complexes may form, which essentially involve the inclusion of a guest dimer into a single host cavity. It is also possible for a long guest dimer to be included by two hosts, giving a 2:2 host:guest complex. While other, even higher stoichiometries might be possible, for example if the guest has multiple (greater than two) binding sites, all of the complexes to be discussed in this chapter are one of these four types, 1:1, 2:1, 1:2 or 2:2.

1.2. Native and modified cyclodextrins as molecular hosts

In this section, the physical and structural properties of cyclodextrins relevant to their host properties will be briefly discussed. More detailed discussions of these properties can be found elsewhere in this book, as well as in the monographs and reviews mentioned above, particularly Refs. [3–5] and [7].

Cyclodextrins are cyclic oligosaccharides of glucopyranose. The chemical structure of β-cyclodextrin, which contains seven glucopyranose monomers, is shown in Fig. 2, where all the R groups are H.

Figure 2 well illustrates the cyclic nature of β-cyclodextrin, and the large, well-defined cavity. A very important feature is the presence of the 7 primary hydroxyl groups on one side of the monomer units, and the 14 secondary hydroxyl groups on the other side. In aqueous solution, the hydroxyl groups on each side interact through hydrogen bonding. Because there are twice as many secondary as primary hydroxyls, the molecule adopts a "truncated cone" shape, with the larger rim lined by the secondary hydroxyls and the narrower rim by the primary ones. On the upper rim, the hydroxyl in the C-2 position will interact with the neighboring hydroxyl on the C-3 position [3]. This shape resembles a "molecular bucket", as mentioned earlier; this vivid description is

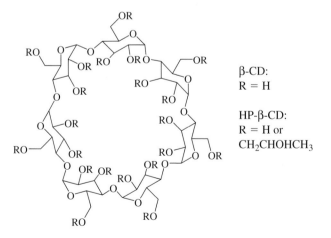

β-CD:
R = H

HP-β-CD:
R = H or
CH$_2$CHOHCH$_3$

Fig. 2. The chemical structure of β-cyclodextrin and HP-β-cyclodextrin.

an appropriate one, as it gives a good visualization of the overall shape of cyclodextrin hosts, and also the visualization of the inclusion process as putting something in a bucket. The hydrogen bonding is responsible for the relatively rigid cavities in these hosts; however, since there is a single covalent bridge between each, there is a certain amount of flexibility.

There are two other "native cyclodextrins", α (six monomers) and γ (eight monomers), so called because they can be produced from enzymatic reactions of starch and other sugar sources [3]. The existence of these three different oligomers is one of the reasons for the widespread use of CD as hosts, since these three have very different cavity sizes (upper rim diameters of 5.7, 7.8, and 9.5 Å), allowing for optimal inclusion of different sized guests, and therefore a high degree of selectivity, which can be tuned by choosing the cavity size.

In terms of the host properties of CD, one of their most important properties is the local polarity within the cavities. This cavity provides a relatively non-polar environment, at least compared to aqueous solution. This of course leads to the preference of hydrophobic guests to become included within the cavity, instead of free in solution. This "hydrophobic effect" is often the main driving force for this inclusion phenomenon. It is of course the strong ability of cyclo-dextrins to form inclusion complexes with a wide range of types of molecular guests which have made them so widely studied and applied. A number of groups have used the effect of cyclodextrins on the spectroscopic properties of included guests to determine a numerical value for the polarity of the local environment with the cavity, mostly for the specific case of β-CD. In such cases, a direct correlation between the fluorescence band maximum and specific sol-vent polarity parameters, such as dielectric constant ε, is observed. There has been a rather wide range of polarities reported for β-CD, ranging from similar to ethyl acetate ($\varepsilon \approx 6$) [37] to $\varepsilon \approx 48$ [38], however many studies put the polarity in the range of that of an alcohol solvent. For example, Hamai estimated the polarity inside the cavity as similar to 1-propanol ($\varepsilon \approx 20$) [39], while Cox *et al.* have reported a polarity similar to ethanol ($\varepsilon \approx 24$) [40] or to t-butylalcohol [41], based on two different included probes. Heredia *et al.* also reported that the polarity inside β-CD is similar to that of ethanol [42].

Liu and Guo [43] have written an excellent recent review of the detailed driving forces involved in the formation of cyclodextrin inclusion complexes. They found that electrostatic interactions, van der Waals interactions, the hydrophobic effect, hydrogen bonding, and charge transfer interactions all can contribute to the formation of these complexes. The relative importance of each of these contributions depends on the properties of the specific guest being included. For example, electrostatic interactions may dominate for charged species, and hy-drogen bonding for species which are hydrogen bond acceptors. Two driving forces which are often invoked for host–guest inclusion phenomena, which Liu and Guo found were not important in the formation of cyclodextrin inclusion complexes, are the expulsion of high-energy water molecules from the cyclodextrin cavities and the release of conformational strain of the cyclodextrin host upon complexation. The authors made a further very important conclusion, namely that the determination of the enthalpy and entropy of inclusion into cyclodextrins is not a useful indication of the contribution of these various driving forces, mainly due to the observation of enthalpy–entropy compensation. This phenomenon for cyclodextrin complexation was also reviewed by Rekharsky and Inoue [31].

For most of the history of cyclodextrin chemistry, the three native cyclodextrins were the only ones studied. However, in the past 20 years or so, the use of chemically modified cyclodextrins has greatly increased. These offer a wide range of improvements over their native parents, such as extended cavity size, reduced effective cavity polarity, and increased aqueous solubility. The three hydroxyls per glucopyranose monomer unit offer ideal sites for chemical modification, by a wide variety of synthetic methodologies [17]. Today, there are a wide range of commercially available modified cyclodextrins, such as hydroxypropyl- and methyl-substituted α-, β-, and γ-CD. The structure of hydroxypropyl-β-cyclodextrin (HP-β) is shown in Fig. 2. As will be illustrated in Section 4.1, these two examples in particular show significantly improved host properties compared to their unmodified parents. One significant disadvantage with these commercial modified CD, however, is that they are not homogeneous. In general, only a relative substitution number is known, usually expressed in terms of how many hydroxyls per monomer unit have been substituted on average (maximum of 3). For example, with a relative substitution of 1.0, only 7 of the 21 hydroxyls have been substituted, and this is not necessarily one on each monomer, or even the same hydroxyl on each monomer (primary substitution is more likely than secondary, due to higher reactivity). Thus, a commercial modified CD sample is a mixture of specific substituted cyclodextrins. This should be kept in mind whenever these commercial modified CD are used. If overall effects on the guest are of interest, then there is no problem with this (except for the possibility of variations in overall composition from bottle to bottle). However, if time-resolved fluorescence is being used, then this leads to an inherent complication in the measured fluorescence decay curve of different lifetimes arising from complexes of different specific CD.

2. Overview of fluorescence spectroscopic techniques and data analysis

Luminescence is the general term for the emission of light from electronically excited molecules. It involves a *relaxation process*, in which a molecule loses excess electronic energy in the form of an emitted photon and thereby *relaxes* to a lower energy electronic state. There must be a source of energy for the initial electronic excitation; this can occur in various ways, including electrical energy (as in the case of fluorescent lights), chemical reactions (chemiluminescence), biological processes (bioluminescence), or the absorption of light (as in the case of fluorescent paint). Fluorescence refers to a specific type of luminescence, namely the emission of light during a transition from a higher to a lower (usually ground) electronic state of the same multiplicity. This usually involves two singlet states, but fluorescence between (for example) triplet states is also possible. By contrast, a luminescence process involving a transition between states of differing multiplicity is referred to as phosphorescence. This usually involves the transition from the first triplet excited state to the ground singlet state.

Although the electronic states of specific molecules can be described using term symbols, based on the point group of the molecule and the corresponding symmetry of the electronic state, a general, iterative notation is more convenient for this chapter. In this scheme, the singlet states are labeled S_0, S_1, S_2, ... in order of increasing energy, and the triplet states are labeled T_1, T_2, ... , again in order of increasing energy. This scheme applies to all molecules in which the ground state is a singlet (S_0).

Absorption (A), fluorescence (F), and phosphorescence (P) are collectively referred to as *radiative transitions*, since they involve the absorption or emission of photons. However, it is also possible to have transitions between electronic states that do involve the absorption or emission of photons; these are referred to as *non-radiative transitions*. Most important to our discussion of fluorescence are the two competing relaxation pathways of *internal conversion* (IC) and *intersystem crossing* (ISC), in which the molecule relaxes from a higher to a lower electronic energy state by releasing the excess electronic energy as heat. Internal conversion involves relaxation to a state of the same multiplicity, whereas intersystem crossing involves relaxation to a state of a different multiplicity; these are thus the non-radiative analogues of fluorescence and phosphorescence, respectively. Because phosphorescence and intersystem crossing require a change in state multiplicity, they are *spin forbidden*, and hence tend to have rate constants which are orders of magnitude smaller than those for fluorescence and internal conversion. Thus, for most fluorescence probes, internal conversion is the most significant competing pathway to fluorescence. However, the presence of heavy atom substituents, such as Br, can greatly increase the rate of intersystem crossing and phosphorescence, making them significant for such probes.

These various radiative and non-radiative pathways discussed are commonly represented on a Jablonski diagram, as illustrated in Fig. 3. This diagram shows the ground (S_0) and first excited singlet (S_1) and triplet (T_1) states with their first few lowest vibrational levels, as well as the transitions between them, with radiative transitions by tradition represented by straight arrows, and non-radiative transitions represented by wavy arrows. This diagram also illustrates a third non-radiative mechanism, *vibrational relaxation* (VR). This is the very

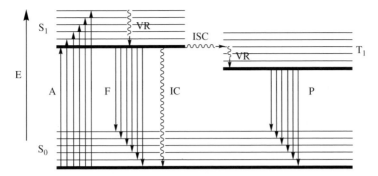

Fig. 3. A Jablonski diagram showing the major photophysical pathways for a typical fluorescent molecule.

rapid (fs to ps timescale) relaxation of vibrational excitation within an electronic level, and has the important consequence that all electronic transitions occur from the lowest vibrational level of an electronic state. One very important consequence of this is that fluorescence always occurs at lower energy (or at best at the same energy in the case of excitation from S_0 ($v = 0$) to S_1 ($v' = 0$)). Vibrational relaxation is also involved in both internal conversion and inter-system crossing, which are in fact isoenergetic processes between vibrational levels of different electronic states, followed by rapid vibrational relaxation to the lowest vibrational level of the lower excited state. This is in fact what is indicated on the diagram for ISC, but this two-step process is shown as a single step for IC, merely for convenience and to reduce congestion on the diagram.

All of these processes indicated on Fig. 3 are first-order processes, and the rate for each process is simply the product of a rate constant and the concentration of the initial electronic state. The mechanism for fluorescence emission can thus be given by the following set of equations:

$$S_0 \rightarrow S_1 \qquad \text{rate} = k_A[S_0] \qquad (3)$$

$$S_1 \rightarrow S_0 + h\nu_F \qquad \text{rate} = k_F[S_1] \qquad (4)$$

$$S_1 \rightarrow S_0 + \text{heat} \qquad \text{rate} = k_{IC}[S_1] \qquad (5)$$

$$S_1 \rightarrow T_1 + \text{heat} \qquad \text{rate} = k_{ISC}[S_1] \qquad (6)$$

The resulting fluorescence can be measured experimentally by two distinct experiments: steady-state and time-resolved spectroscopy; the former will be described in detail.

2.1. Steady-state techniques

Steady-state fluorescence spectroscopy refers to the measurement of the fluorescence intensity of a sample under the condition of constant illumination (excitation) of the sample. This results in a constant rate k_A of absorption, and hence a constant rate of formation of the first excited singlet state, S_1, as given by Eq. 3. This leads to the establishment of steady-state conditions, in which the rate of relaxation (decay) of the S_1 population is exactly the same as the rate of its formation. Thus, a constant S_1 population is established. Under these conditions, the rate of Eq. 4 ($= k_F[S_1]$), i.e. the rate of fluorescence emission, is constant. Since intensity (I_F) is defined as the rate of photon emission per unit time (usually expressed in counts per second, cps), the measured fluorescence intensity is therefore constant with time.

The primary measurement in steady-state spectroscopy is the fluorescence intensity as a function of emission wavelength, I_F versus λ_F, i.e. the fluorescence spectrum. This is done quite simply using a steady-state lamp, and scanning the constant emission using a monochrometer. In general, fluorescence spectra tend to be broad and featureless. This is true of all of the fluorescent guests described in this chapter, with the exception of pyrene, which shows vibronic resolution (a result of its highly symmetric structure). There are three important parameters which are obtained from the spectrum, and which are used to characterize the observed fluorescence. The first is the wavelength at which maximum emission occurs, $\lambda_{F,max}$. For many polarity-sensitive fluorophores, this can provide a direct indication of the polarity experienced by the fluorophore. The second is the fluorescence intensity, which depends on many factors, including the fluorophore concentration, the excitation wavelength, the lamp intensity, and the monochrometer slit widths. Thus, the absolute value of the measured intensity has little significance, and can vary from experiment to experiment and between different instruments. However, the *relative* intensity is very important, e.g. the intensity in the presence relative to the absence of added CD. The third parameter obtainable from the spectrum is the *fluorescence quantum yield*, ϕ_F. This is defined as the fraction of excited molecules, which decay from S_1 to S_0 via fluorescence. Considering Eq. 3–6, this can be calculated by the following equation:

$$\phi_F = k_F/(k_F + k_{IC} + k_{ISC}) \tag{7}$$

This is a measure of the efficiency of fluorescence. The quantum yield is determined by integrating the area under the sample fluorescence spectrum (F_S) expressed as a function of frequency v (or wavenumber): $F = \int I_F(v)dv$. This area is then compared to that of a reference fluorescent probe (F_R) measured under the identical experimental conditions (excitation wavelength, monochrometer slit widths, etc.). The quantum yield of the sample compound, $\phi_{F,S}$, can be obtained from the known value of the quantum yield for the reference, $\phi_{F,S}$, using Eq. 8 [44]:

$$\phi_{F,S} = \phi_{F,R} \times (F_S/F_R) \times (A_R/A_S) \times (n_S/n_R)^2 \tag{8}$$

where A is the absorbance at the excitation wavelength and n is the solution refractive index. All three of these parameters $\lambda_{F,max}$, I_F, and ϕ_F can be significantly affected by inclusion of the probe into a CD cavity, as will be described in Section 3.

Fluorescence spectrometers are a very well established class of commercial scientific instrument, and need not be further described here, as they are well described elsewhere. For example, a schematic diagram of the basic components of a steady-state fluorimeter is given in Ref. [45], and a review of fluorescence instrumentation is also given in Ref. [46].

Although fluorescence spectroscopy is widely used, it is useful in this chapter to discuss some important practical considerations which are not often discussed in the literature (an important exception is Ref. [47]), but which need to be taken into account when the technique is applied to the study of cyclodextrin inclusion complexes (and any other supramolecular system). Although fluorescence is always collected at right angles to the incident excitation light path, Rayleigh scattering (i.e. at the excitation wavelength) still occurs; this can be significant for weakly fluorescent samples. This means that the fluorescence spectrum scan must be started at a wavelength somewhat higher than the excitation wavelength. In our lab, we start 10 nm above the excitation wavelength as a rule of thumb. However, this may require care in the selection of the excitation wavelength: if it is too high, then the blue end of the fluorescence spectrum may be missed. In general, we choose an excitation wavelength low enough to ensure that there is some baseline measurement before the fluorescence spectrum is measured (i.e. the intensity begins to rise). However, if it is too low, then the CD host may begin to absorb (this generally happens below 300 nm for most CD); this must be avoided, as this CD concentration will be changed during the experiment. Once the proper excitation wavelength has been selected, the guest solution concentration is chosen to give an absorbance at the excitation wavelength in the range of 0.20–0.40 (any lower and the measured fluorescence will be weak; any higher and non-linear effects and complete absorption at the front of the fluorescence cuvette may occur).

Raman scattering is another possible complication which needs to be considered, as it can occur at wavelengths throughout the fluorescence spectrum of interest, depending on the solvent and excitation wavelength chosen. Since this is a relatively weak phenomenon, it tends to be solvent Raman bands which are observed. The observation of narrow, symmetric emission bands in the fluorescence spectrum should be checked by running a solvent blank — if the bands are still present, then they are indeed solvent Raman bands. In such cases, it may be possible to move the bands out of the wavelength region of interest by changing the excitation wavelength, otherwise the solvent blank spectrum will need to be subtracted from the fluorescence spectrum.

An additional decay path also needs to be considered, namely bimolecular quenching of the S_1 state by quencher molecules. Unless quenching experiments are being performed, the only potential quencher molecule which should be present in carefully prepared guest/CD solutions is molecular oxygen, O_2. In aqueous solutions equilibrated with the atmosphere, O_2 will be present at a concentration of 2.4×10^{-4} M. It is a highly efficient quencher, with a diffusion-controlled quenching rate constant. Thus, except for fluorophores that have relatively short fluorescent lifetimes, the presence of O_2 can significantly reduce the measured intensity. It can be removed from solutions in two ways. The first is the freeze-pump-thaw technique, which requires a fluorescence cell with a stop-cock that can be attached to a vacuum system. The solution in the cell is frozen in liquid nitrogen, the cell is pumped out, then the stopcock is closed and the solution allowed to thaw. This is repeated three or four times, and results in a solution completely free of dissolved gases. The second approach, purging, is much more simple, and involves the bubbling of an inert gas (usually N_2 or Ar) through the solution using syringe needles (one for the purge gas and one as a bleed) through a septum over the fluorescence cell stem. As a result of this possible effect of O_2, it is recommended in all cases that the fluorescence of the guest in the absence of CD in purged and unpurged solutions be measured. If the effect is negligible (we use less than 5% change in intensity as a rule of thumb), then this purging step can be omitted, making the experiment easier to perform. If O_2 has a significant effect, however, then all solutions need to be purged. This does present an additional complication, as CD solutions display surfactant behavior, and generate a high amount of foaming when bubbled. Great care must therefore be taken when these CD solutions are purged, to make sure that no solution is lost.

Application of steady-state fluorescence to the study of CD inclusion involves a basic experimental approach referred to as a *fluorescence titration*. This involves measuring the fluorescence spectrum of the guest (at fixed concentration) as a function of CD concentration. This is done by using a stock solution of the guest at the appropriate concentration (as determined above), weighing out the appropriate amount of CD in a vial or volumetric flask, then adding the appro-priate (fixed) volume of the guest stock solution to dissolve the CD. The fluorescence spectrum of each of these solutions is then measured using *identical* experimental conditions (excitation wavelength, monochrometer slit widths, etc.).

2.2. Extraction of binding constants from steady-state fluorescence data

The binding constant K can be extracted from steady-state fluorescence data in the following way. The results of the fluorescence titration experiment described above are quantified by determining F, the integrated area under the measured fluorescence spectrum (also referred to as the *total* fluorescence) of the guest in the presence of CD, and dividing by F_0, the integrated area of the guest *in the absence* of CD. This ratio gives a direct measurement of the effect of the CD on the guest fluorescence intensity (the mechanisms for this will be discussed in the next section). As will be seen, in most cases, the CD *increases* the guest fluorescence, so the ratio F/F_0 is referred to as the *fluorescence enhancement*. For a simple 1:1 host:guest complex (K defined in Eq. 1), F/F_0 will vary with the added CD concentration ($[CD]_0$) according to [48,49]

$$F/F_0 = 1 + (F_{max}/F_0 - 1)\frac{[CD]_0 K}{1 + [CD]_0 K} \qquad (9)$$

where F_{max}/F_0 is the maximum enhancement, i.e. when 100% of the guest is included in the CD. A plot of F/F_0 versus $[CD]_0$ can be fit to Eq. 9 using non-linear least squares fitting programs (we have written one for use in our lab, but many commercial software plotting and data analysis packages can be used to fit this equation). The quality of the fit can be determined by the χ^2 values, as well by plotting Eq. 9 with the two recovered parameters (F_∞/F_0 and K) and comparing to the experimental data (this can be made quantitative using a residual plot). Since Eq. 9 is derived based on a 1:1 complex, it will not fit the data if higher-order complexes form. A very easy way to test this is to plot a double-reciprocal plot of $1/(F/F_0-1)$ versus $1/[CD]_0$; a non-linear plot indicates higher-order complexes, and invalidates the use of Eq. 9. Note that this is basically the Benesi–Hildebrand method [50], and is sometimes used to extract K using linear regression. However, this approach is not recommended, as it emphasizes the data at low CD concentration, which in fact is the data with the highest relative errors. Hoenigman and Evans [51] have suggested the use of an iterative approach to correct the errors inherent in the Benesi–Hildebrand method, and demonstrate that this makes this a highly accurate approach.

In the case of higher-order complexes, other equations must be used. In the case of 2:1 complexation, the following equation can be obtained based on stepwise binding of a guest by two hosts [52]:

$$F/F_0 = \frac{1 + F_A K_1[CD]_0 + F_B K_1 K_2[CD]_0^2}{1 + K_1[CD]_0 + K_1 K_2[CD]_0^2} \qquad (10)$$

where F_A and F_B are the relative fluorescence of the complexed versus free guest in the 1:1 and 2:1 complexes, respectively, and K_1 and K_2 are the equilibrium constants for sequential binding by the first followed by the second CD, respectively. In addition, Hamai *et al.* have reported equations for treating 2:2 complexation (which will not be used in this chapter) [53].

It is possible to determine the stoichiometry of a CD complex using a Job plot, in which the concentration ratio of host:guest is varied from 1 to 0; a maximum in the measured signal will occur at the mole fraction corresponding to the complex stoichiometry. For example, a 1:1 complex will show a maximum at 0.50. Hirose has published a comprehensive review article entitled "A Practical Guide for the Determination of Binding Constants" [54], which includes a good description of the use of a Job plot. However, while this approach works well for NMR data, it is not practical for fluorescence data, due to the very low concentration of fluorescent guests which are used in fluorescence titrations. If a Job plot were attempted, the absorption of the guest at high concentration would be too large to allow for proper measurement of the fluorescence spectrum. Thus, in practical terms, if a non-linear double reciprocal plot is obtained, then the specific higher-order stoichiometry in fluorescence experiments is determined by applying the fit equations arising from the various complex models, to obtain the best fit to the data. However, Loukas [55] has published an alternative approach, using continuous variation methods to determine the complex stoichiometry.

Some researchers use I_F, the intensity at a specific wavelength (often the spectral peak), or ΔI_F, the change in intensity at a given wavelength, as the dependent variable in fluorescence titration experiments. These intensity or change in intensity values are then plotted as a function of [CD], and the data fit to extract K in a similar manner to that described above. However, the use of the fluorescence enhancement F/F_0 has two major advantages over the use of intensities. First, the use of the ratio of the total fluorescence (integrated spectrum) in the presence and absence of CD means that F/F_0 is a *direct* measure of the CD-induced change in ϕ_F. This can be seen from Eq. 8, making the assumptions that CD inclusion does not significantly change the guest absorbance (which is easy to check) and that the CD solution has the same refractive index as pure water (which should be valid at typical CD concentrations). Thus, unlike I_F or ΔI_F, F/F_0 has a direct physical significance. Second, because F/F_0 is a ratio, it does not depend on experimental conditions such as guest concentration, or instrument parameters such as slit widths or instrument sensitivity. Thus, again unlike I_F or ΔI_F, the value of F/F_0 for a given [CD] measured on different instruments and even in different laboratories should be the same. This allows for comparisons of fluorescence enhancements for different CD, or even for different guests as a measure of the relative effect of inclusion.

3. General effects of cyclodextrin inclusion on guest fluorescence

Inclusion of a polarity-sensitive fluorescent probe molecule into the cavity of a cyclodextrin in aqueous solution can have significant effects on its florescence spectrum, including its wavelength maximum (fluorescent color), intensity, and quantum yield. Most commonly, CD inclusion results in enhanced, blue-shifted emission, but this is not always the case. In fact, the specific effect of CD inclusion, and the factors which are responsible for these effects, vary significantly depending on the specific guest.

In general, there are five factors which can be responsible for these observed changes:

1. The local polarity experienced by the probe inside the cavity is much lower than that of aqueous solution.
2. Inclusion within the cavity results in restricted intramolecular rotational freedom in the guest.
3. Specific guest–water interactions are prevented upon inclusion.
4. The included guest is protected from quenchers, including the ubiquitous O_2.
5. Inclusion occurs as dimers, thus resulting in excimer emission or self-quenching.

Each of these factors can be significant for specific CD complexes, and will be discussed in the following. By far the largest and the most general effect is that of polarity, which will be described in detail. The effect of polarity on probe fluorescence is well known, and is a major factor in *solvatochromism*, the shifting of the fluorescence spectrum of a fluorophore to different wavelength maxima in different solvents [56]. (The other possible factor in solvatochromism is specific solvent–solute interactions.) As mentioned, most polarity-sensitive fluorescent probes show an increase in intensity and a blue-shift in their spectrum upon CD inclusion; this is because the S_1 electronic state of most fluorophores is more polar than the ground S_0 state. When an aqueous fluorescent guest becomes included within a CD cavity, it experiences a significantly less polar microenvironment. As mentioned in Section 1.2, the polarity within the interior of the β-CD cavity has been widely reported to be similar to that of alkyl alcohols, such as ethanol. This greatly reduced polarity of the medium results in a significant destabilization of the relatively polar S_1 ground state, but a much smaller destabilization of the less polar S_0 ground state (or in the case of a non-polar guest such as a polynuclear aromatic hydrocarbon, stabilization might in fact occur). This is illustrated in Fig. 4.

The effect of this greater inclusion-induced destabilization of S_1 relative to S_0 is a significantly larger S_1–S_0 energy gap, ΔE_{10}, for the included probe, as can be seen from Fig. 4. This has two significant results. First, since the energy of the fluorescence photon is determined by the energy gap ΔE_{10} ($\Delta E_{10} = h\nu_F = hc/\lambda_F$), the fluorescence of such a probe is blue-shifted, i.e. the fluorescence maximum shifts to higher frequency (or lower wavelength). (This shift in $\lambda_{F,max}$ was in fact exploited to determine the polarity inside the β-CD cavity in the studies described in Section 1.2; this will also be discussed in Section 4.1). Second, the increased energy gap results in a significantly reduced rate of internal conversion, i.e. a significantly smaller k_{IC}, according to the energy gap law [57]. Since k_{IC} is smaller for the included guest, a larger proportion of guests decay by

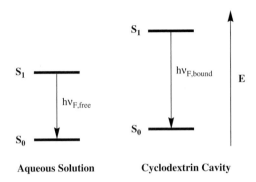

Fig. 4. The relative S_1–S_0 energy gap of a typical polarity-sensitive fluorescent guest in aqueous solution and inside a cyclodextrin cavity.

fluorescence, i.e. the fluorescence quantum yield ϕ_F increases, as can be seen from Eq. 7. This results in an increase in guest fluorescence intensity, or fluorescence enhancement. It is this second effect which is exploited to extract the equilibrium constant K, as described in Section 2.2.

Obviously, the degree to which the fluorescence of a probe is enhanced by CD inclusion will vary from probe to probe, since the polarity sensitivity is different. In order to quantify the degree to which the fluorescence of a given molecule is sensitive to the local polarity, we developed the PSF, or *polarity sensitivity factor* [58,59]. This PSF is simply the ratio of the integrated fluorescence spectrum (F, or total fluorescence) of a probe in ethanol compared to that in water, corrected for the difference in absorbance of the two solutions:

$$PSF = (F_{EtOH}/F_{H_2O}) \times (A_{H_2O}/A_{EtOH}) \tag{11}$$

The PSF is basically the fluorescence enhancement of the probe in ethanol relative to water, and is related (but not equivalent) to the relative quantum yield in ethanol versus water. The greater the sensitivity of a given probe, the larger will be its PSF. Also, by expressing the maximum enhancement observed upon CD inclusion to the PSF of the probe, an indication of the degree of inclusion can be obtained, especially in a comparison between different probes in the same CD. The PSF of specific probes studied in our lab will be given throughout Section 4. Furthermore, an example of the comparison of fluorescence enhancement relative to PSF for two probes will be discussed in Section 4.1.

In contrast to the discussion above, decreased polarity can also result in decreased fluorescence. In the case of 7-methoxycoumarin, ISC from S_1 to T_1 competes efficiently with fluorescence in a non-polar medium, but this ISC efficiency is reduced in a polar medium. This results in a greatly reduced quantum yield and intensity in a less polar environment (cf. Section 4.4 for details).

Thus, the same polarity effect on S_1 (i.e. the lowering of its energy) results in a very different effect on the observed fluorescence.

The effect of restricted rotational motion of the guest inside the cavity is particularly relevant for guests which exhibit twisted intramolecular rotation, resulting in a twisted intramolecular charge transfer (TICT) state. This phenomenon can happen when there are electronic donor and acceptor moieties within the same molecule. In specific relative orientations, charge transfer between donor and acceptor moieties will occur in the excited state; this provides a competing decay pathway for the S_1 state, reducing the fluorescence quantum yield. However, this charge transfer can only occur with the proper orientation of the molecular orbitals of these two parts of the molecule. Since the lowest energy conformation of the ground state is usually different from the optimum orientation for charge transfer in the excited state, this requires an intramolecular rotation, or twist. Inside a CD cavity, this intramolecular rotation may be severely hindered, resulting in a significant decrease in the TICT rate, and hence enhanced fluorescence relative to the guest in free solution. This has been reported by Kim *et al.* [60], for example, who observed a significant and different effect of α-CD and β-CD cavities on the TICT formation in *p*-(*N*,*N*-dimethylamino)benzoic acid. Similarly, Turro *et al.* [41] proposed an effect of hindered rotation by CD inclusion on the TICT efficiency in dialkylaminobenzonitrile guests.

There also is a large role of specific water-solute interactions for guests which can undergo TICT. In fact, the report by Kim *et al.* mentioned above [60] attributes much of the effect of CD inclusion on TICT formation in *p*-(*N*,*N*-dimethylamino)benzoic acid to specific hydrogen bonding between the carbonyl group of this molecule and water; this interaction is prevented upon inclusion into β-CD. Samanta *et al.* also report the effect of CD inclusion in a TICT fluorophore as a result of the prevention of specific guest–water interactions [61]. In this case, the fluorescence *arises* from a CT state, and elimination of water from the probe surroundings leads to enhanced CT fluorescence.

The effect of CD inclusion by prevention of quenching is fairly straightforward. Most quenchers involved a contact deactivation mechanism, and inclusion in a CD cavity effectively prevents this. Even in the case of Förster (long-range) energy transfer, the efficiency of quenching is highly dependent on the distance between the fluorophore and donor, so again inclusion into a CD cavity increases the minimum distance. For example, the protection of pyrene from quenchers by CD inclusion will be discussed in Section 4.2.

Preferential inclusion of some guests as dimers will have a large effect on the observed fluorescence properties, in cases where the free guests exist as monomers. A striking example of the effect of dimer inclusion will be discussed in Section 4.3 for the case of Nile Red. This is also observed in studies of the complexes of pyrene in γ-CD, which has a large enough cavity to accommodate two guest molecules; this will be discussed in Section 4.2.

Eaton [62] reports a variety of effects of CD inclusion on the photophysical properties of a number of guests, including the effects of restricted intramolecular rotation, the inclusion of dimers, and the prevention of quenching by oxygen.

4. Steady-state fluorescence studies of the inclusion of specific polarity-sensitive guests in cyclodextrins

As discussed in the introduction, steady-state fluorescence is an excellent method for studying inclusion of polarity-sensitive guests into cyclodextrins, due to the high sensitivity of fluorescence and the large changes in guest fluorescence intensity that can be observed in many cases. In this section, the use of steady-state fluorescence in the study of some specific fluorescent guests which show high polarity sensitivity will be discussed. The specific guests chosen are mainly reflection of studies done in our laboratory, with the exception of pyrene, and have been chosen to illustrate the five mechanisms whereby inclusion into CD can affect guest fluorescence, listed and described in the previous section.

4.1. Anilinonaphthalene sulfonates

1-Anilino-8-naphthalene sulfonate (1,8-ANS) **1** is one of the most widely used polarity-sensitive fluorescent probes, and is in fact the first guest to be studied in a CD inclusion complex using fluorescence spectroscopy. Cramer *et al.* reported the significant increase in fluorescence intensity of 1,8-ANS in the presence of β-CD way back in 1967 [63]. Since then, other groups have also reported on the large fluorescence enhancement of 1,8-ANS by CD using fluorescence [51,64–66], including a report using a modified CD [67]. More recently, work has also been done with 2,6-ANS **2**, an isomer of 1,8-ANS which also exhibits significant polarity sensitivity, but with a very different shape [68].

1   2

Our work [58,69,70] showed the much higher fluorescence enhancement and binding constants obtained in modified relative to parent CD. In the case of 1,8-ANS [69], we found that by using HP-β-CD, the fluorescence of 1,8-ANS could be enhanced by as much as a factor of 180, compared to an enhancement of a factor of 8.4 in the case of β-CD itself. This is one of the largest enhancements ever reported. It is so large in fact that it is clearly visible to the naked eye, and we developed a fluorescence demonstration of host–guest inclusion phenomena based on this host–guest pair [70]. The relative fluorescence spectrum of 1,8-ANS in water and in various 10 mM aqueous CD solutions is shown in Fig. 5.

Fluorescence titrations were performed to obtain the binding constants. The experimental results for 1,8-ANS in Me-β-CD fit to Eq. 9 for 1:1 complexation is shown in Fig. 6. (The double-reciprocal plots were found to be linear, confirming simple 1:1 complexes.) Once again the results showed the much greater host abilities of modified CD; for example, Me-β-CD was found to bind 1,8-ANS with a binding constant $K = 370 \, \text{M}^{-1}$, compared to $K = 85 \, \text{M}^{-1}$ [65] for β-CD. We also observed a significant blue-shift in the 1,8-ANS emission spectrum upon inclusion into the modified CD, from 520 nm as the free probe in aqueous solution to 465 nm in 10 mM HP-β-CD solution, compared to 492 nm in β-CD. Thus, 1,8-ANS experiences a much lower cavity polarity in HP-β-CD than in β-CD. The position of the 1,8-ANS spectral maximum (in v) in various solvents was found to correlate well to the dielectric constant of the solvent; the spectral maximum obtained in HP-β-CD and β-CD corresponded to local

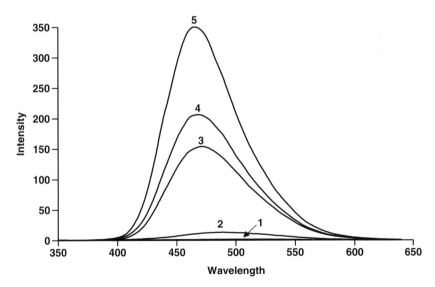

Fig. 5. Relative fluorescence spectra of 1,8-ANS in various cyclodextrin solutions (10 mM in phosphate buffer): (1) no CD; (2) β-CD; (3) HP-β-1; (4) Me-β; and (5) HP-β-2. Reprinted from Ref. [69].

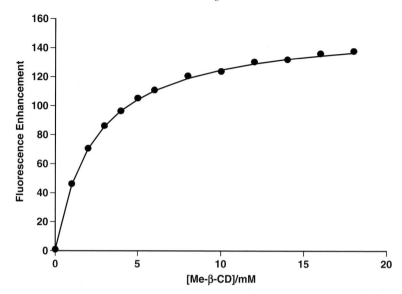

Fig. 6. Observed fluorescence enhancement of 1,8-ANS as a function of Me-β concentration. Solid line: best fit to Eq. 9: $K = 370\,\mathrm{M}^{-1}$. Reprinted from Ref. [69].

polarities with dielectric constants of 25 and 54, respectively. Both the higher enhancement and the better binding of the modified compared to parent CD were proposed to be a result of the extension of the size of the non-polar CD cavity by the presence of the alkyl modifying groups.

In a subsequent paper, we compared the binding of 1,8- and 2,6-ANS in these parent and modified CD [58]. We found that the more streamlined shape of 2,6-ANS results in a better fit to the β-CD cavities, and hence found much greater binding constants for 2,6-ANS. For example, in HP-β-CD, the binding constant K was found to be $7200\,\mathrm{M}^{-1}$ for 2,6-ANS, compared to $480\,\mathrm{M}^{-1}$ for 1,8-ANS. In all cases, the parent or modified β-CD cavity gave the best results, compared to the smaller α-CD or larger γ-CD cavities. This illustrates the importance of the fit match between the size and shape of the guest and the size and shape of the host cavity. We propose that this difference in host–guest matching can be explained by the differences in the pattern of substitution on these two isomeric probes. In 1,8-ANS, the two substituents are both on the same long side of the molecule. As shown in Fig. 7a, this prevents inclusion along the long axis of the naphthalene, and necessitates inclusion along the short naphthalene axis, i.e. inclusion of the opposite long side of naphthalene into the cavity, or *equatorial* inclusion. However, in the case of 2,6-ANS, having the substituents on the two ends of the molecule allows for inclusion along the long naphthalene axis, i.e. *axial* inclusion, as shown in Fig. 7b. As can be seen in Fig. 7, axial inclusion is favoured, as it will allow for deeper penetration and maximize host–guest interactions. Once again, the binding by modified CD was much greater than for

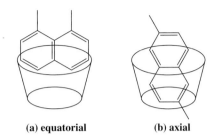

(a) equatorial   (b) axial

Fig. 7. A comparison of (a) equatorial and (b) axial binding modes for 1,8- and 2,6- substituted naphthalene derivatives, respectively.

the parent; a value of K of $1350\,\text{M}^{-1}$ was found for 2,6-ANS in β-CD, compared to the value reported above of $7200\,\text{M}^{-1}$ for 2,6-ANS in HP-β-CD.

Interestingly, although 2,6-ANS was found to bind more strongly than 1,8-ANS in all of the CD studied, 1,8-ANS actually exhibited a larger fluorescence enhancement. For example, in HP-β-CD, maximum fluorescence enhancements of 125 and 81 were obtained for 1,8- and 2,6-ANS, respectively. This illustrates the different dependencies of the magnitude of the fluorescence enhancement and that of the binding constant. The binding constant K depends on how well the guest fits into the host, as well as the polarity inside the cavity. However, the enhancement depends *only* on the polarity, and the degree of sensitivity is different for different guests. Thus, if a guest has a low polarity sensitivity, it might be bound very tight, but have only a small fluorescence enhancement. As defined in Section 3, we developed the *polarity sensitivity factor* (PSF) to take into account the effect of probe sensitivity. We measured the PSF of 1,8- and 2,6-ANS to be 197 and 120, respectively. Thus, 1,8-ANS is significantly more polarity sensitive than 2,6-ANS, as expected from the fluorescence enhancement results. When the maximum fluorescence enhancement of these two guests in HP-β-CD were scaled using the PSF values, enhancements of 64 and 68 were obtained for 1,8- and 2,6-ANS, respectively (out of a maximum of 100 being the emission observed in ethanol). Thus, the two probes experience a similar polarity cavity.

We proposed the use of these two structural isomers, with enhancement results scaled using these PSFs, for the diagnostic study of inclusion in various hosts, such as other modified CD. The significant difference in shape, and the large polarity sensitivities, make this pair ideal for probing the cavity properties of new hosts.

There have also been a number of studies of the CD inclusion of 6-*p*-toluidinylnaphthalene-2-sulfonate [71]. This is the derivative of 2,6-ANS with a *p*-methyl group on the anilino phenyl moiety. This probe is even more polarity sensitive than either 1,8- or 2,6-TNS, with a PSF measured in our lab of over 200, and shows high fluorescence enhancement and higher-order complexes.

4.2. Pyrene

Although not studied in my work on cyclodextrins, pyrene **3** has been included in this chapter because it has a very different polarity-sensitive response than the anilinonaphthalene sulfonates described above. Instead of showing an overall increase in intensity in less polar media, pyrene exhibits a change in the *relative* intensity of two resolved vibronic bands, namely the first and third [72]. Thus, the value of I_{III}/I_I changes according to the local polarity. For example, $I_{III}/I_I = 1.68$ in cyclohexane compared to 0.63 in water [72].

3

This vibronic band ratio has been shown to change significantly upon inclusion into CD, and a number of groups have performed fluorescence tractions based on this change in fluorescence spectrum shape, fitting plots of I_{III}/I_I versus [CD] to extract the binding constants. Hamai [73,74] was the first to do this, and found ternary complex formation in the presence of added alcohol [73], and 1:1, 1:2, and 2:2 complexes (the latter exhibiting excimer fluorescence, see below) in the absence of added alcohol [74]. Warner *et al.* also used this I_{III}/I_I method extensively [75,76], and further studied the range of stoichiometries obtained in β-CD and γ-CD [75], as well as ternary inclusion complexes with alcohols [76]. Eddaoudi *et al.* used the I_{III}/I_I ratios of pyrene in per-6-O-*tert*-butyldemethylsilyl α-, β-, and γ-CD in a different way, namely to determine the cavity polarity in this interesting set of modified CD [77]. They showed that the modified γ-CD had a more polar cavity than either the modified α- or β-CD.

Two other effects of CD inclusion on guest fluorescence have been exploited for the study of pyrene CD complexes. A number of researchers have studied pyrene complexation into CD by measuring the effect of added CD on bimolecular quenching [78–80]. In some cases, CD were found to increase the quenching efficiency, due to the formation of ternary complexes which served to bring the quencher and pyrene together [78]. In other cases, however, inclusion of pyrene into CD dramatically decreased the quenching efficiency [80], due to protection of the pyrene from contact by the quenchers, as discussed in Section 3. This difference was shown by Kano *et al.* to depend on the size of the quencher [79]; in a series of experiments with aliphatic and aromatic amine quenchers, they found that the CD accelerated the quenching of pyrene in the case of small quenchers (which could form ternary complexes), but decreased the quenching by larger amine quenchers.

Alternatively, other researchers have used the tendency of pyrene to form excimers to study its CD inclusion complexes [74,81]. In fact, pyrene in γ-CD was one of the first systems in which 2:2 complexation was reported to occur; this results in a very different broad, red-shifted excimer fluorescence, and the ratio of excimer to monomer fluorescence as a function of γ-CD concentration has been used to determine the binding constants.

4.3. Nile Red

Nile Red **4** is another example, like ANS, of a highly polarity-sensitive fluorescent probe. Unlike ANS, however, Nile Red is a neutral molecule, and has a very low aqueous solubility. The major source of the high polarity sensitivity of this probe is the formation of a TICT state, which depends on the polarity of the medium, and is accelerated in a high-polarity medium (which supports the charge transfer) [82]. Since this process competes with fluorescence, polar media strongly reduce Nile Red fluorescence.

H₃CH₂C—N—CH₂CH₃

4

Srivatsavoy published a study on the inclusion of Nile Red in β- and γ-CD, and found that its fluorescence is strongly reduced in γ-CD, but not in β-CD, and that multiple types of complexes were formed in both cases [83]. He rationalized these observations in terms of the effects of inclusion in these two different sized cavities on the TICT state, and concluded that β-CD has no effect on the formation of this state. However, a mechanism for *enhancement* of TICT formation by γ-CD, and the complex stoichiometries, were not reported. Our group decided to further investigate the effect of β- and γ-CD on Nile Red, as well as HP-β-CD and HP-γ-CD [84], the latter due to the much improved host properties we found for these modified CD in the case of ANS probes. We also found that γ-CD strongly suppressed Nile Red fluorescence, but β-CD enhanced it; furthermore, a significant blue-shift was observed in the absorption spectrum in both cases, indicating that Nile Red is experiencing a less polar cavity in both CD. Furthermore, it was found that if the Nile Red concentration was reduced by addition of more γ-CD solution, or if the solution was prepared using a saturated Nile Red solution (rather than using γ-CD to help increase the Nile Red solubility, as was done in the first case), then γ-CD showed stronger enhancement than β-CD! We concluded that the fluorescence suppression of Nile Red by γ-CD was not a result of its effect on formation of the TICT (which in fact would enhance the fluorescence due to the lower cavity polarity), but rather a result of the inclusion of Nile Red as dimers (which is not possible in the much smaller β-CD cavity), giving 1:2 host:guest complexes. Self-quenching of the dimer explains the loss of fluorescence. At higher host:guest ratios, we in fact observed 2:1 γ-CD:Nile Red complexes, which showed very high enhancement. Similar results were found with HP-β-CD and HP-γ-CD, and the enhancements were significantly larger for the modified as compared to the parent CD. In contrast to the case of ANS, γ-CD was found to be a better size match for this larger guest than β-CD. We also did electrospray mass spectroscopy and molecular modeling studies of these γ-CD:Nile Red complexes, both of which suggest that Nile Red is not completely inserted in the cavities, but that the complexation involves more of a capping process, with only partial insertion into the cavity.

Sarkar *et al.* have published a subsequent study of Nile Red in CD [85], in which they determined quantitatively the effect of β- and γ-CD on the TICT process. They found that TICT formation is suppressed by a factor of 2.5 in β-CD, but shows only a "little retardation" in γ-CD, consistent with our work. They explain this in terms of different types of complexation with these two CD, and supported our proposal of capped complexes with γ-CD.

4.4. Coumarins

Coumarins are another family of polarity-sensitive fluorescent probes, which have been widely studied in CD, typically showing significant fluorescence enhancement upon CD inclusion [86]. In this section, the effect of CD inclusion on one particular member of this family, 7-methoxycoumarin (7-MC, **5**), will be discussed, because it exhibits the *reverse* polarity effect of other probes such as ANS, Nile Red, and other coumarins: it becomes *less* fluorescent in lower polarity media [87]. This has been explained in the following way [87]: the fluorescence of 7-alkoxycoumarins originates from the S_1 $\pi\pi^*$ state, but this is in strong competition to an intersystem crossing pathway from this $\pi\pi^*$ state to the closely lying T_1 $n\pi^*$ state. In polar solvents, the S_1 energy level is lowered relative to the T_1 state, reducing the ISC efficiency, and thus resulting in a larger fluorescence quantum yield.

H₃CO

5

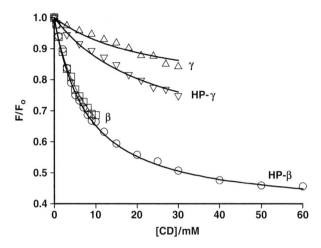

Fig. 8. The effect of cyclodextrin concentration on the relative fluorescence (F/F_0) of 7-methoxycoumarin for various cyclodextrins. Experimental data: $\square$ β-CD, $\bigcirc$ HP-β-CD, $\triangle$ γ-CD, ∇ HP-γ-CD; — fit to Eq. 9. Reprinted from Ref. [88] (with permission from Springer Science and Business Media).

We studied the effect of the inclusion of 7-MC into α-, β-, and γ-CD, as well as their HP-modified counterparts [88]. We did indeed observe extremely strong reduction in the 7-MC fluorescence intensity in β-, γ-, HP-β-, and HP-γ-CD. No effect was observed with α-CD or HP-α-CD; these cavity sizes are obviously too small to accommodate the rather large 7-MC molecule. Figure 8 shows the fluorescence titration results for these four CD; F/F_0 values less than 1 are observed in all cases. However, we showed that Eq. 9 derived for the case of fluorescence enhancement by 1:1 host–guest complexation works equally well for this case of reduced fluorescence. Linear double-reciprocal plots were obtained, confirming simple 1:1 complexation, as were excellent fits to Eq. 9, as can be seen in Fig. 8.

Both the decrease in fluorescence and the binding constants K were larger for the β-CD relative to the γ-CD cavities, indicating the better size match with the former cavity size. However, in contrast to all of our other experimental results for modified CD, there was essentially no difference in the results *either* in terms of fluorescence reduction or binding constant when comparing the parent to the modified CD. The binding constant K (in units of M^{-1}) was found to be 128 ± 32 and 120 ± 20 in β-CD and HP-β-CD, respectively, and 41 ± 8 and 42 ± 6 in γ-CD and HP-γ-CD, respectively. We proposed this to be the result of the 7-MC guest being completely included within the CD cavities, thus the presence of side groups around the CD rims would have minimal effect on the environment experienced by the fluorophore. This is in contrast to the ANS probes, where a substantial portion of the guest would stick outside the cavity, making the presence of side chains around the rim important.

Although reduced guest fluorescence had been reported in previous studies, it was always described and explained in terms of bimolecular quenching by the CD hosts. To emphasize that the effect of CD on 7-MC is an environmental polarity effect, and does not involve energy transfer to the CD as quenchers, we proposed the term *fluorescence suppression* to describe this observed effect. This corresponds nicely to the term *fluorescence enhancement* commonly used to describe the opposite (and much more common) polarity effect.

4.5. Curcumin

Curcumin **6** is the main component of the Indian spice turmeric, and is thus found in large amounts in curry powders and sauces. This is a very interesting compound, as it has been shown to exhibit a wide array of beneficial health effects, including antibacterial, antioxidant, and anticancer properties [89]. It is sold in health food stores in pill form as a dietary supplement. It is also a potential photosensitizer for photodynamic therapy for oral lesions and cancers [90]. This fascinating molecule is of particular interest to my work for two reasons: it has a very low aqueous solubility, which might be enhanced by CD inclusion, and it exhibits highly polarity-sensitive fluorescence [91], which gives us an excellent way to study its CD inclusion.

There were two studies of the inclusion of curcumin into CD reported prior to our work, neither of which used fluorescence spectroscopy. Tang *et al.* [92] reported the formation of a 2:1 host:guest complex of curcumin with β-CD, studied using absorption spectroscopy. Tonnesen *et al.* [93] investigated the inclusion of curcumin into the three parent CD as well as a number of modified CD, using the effect of inclusion on curcumin hydrolysis rates, and reported solubility enhancements. However, they assumed 1:1 host:guest complexation, which is not an obvious stoichiometry based on the length of the molecule (which is much greater than the CD cavity depth) and the presence of identical phenyl moieties at each end of the molecule.

Our work exploited the reported polarity-sensitive fluorescence of curcumin to allow for accurate determinations of host–guest stoichiometries, binding constants, and solubility enhancements for its inclusion into the three parent CD as well as their HP-modified derivatives [94]. We were able to demonstrate

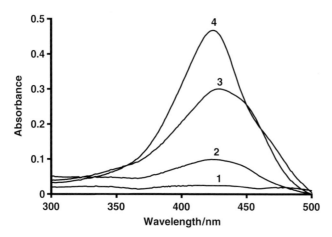

Fig. 9. Absorption spectra of saturated solutions of curcumin in various aqueous solutions: (1) no CD; (2) 10 mM β-CD; (3) 10 mM HP-β-CD; and (4) 10 mM HP-γ-CD. Reprinted from Ref. [94].

quite clearly the CD-induced enhancement of curcumin solubility, using UV-visible absorption spectroscopy. Figure 9 shows the absorption of saturated aqueous solutions of curcumin containing 10 mM concentrations of various CD; HP-γ-CD clearly gives the best solubility enhancement. These results were quantified in the case of HP-β-CD and HP-γ-CD by determining the extinction coefficients of curcumin in each CD using solutions of known curcumin concentration; this yielded solubilities of 5.2×10^{-5} and 1.4×10^{-5} M, respectively. These correspond to an increase in solubility relative to pure aqueous solution of factors of 1700 and 4700, respectively; this result is of great potential application for this pharmaceutically active compound.

We also investigated the polarity sensitivity of curcumin, and found it to have a PSF of 39. Thus, while not as polarity sensitive as 1,8- or 2,6-ANS, it is highly polarity sensitive, and should show significant effects upon CD inclusion. This was in fact the case, a maximum fluorescence enhancement of 7.1 at 30 mM HP-β-CD was observed. Fluorescence titration experiments were performed in all six CD. The titration results for the case of HP-β-CD are shown in Fig. 10.

The double-reciprocal plot was non-linear, indicating higher-order complexation. The data were found to fit extremely well to Eq. 10, for 2:1 host:guest complexation, as shown by the solid line in Fig. 10. The equilibrium constants K_1 and K_2 were determined to be 3400 and 120 M^{-1} for this case, for an overall binding constant (K_1K_2) of $4.1 \times 10^5\,M^{-2}$. Figure 11 shows our proposed structure for this 2:1 complex, in which a CD host encapsulates each end of the curcumin molecule. It is interesting that $K_1 >> K_2$; this indicates that despite the long length of the guest, encapsulation of one end by an HP-β-CD inhibits binding of the other end; this is most likely a steric crowding effect.

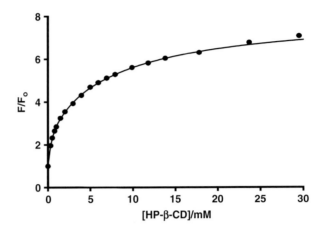

Fig. 10. The fluorescence enhancement, F/F_0, for 1.00×10^{-5} M curcumin in 1% methanol/water as a function of HP-β-CD concentration. The solid line shows the fit of the data to Eq. 10. Adapted from Ref. [94].

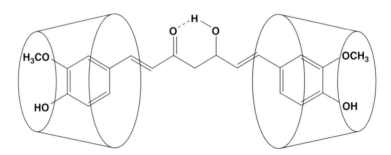

Fig. 11. The proposed structure of the 2:1 CD:curcumin complex. Note that this is a cartoon depiction only, and that the relative size of curcumin and β-CD cavity is indicated. Reprinted from Ref. [94].

Similar complexation occurred with α- and HP-α-CD. In fact, even larger K_1 values were obtained for α- and HP-α-CD relative to β- and HP-β-CD, indicating that the size of the α-CD cavity gives the best match to the size of the ends of the curcumin molecule. However, a very different result was observed with γ- and HP-γ-CD: in both cases, fluorescence suppression was observed! We proposed that this was a result of the curcumin forming 1:1 complexes with this much larger cavity, but as a folded molecule, so that the two phenyl ends of the molecule are included essentially as a dimer pair. This would require the curcumin to be in its diketone form, instead of the enol form depicted in Fig. 11. This would result in the breaking of the conjugation along the length of the molecule, which would indeed result in a significant reduction in the curcumin fluorescence quantum yield. At higher HP-γ-CD concentrations, a recovery of this fluorescence is observed, presumably as a result of the formation of 2:1 complexes involving the unfolded curcumin as CD concentration becomes very high.

These results illustrate the great range of fluorescent effects which can be observed with a single guest by varying the CD cavity size and modifications.

4.6. OPA-derived isoindoles

The final example of a guest studied by steady-state fluorescence is the OPA-derived isoindoles of amino acids. These highly fluorescent derivatives **7** are formed by the reaction of o-phthalaldehyde with an amino acid R-NH$_2$ in the presence of a thiol R'-SH [95]. This has the desirable effect of transforming an alkyl amino acid with no chromophore (and hence no spectroscopic label) to a highly fluorescent derivative. This has been widely used in various amino acid assays, including the separation and detection of amino acids in high-performance liquid chromatography [96]. A major problem with this method, however, is that the isoindoles produced are unstable, with lifetimes on the order of minutes to hours; this can create problems using this derivatization in trace analysis methods.

7

A number of groups have investigated the use of CD to stabilize and enhance the fluorescence of these isoindoles. For example, Baeyens *et al.* [97] investigated both β-CD and micelles, and found strong fluorescence enhancement by micelles, but only very modest ones (less than a factor of 2) with β-CD. Khokhar and Miller [98] reported 2–10 fold fluorescence enhancement of the isoindoles of lysine in β- and γ-CD, and serine in γ-CD, but no enhancement of the derivatives of any other amino acids. This obviously is useful for the application of this method in HPLC and other analysis techniques.

In light of the improved host properties we found for modified CD, we investigated the potential effects of HP-β-CD on a wide array of amino acid OPA-derived isoindoles, to see if significant fluorescence enhancements could be observed, and in addition, whether a stabilization of the isoindoles could be obtained (this was not observed in the above-mentioned previous studies) [99]. Our measurements showed modest fluorescence enhancements for some of the derivatives, including lysine and glycine, with values between 2 and 3. However,

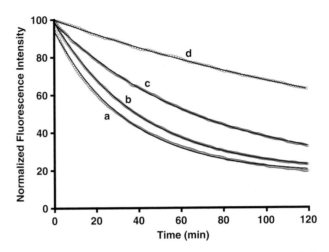

Fig. 12. The normalized decay of the fluorescence intensity at 450 nm of the lysine isoindole under steady-state illumination (open circles) and the first-order kinetic fits (solid lines). (a) No added CD ($\tau = 35.4$ min); (b) 5 mM β-CD ($\tau = 41.6$ min); (c) 5 mM HP-β-CD ($\tau = 71.1$ min); and (d) 20 mM HP-β-CD ($\tau = 208$ min). Reprinted from Ref. [99].

much more importantly, significant stabilization of the isoindoles of these two common amino acids were observed. These were determined by measuring the steady-state emission of the isoindole at a fixed wavelength as a function of real time in minutes (i.e. not using time-resolved fluorescence on the ns scale). The resulting time traces could be fit to first-order kinetics, to give the lifetime of the derivative. Figure 12 shows the results for the isoindole of lysine in the absence and presence of β- and HP-β-CD. As can be seen, HP-β-CD gave much greater stabilization than β-CD. At 20 mM HP-β-CD, the lifetime of this isoindole was increased from 35.4 min to 208 min, an increase by a factor of 5.9. Similar results were obtained for the isoindole of glycine, which showed an increase in lifetime from 16 to 74 min, giving a stabilization factor of 4.6. Simple 1:1 host:guest complexation was observed, but with rather low binding constants of 44 and 130 M^{-1} for glycine and lysine, respectively. This significant stabilization of these isoindole derivatives using HP-β-CD has the potential to be beneficial in the detection and measurement of amino acids in solution.

5. Concluding Remarks

Fluorescence spectroscopy is an excellent experimental technique for the study of cyclodextrin host–guest inclusion. This is especially true in the case of highly polarity sensitive probes, such as 1,8-ANS, 2,6-ANS, Nile Red, and curcumin as described in this chapter, in which very large effects on guest fluorescence can be observed upon cyclodextrin inclusion. Furthermore, very low concentration of guests are required, due to the high sensitivity of fluorescence. The various guests described in this chapter well demonstrate the range of effects and behavior that can be observed using fluorescence. Both ANS isomers, Nile Red and curcumin show large enhancements of their fluorescence upon inclusion, whereas 7-methoxycoumarin actually shows strongly suppressed fluorescence. Pyrene shows a change in the relative intensity of two of its vibronic bands. In the case of isoindoles, CD inclusion provided significant guest stabilization, while in the cases of Nile Red and curcumin, CD inclusion greatly increased the guest solubility. Furthermore, the change in fluorescence intensity can have practical applications, for example in trace analysis and fluorescent sensor technology, making the study of CD inclusion complexes by fluorescence spectroscopy even more relevant and compelling.

References

[1] A. Villiers, Compt. Rend. 112 (1891) 536.
[2] F. Schardinger, Z. Unters. Nahr. u. Genussm. 6 (1903) 865.
[3] J. Szejtli, Chem. Rev. 98 (1998) 1743.
[4] W. Saenger, Angew. Chem. Int. Ed. Engl. 19 (1980) 344.
[5] J. Szejtli, Cyclodextrins and Their Inclusion Complexes, Akadémiai Kiadó, Budapest, 1982.
[6] A.P. Croft, R.A. Bartsch, Tetrahedron 39 (1983) 1417.
[7] J.S. Pagington, Chemistry in Britain 23 (1987) 455.
[8] V. Ramamurthy, D.F. Eaton, Acc. Chem. Res. 21 (1988) 300.
[9] D. Duchêne (Ed.), New Trends in Cyclodextrins and Derivatives, Editions de Santé, Paris, 1991.
[10] S. Li, W.C. Purdy, Chem. Rev. 92 (1992) 1457.
[11] J.F. Stoddart, Angew. Chem. Int. Ed. Engl. 31 (1992) 846.
[12] G. Wenz, Angew. Chem. Int. Ed. Engl. 33 (1994) 803.
[13] P. Bortolus, S. Monti, Photochemistry in Cyclodextrin Cavities, in: Advances in Photochemistry, D.C. Neckers, D.H. Volman, G. von Bünau (Eds), Vol. 21, Wiley, London, 1996, p. 1.
[14] C.J. Easton, S.F. Lincoln, Chem. Soc. Rev. 25 (1996) 163.
[15] K.A. Connors, Chem. Rev. 97 (1997) 1325.
[16] L. Szente, J. Szejtli, Analyst 123 (1998) 735.
[17] C.J. Easton, S.F. Lincoln, Modified Cyclodextrins: Scaffolds and Templates for Supramolecular Chemistry, Imperial College Press, London, 1999.
[18] C. Bohne, The Spectrum 12 (2000) 14.
[19] A. Harada, Acc. Chem. Res. 34 (2001) 456.
[20] E. Engeldinger, D. Armspach, D. Matt, Chem. Rev. 103 (2003) 4147.
[21] S. Fery-Forgues, R. Dondon, F. Bertorelle, Chapter 3 in Handbook of Photochemistry and Photobiology, in: H. Nalwa (Ed.), Supramolecular Photochemistry, Vol. 3, American Scientific Publishers, Stevenson Ranch, CA, 2003, p. 121.
[22] A. Douhal, Chem. Rev. 104 (2004) 1955.
[23] K. Uekama, F. Hirayama, T. Irie, Chem. Rev. 98 (1998) 2045.

[24] J. Villaverde, C. Maqueda, E. Morillo, J. Agric. Food. Chem. 53 (2005) 5366.
[25] D.A. Lerner, B. Del Castillo, S. Muñoz-Botella, Anal. Chim. Acta 227 (1989) 297.
[26] A.R. Hedges, Chem. Rev. 98 (1998) 2035.
[27] M. Narita, S. Koshizaka, F. Hamada, J. Incl. Phenom. Macro. Chem. 35 (1999) 605.
[28] K. Takahashi, Chem. Rev. 98 (1998) 2013.
[29] K.W. Busch, I.M. Swamidoss, S.O. Fakayode, M.A. Busch, J. Am. Chem. Soc. 125 (2003) 1690.
[30] H.-J. Schneider, F. Hacket, V. Rüdiger, H. Ikeda, Chem. Rev. 98 (1998) 1755.
[31] M.V. Rekharsky, Y. Inoue, Chem. Rev. 98 (1998) 1875.
[32] H. Bakirci, X. Zhang, W.M. Nau, J. Org. Chem. 70 (2005) 39.
[33] K. Harata, Chem. Rev. 98 (1998) 1803.
[34] B. Wagner, Chapter 1 in Handbook of Photochemistry and Photobiology, in: H. Nalwa (Ed.), Supramolecular Photochemistry, Vol. 3, American Scientific Publishers, Stevenson Ranch, CA, 2003, p. 1.
[35] K.B. Lipkowitz, Chem. Rev. 98 (1998) 1829.
[36] J.-M. Lehn, Angew. Chem. Int. Ed. Engl. 27 (1988) 89.
[37] S. Hamai, J. Phys. Chem. 92 (1988) 6140.
[38] K.W. Street Jr., W.E. Acree Jr., Appl. Spectrosc. 42 (1988) 1315.
[39] S. Hamai, J. Phys. Chem. 94 (1990) 2595.
[40] G.S. Cox, N.J. Turro, N.C. Yang, M.J. Chang, J. Am. Chem. Soc. 106 (1984) 422.
[41] G.S. Cox, P.J. Hauptman, N.J. Turro, Photochem. Photobiol. 39 (1984) 597.
[42] A. Heredia, G. Requena and F. García-Sánchez, J. Chem. Soc., Chem. Commun. (1985) 1814.
[43] L. Liu, Q.-X. Guo, J. Incl. Phenom. Macro. Chem. 42 (2002) 1.
[44] D.F. Eaton, J. Photochem. Photobiol. B 2 (1988) 523.
[45] L.J. Johnston, B.D. Wagner, Chapter 13 in Physical Methods in Supramolecular Chemistry, in: J.E.D. Davies, J. Ripmeester (Eds), Comprehensive Supramolecular Chemistry, Vol. 8, Pergamon, Oxford, 1996.
[46] E.L. Wehry, Anal. Chem. 54 (1982) 131R.
[47] G. Patonay, M.E. Rollie, I.M. Warner, Anal. Chem. 57 (1985) 569.
[48] A. Muñoz de la Peña, F. Salinas, M.J. Gómez, M.A. Acedo, M. Sánchez Peña, J. Incl. Phenom. Mol. Rec. Chem. 15 (1993) 131.
[49] C.N. Sanramé, R.H. de Rossi, G.A. Argüello, J. Phys. Chem. 100 (1996) 8151.
[50] H.A. Benesi, H. Hildebrand, J. Am. Chem. Soc. 71 (1949) 2703.
[51] S.M. Hoenigman, C.E. Evans, Anal. Chem. 68 (1996) 3274.
[52] S. Nigam, G. Durocher, J. Phys. Chem. 100 (1996) 7135.
[53] S. Hamai, A. Hatamiya, Bull. Chem. Soc. Jpn. 69 (1996) 2469.
[54] K. Hirose, J. Incl. Phenom. Macro. Chem. 39 (2001) 193.
[55] Y.L. Loukas, J. Phys. Chem. B 101 (1997) 4863.
[56] P. Suppan, J. Photochem. Photobiol. A: Chem. 50 (1990) 293.
[57] R. Englman, J. Jortner, Mol. Phys. 18 (1970) 145.
[58] B.D. Wagner, S.J. Fitzpatrick, J. Incl. Phenom. Macro. Chem. 38 (2000) 467.
[59] M.A. Rankin, B.D. Wagner, Supramol. Chem. 16 (2004) 513.
[60] Y.H. Kim, D.W. Cho, M. Yoon, D. Kim, J. Phys. Chem. 100 (1996) 15670.
[61] T. Soujanya, T.S.R. Krishna, A. Samanta, J. Photochem. Photobiol. A: Chem. 66 (1992) 185.
[62] D.F. Eaton, Tetrahedron 43 (1987) 1551.
[63] F. Cramer, W. Saenger, H.-Ch. Spatz, J. Am. Chem. Soc. 89 (1967) 14.

[64] J. Franke, T. Merz, H.-W. Losensky, W.M. Müller, U. Werner, F. Vögtle, J. Incl. Phenom. 3 (1985) 471.
[65] J.W. Park, H.J. Song, J. Phys. Chem. 93 (1989) 6454.
[66] H.-J. Schneider, T. Blatter, S. Simova, J. Am. Chem. Soc. 113 (1991) 1996.
[67] O.S. Tee, T.A. Gadosy, J.B. Giorgi, Can. J. Chem. 74 (1996) 736.
[68] S.G. Penn, R.W. Chiu, C.A. Monnig, J. Chromatogr. A 680 (1994) 233.
[69] B.D. Wagner, P.J. MacDonald, Photochem. Photobiol. A: Chem. 114 (1998) 151.
[70] B.D. Wagner, P.J. MacDonald, M. Wagner, J. Chem. Ed. 77 (2000) 178.
[71] H. Kondo, H. Nakatani, K. Hiromi, J. Biochem. 79 (1976) 393.
[72] K. Kalyanasundaram, J.K. Thomas, J. Am. Chem. Soc. 99 (1977) 2039.
[73] S. Hamai, J. Phys. Chem. 93 (1989) 2074.
[74] S. Hamai, J. Phys. Chem. 93 (1989) 6527.
[75] A. Muñoz de la Peña, T. Ndou, J.B. Zung, I.M. Warner, J. Phys. Chem. 95 (1991) 3330.
[76] A. Muñoz de la Peña, T.T. Ndou, J.B. Zung, K.L. Greene, D.H. Live, I.M. Warner, J. Am. Chem. Soc. 113 (1991) 1572.
[77] M. Eddaoudi, A.W. Coleman, P. Prognon, P. Lopez-Mahia, J. Chem. Soc., Perkin Trans. 2 (1996) 955.
[78] K. Kano, I. Takenoshita, T. Ogawa, Chem. Lett. (1980) 1035.
[79] K. Kano, I. Takenoshita, T. Ogawa, J. Phys. Chem. 86 (1982) 1833.
[80] G. Patonay, A. Shapira, P. Diamond, I.M. Warner, J. Phys. Chem. 90 (1986) 1963.
[81] N. Kobayashi, R. Saito, H. Hino, Y. Hino, A. Ueno, T. Osa, J Chem Soc., Perkin Trans. II (1983) 1031.
[82] A.K. Dutta, K. Kamada, K. Ohta, J. Photochem. Photobiol. A: Chem. 93 (1996) 57.
[83] V.J.P. Srivatsavoy, J. Lumin. 82 (1999) 17.
[84] B.D. Wagner, N. Stojoanovic, G. LeClair, C.K. Jankowski, J. Incl. Macro. Chem. 45 (2003) 275.
[85] P. Hazra, D. Chakrabarty, A. Chakraborty, N. Sarkar, Chem. Phys. Lett. 388 (2004) 150.
[86] S. Scypinski, J.M. Drake, J. Phys. Chem. 89 (1985) 2432.
[87] K. Muthuramu, V. Ramamurthy, J. Photochem. 26 (1984) 57.
[88] B.D. Wagner, S.J. Fitzpatrick, G.J. McManus, J. Incl. Phenom. Macro. Chem. 47 (2003) 187.
[89] T.A. Dahl, P. Bilski, K.J. Reszka, C.F. Chignell, Photochem. Photobiol. 59 (1994) 290.
[90] E.M. Bruzell, E. Morisbak, H.H. Tønnesen, Photochem. Photobiol. Sci. 4 (2005) 523.
[91] P.-H. Bong, Bull. Korean Chem. Soc. 21 (2000) 81.
[92] B. Tang, L. Ma, H.-Y. Wang, G.-Y. Zhang, J. Agric. Food Chem. 50 (2002) 1355.
[93] H.H. Tonnesen, M. Másson, T. Loftsson, Int. J. Pharm. 244 (2002) 127.
[94] K.N. Baglole, P.G. Boland, B.D. Wagner, J. Photochem. Photobiol. A: Chem. 173 (2005) 230.
[95] M.C. García Alvarez-Coque, M.J. Medina Hernández, R.M. Villanueva Camañas, C. Mongay Fernández, Anal. Biochem. 178 (1989) 1.
[96] D. Tsikas, J. Sandmann, D. Holzberg, P. Pantazis, M. Raida, J.C. Frölich, Anal. Biochem. 273 (1999) 32.
[97] W.R.G. Baeyens, B.L. Ling, V. Corbisier, A. Raemdonck, Anal. Chim. Acta 234 (1990) 187.
[98] M.Y. Khokhar, J.N. Miller, J. Chem. Soc. Pak. 17 (1995) 226.
[99] B.D. Wagner, G.J. McManus, Anal. Biochem. 317 (2003) 233.

Cyclodextrin Materials Photochemistry, Photophysics and Photobiology
Abderrazzak Douhal (Editor)
DOI 10.1016/S1872-1400(06)01003-X

Chapter 3

Measurement of Chiral Recognition in Cyclodextrins by Fluorescence Anisotropy

Matthew E. McCarroll, Irene W. Kimaru, Yafei Xu

Department of Chemistry and Biochemistry, Southern Illinois University, Carbondale, IL 62901, USA

1. Chiral recognition

Chiral recognition can be described as the discrimination between the two enantiomers of a chiral molecule. Because the physical properties that are typically used to separate molecular species are identical in the case of enantiomers, it is difficult to separate the two species. It is only through the interactions with a discriminating secondary species (i.e. chiral selector) that physical differences can be observed. Despite the subtle differences between enantiomers, the biological effects of chiral drugs can be radically distinct for each enantiomer. In some cases, such as the anti-inflammatory ibuprofen, one enantiomer provides safe therapeutic activity and the other enantiomer is simply inactive. In other cases, such as thalidomide, one enantiomer of the otherwise safe drug can have significant deleterious effects. Given that molecular building blocks such as sugars and amino acids are chiral in nature, chiral recognition plays a fundamental role in the biological activity of living systems. In addition, the enantiomers of chiral drugs may exhibit completely different pharmacokinetic properties and can be metabolized by different pathways.

Following a 1992 mandate by the Federal Food and Drug Administration, producers of pharmaceuticals are required to evaluate the effects of individual enantiomers and to verify the enantiomeric purity of chiral drugs that are produced [1]. The pharmaceutical properties and toxicity must be established independently for both enantiomers even if the drug is to be marketed as a single enantiomer. This requirement has resulted in an increased need for techniques to evaluate chiral interactions and methods for determining enantiomeric purity, as well as preparative-scale enantiomeric separations. Thus, the ability to separate chiral compounds is of utmost importance, especially considering the fact that currently more than one-third of the drugs are marketed as single enantiomers [2]. Simply stated, the need to produce pure enantiomers of chiral compounds is becoming synonymous with the pharmaceutical industry.

The simplest case of chiral separation was first demonstrated by Pasteur in 1848 when he physically separated the enantiomers of sodium ammonium tartrate [3]. Unfortunately, most chiral separations are more tedious and a range of techniques has been developed for separating pure enantiomers, the most common being various chromatographic methods that are dependent on the formation of a transient diastereomeric complex of distinct physical properties. The formation of this complex is governed and fundamentally limited by the selectivity of the chiral recognition process. It is generally accepted that for chiral recognition to occur between two molecules, a minimum of three simultaneous interactions must be present, and at least one must be dependent on stereochemistry (i.e. three-point rule) [4,5]. As illustrated in Fig. 1, only one enantiomer is capable of forming three points of interaction, resulting in the formation of two distinct diastereomeric complexes with differing free energies of formation (i.e. $\Delta\Delta G_{R,S}$). This model has been exploited for many years and has resulted in the development of a wide array of chiral stationary phases used in gas and liquid chromatography [6–9]. Despite significant advances that have been made, a complete model of chiral recognition that is of predictive utility is yet to be materialized and much research has gone

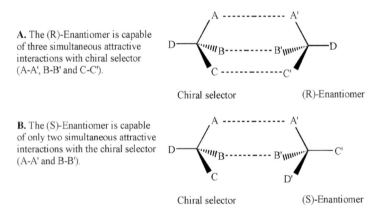

A. The (R)-Enantiomer is capable of three simultaneous attractive interactions with chiral selector (A-A', B-B' and C-C').

Chiral selector   (R)-Enantiomer

B. The (S)-Enantiomer is capable of only two simultaneous attractive interactions with the chiral selector (A-A' and B-B').

Chiral selector   (S)-Enantiomer

Fig. 1. Three-point model of chiral discrimination. (A) The R-enantiomer is shown to be capable of three simultaneous interactions with the chiral selector and (B) the S-enantiomer is only capable of two simultaneous interactions (A–A′ and B–B′).

into developing various methods for the qualitative and quantitative examination of chiral interactions.

1.1. Measurement of chiral recognition

In order to understand and optimize various processes based on chiral recognition, it is necessary to accurately measure the phenomenon of chiral recognition. Common methods for the examination of chiral interactions include separation-based methods such as high-performance liquid chromatography (HPLC), gas chromatography, and capillary electrophoresis (CE). Spectroscopic methods, such as circular dichroism and the use of NMR chiral shift reagents have also been used. In addition, microcalorimetry and mass spectrometry [10–13] have also been used to study the thermodynamic parameters of enantioselective binding interactions [9]. There are, however, relatively few examples of enantioselective photophysical behavior [14]. The majority of these examples are based on enantioselective quenching [15–20] and excimer formation, [21,22] or spectral shifts [23,24] upon host/guest complexation. While these advances are significant, a limitation of these approaches is that they are dependent on specific photophysical properties of the analyte, which are typically not broadly applicable. New advances are needed and work is ongoing in this area of research.

2. Fluorescence anisotropy

Fluorescence anisotropy is a polarization-based spectroscopic technique that has been widely used to study molecular interactions, particularly in biological systems [25]. The measurement of fluorescence anisotropy (r) is straightforward, involving the determination of the fluorescence intensity of the parallel ($I_{\parallel}$) and perpendicular ($I_{\perp}$) components of the fluorescence emission, following excitation with vertically polarized light (Eq. 1) (Figure of experimental layout, Fig. 2).

$$r = \frac{I_{\parallel} - I_{\perp}}{I_{\parallel} + 2I_{\perp}} \tag{1}$$

Fluorescence polarization is frequently expressed in terms of anisotropy (r) due to the comparatively simple equations describing rotational depolarization. The dependence of fluorescence anisotropy on molecular rotation can be described quantitatively with the well-known Perrin equation,

$$\frac{r_0}{r} = 1 + \frac{\tau}{\phi} \tag{2}$$

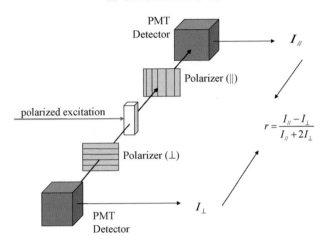

Fig. 2. Depiction of the experimental layout for the measurement of steady-state fluorescence anisotropy.

where r_0 is the intrinsic anisotropy, τ the fluorescence lifetime, and ϕ the rotational correlation time. Because the magnitude of the anisotropy depends on the rotational rate of the fluorophore, the technique has found wide use as a method to study molecular interactions [25, 26]. Fluorescence polarization is frequently expressed in terms of anisotropy (r) due to the comparatively simple equations describing rotational depolarization. Consider the case of a fluorophore that interacts with cyclodextrin (CD), such that the fluorophore can exist as the free species or in the bound state. Based on the additive property of polarization, the total anisotropy is given by the weighted average of the anisotropy values of the bound species and free analyte [26],

$$r_{\text{avg}} = f_f r_f + f_b r_b \tag{3}$$

where $f_b, f_f, r_b,$ and r_f are the fractions and anisotropy values of the bound and free species, respectively. The anisotropy of a small fluorophore rotating freely in solution typically approaches zero. Under these circumstances, the total anisotropy of the system mainly depends on the anisotropy of the bound species and the fractional distribution. As a result, the measured anisotropy can be taken as a measure of the fraction of host-bound analyte, which in turn is indicative of the stability of the host–guest complex (i.e. binding constant) (model of CD binding). Furthermore, a bound fluorophore may also undergo segmental motion that is independent of overall rotational diffusion. Therefore, the experimentally measured anisotropy is actually a weighted average of contributions from distributional effects as well as any segmental motions of the bound fluorophore. The contributions of the segmental motion to the measured anisotropy mainly depend on the binding strength of the fluorophore to the host molecule.

2.1. Measurement of chiral recognition via fluorescence anisotropy

There are relatively few reports of fluorescence anisotropy being used to examine chiral interactions. In 1999, Al Rabaa, Tfibel, Pernot, and Fontaine-Aupart reported [27] the spectroscopic characterization of a chiral anthryl probe and its interactions with DNA, and in 2001, McCarroll, Haddadian, and Warner [28] reported the use of fluorescence polarization to specifically study chiral interactions of a pseudostationary phase. These two papers represent the first efforts to use anisotropy as a measure of chiral recognition. Subsequently, Billiot, McCarroll, Billiot, and Warner applied the method toward character-izing a chiral separation system [29]. In the past several years, the McCarroll group has developed a theoretical framework from fundamental principles to relate steady-state fluorescence anisotropy measurements to enantioselective binding (i.e. chiral recognition) (Fig. 3) [30]. The remainder of this chapter describes and discusses the results of this work, with a particular emphasis on the measurement of chiral recognition in cyclodextrins.

Results from the first investigation of this phenomenon are reproduced in Fig. 4, which illustrates the chiral selectivity of the enantiomers of binaphthyl phosphate (BNP) by an amino acid-based molecular micelle. This observation not only indicated that the anisotropy could actually be sensitive to chiral recognition, but led us to posit that the magnitude of the difference could be a quantitative measure of the selectivity [28].

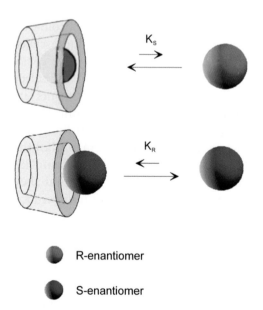

Fig. 3. Representation of enantioselective binding with the chiral selector β-cyclodextrin.

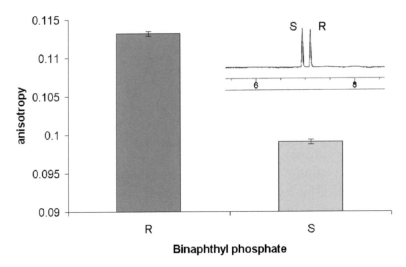

Fig. 4. Fluorescence anisotropy of the enantiomers of binaphthyl phosphate in the presence of a micelle polymer. Capillary electrophoresis measurements confirmed the enantiomer with the greatest interaction also had the greatest anisotropy value (inset). Reproduced with permission from Ref. [28]. Copyright 2001, Am. Chem. Soc.

Following the initial investigation, studies in the McCarroll group examined the empirical correlation between the anisotropy ratio and chiral selectivity. For example, Figs. 5–7 show the clear correlations that were observed between the fluorescence anisotropy data and published chromatographic data [31, 32]. Figure 5a shows the effect of pH on the fluorescence anisotropy of the enantiomers of mandelic acid in the presence of β-cyclodextrin and Fig. 5b shows the trend observed for the anisotropy ratio and selectivity as a function of pH. The trend predicted from anisotropy experiments, including the reversal, was in fact observed in CE experiments carried out under the same conditions. Figure 6a shows the effect of polymer concentration on the chiral resolution obtained in CE–MS experiments separating binaphthol that were recently published [33] and Fig. 6b shows data generated by mimicking the various separation conditions used in Fig. 6a [33]. It is clear that the optimum separation conditions could have been determined using fluorescence anisotropy prior to performing separations and mass spectrometry.

Another example in Fig. 7 shows the chiral resolution obtained in electrokinetic chromatography using vancomycin as a chiral selector (Fig. 7a), and the chiral selectivity observed in HPLC separations using a sulfated β-cyclodextrin stationary phase (Fig. 7b). In both cases, the differences observed from fluorescence anisotropy follow very similar trends as those obtained from the separation-based methods.

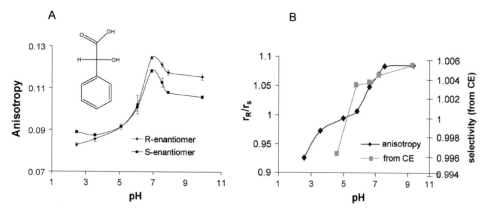

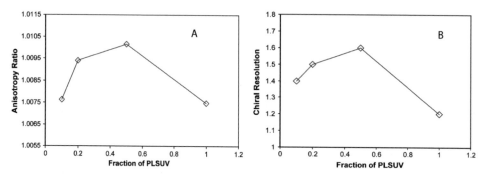

Fig. 5. (A) Anisotropy of mandelic acid in the presence of β-cyclodextrin as a function of solution pH. (B) Ratio of anisotropy values and selectivity as a function of pH. Ratios and selectivities less than one indicate reversal of preferential binding. Reproduced with permission from Ref. [39].

Fig. 6. Comparison of anisotropy ratio (A) and chiral resolution (B) as a function of polymer concentration. The data for (B) were taken from published literature [33]. Reproduced with permission from Ref. [39].

2.2. Theoretical basis

To relate the measured anisotropy to chiral recognition, the following has been derived from fundamental principles [30]. Assuming a 1:1 association stoichiometry, the fractional distribution can be represented in terms of the binding constant, K, and the concentration of free selector, S, Eq. 3 can be given as,

$$r_{avg} = r_b \frac{K[S]}{(K[S]+1)} + r_f \left[1 - \frac{(K[S])}{(K[S]+1)} \right] \qquad (4)$$

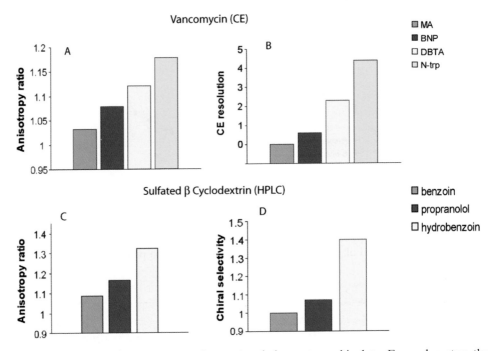

Fig. 7. Comparison of fluorescence anisotropy and chromatographic data. For each system the fluorescence anisotropy (A and C) was measured under conditions mimicking those published in the literature (B and D). Data for (B) and (D) were taken from published literature [30,31]. Reproduced with permission from Ref. [39].

which relates the average anisotropy to the anisotropy of the bound species, the binding constant, and the concentration of the selector. Because r_f usually approaches zero, Eq. 3 can typically be approximated by Eq. 5.

$$r_{avg} = r_b \frac{K[S]}{(K[S] + 1)} \tag{5}$$

Many factors affect the average anisotropy value in a given system, only some of which are enantioselective. Non-enantioselective factors, however, affect both enantiomers equally and do not lead directly to a difference in anisotropy values between two enantiomers. To examine the enantioselectivity, the ratio of anisotropy values can be examined (Eq. 6).

$$\frac{r_{avg,R}}{r_{avg,S}} = \frac{r_{b,R}}{r_{b,S}} \frac{K_R}{K_S} \frac{(K_S[S] + 1)}{(K_R[S] + 1)} \tag{6}$$

The first and last terms of the equation typically approach unity, hence the ratio of the binding constants (i.e. selectivity) is the primary factor leading to differences in the measured anisotropy. It is well established that the Gibbs free energy for complex formation can be related to the binding constant by Eq. 7.

$$\frac{\Delta\Delta G^{\circ}_{R,S}}{-RT} = \ln\left(\frac{K_R}{K_S}\right) = \frac{-\Delta\Delta H^{\circ}}{RT} + \frac{\Delta\Delta S^{\circ}}{R} \tag{7}$$

Evaluating the logarithm of Eq. 6 results in Eq. 8,

$$\ln\left(\frac{r_{avg,R}}{r_{avg,S}}\right) = \ln\left(\frac{K_R}{K_S}\right) + \ln\left(\frac{r_{b,R}}{r_{b,S}}\right) + \ln\left(\frac{(K_S[S]+1)}{(K_R[S]+1)}\right) \tag{8}$$

which allows substitution of the $\Delta\Delta G^{\circ}$ term from Eq. 7 to produce the useful form shown in Eq. 9 [30].

$$\ln\left(\frac{r_{avg,R}}{r_{avg,S}}\right) = \frac{-\Delta\Delta H^{\circ}}{RT} + \frac{\Delta\Delta S^{\circ}}{R} + \ln\left(\frac{r_{b,R}}{r_{b,S}}\right) + \ln\left(\frac{(K_S[S]+1)}{(K_R[S]+1)}\right) \tag{9}$$

The terms, $\ln(r_{b,R}/r_{b,S})$ and $\ln((K_S[S]+1)/(K_R[S]+1))$ typically approach zero and the differences observed in the steady-state anisotropy measurements are primarily a result of the distribution between free and bound species. A plot of $\ln(r_{avg,R}/r_{avg,S})$ vs. $1/T$ results in a response where the slope is equal to $\Delta\Delta H^{\circ}/R$ and the intercept is equal to $\Delta\Delta S^{\circ}/R$ plus a constant that typically approaches zero. Importantly, the selectivity (K_R/K_S) is a reflection of the thermodynamic consequence of chiral recognition and the term, $r_{b,R}/r_{b,S}$, is a measure of the differences in rotational rates of each diastereomeric complex, which represent indirect and direct measures of chiral recognition, respectively.

2.3. Determination of thermodynamic binding parameters in cyclodextrin

The temperature dependence of the fluorescence anisotropy has also been examined, an example of which is shown in Fig. 9. The results have prompted us to focus on the possibility of obtaining thermodynamic parameters from anisotropy measurements. The enantiomeric difference in Gibbs free energy $(\Delta\Delta G^{\circ}_{R,S})$ upon complexation is a good descriptor of enantioselectivity. Additionally, the associated differential enthalpy $(\Delta\Delta H^{\circ}_{R,S})$ and entropy $(\Delta\Delta S^{\circ}_{R,S})$ parameters reflect the mechanism of enantioselectivity. We have begun to fully develop methodology with which the thermodynamic parameters of enantioselective binding can be characterized with great precision, sensitivity, and speed.

Based on the relationship between the anisotropy and the binding constants (Eq. 9), the thermodynamics of the enantioselective binding can be evaluated. A plot of $\ln(r_{avg,R}/r_{avg,S})$ vs. $1/T$ results in a response where the slope is equal to $\Delta\Delta H^{\circ}/R$ and the intercept is equal to $\Delta\Delta S^{\circ}/R$ plus a constant that often approached zero. An example of this type of analysis is shown in Fig. 8, which

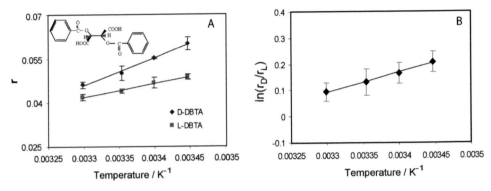

Fig. 8. Temperature dependence of the fluorescence anisotropy of the individual enantiomers of DBTA in the presence of β-cyclodextrin (A) and the linear response predicted from Eq. 9 (B). Reproduced with permission from Ref. [30]. Copyright 2005, Am. Chem. Soc.

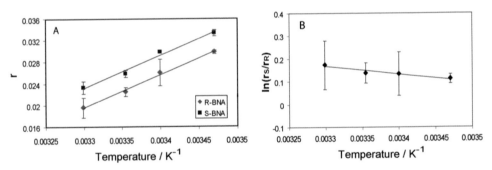

Fig. 9. Temperature dependence of the fluorescence anisotropy of the individual enantiomers of BNA in the presence of β-cyclodextrin (A) and the linear response predicted from Eq. 9 (B). Reproduced with permission from [30]. Copyright 2005, Am. Chem. Soc.

shows the interactions of dibenzoyl tartaric acid (DBTA) in the presence of β-CD. The data shown in Fig. 8A represents the general binding properties of each enantiomer, but contains artifacts that affect the anisotropy. In Fig. 8B, which is a representation of Eq. 9, the non-stereoselective artifacts affect both enantiomers equally and are effectively nullified. The non-stereoselective artifacts include solvent viscosity, temperature, fluorescence quenching, and any other non-stereoselective interactions. The predicted linear response is observed and the slope and intercept can be used to infer the differential entropy and enthalpy of the enantioselective binding. Another example is shown in Fig. 9, which shows the interactions of the enantiomers of binaphthyl amine (BNA) with β-CD. The negative slope observed is indicative of a positive $\Delta\Delta S$ value.

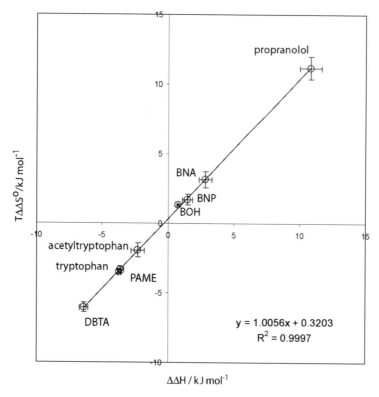

Fig. 10. Enthalpy–entropy compensation plot of several enantiomer pairs interacting with β-CD. Reproduced with permission from Ref. [30]. Copyright 2005, Am. Chem. Soc.

It should be noted that the thermodynamic parameters determined in this technique refer explicitly to the differential values that result from chiral recognition, not the thermodynamics of the actual binding event.

The results of the determination of enantioselective thermodynamic parameters for eight enantiomer pairs with β-CD are shown in the enthalpy–entropy compensation plot [34,35] in Fig. 10 [30]. The plot has excellent linearity, indicating a great degree of self consistency in the data. Furthermore, an interesting observation was that analytes were generally grouped according to their molecular structure. Specifically, the amino acids and amino acid derivatives were grouped together, as were the binaphthyl derivatives. The drug propranolol was isolated on the plot, which is consistent with the fact that the molecular structure is distinct from those of the other analytes examined. Efforts are ongoing to explore the feasibility of using the method to characterize the mode of chiral discrimination in various systems exhibiting chiral selectivity.

2.4. Determination of enantiomeric composition

The significant reliance of the pharmaceutical industry on the production of pure enantiomers has lead to an increased need for the rapid determination of enantiomeric composition, often represented as the enantiomeric excess (e.e.). As a result of this need, great efforts have been made to improve existing methods for determination of enantiomeric excess. The sensitivity of fluorescence-based methods have lead to various approaches for examining enantiomeric composition [15,16,36,37]. Recently, the use of fluorescence anisotropy to determine enantiomeric composition has been reported [38] and is detailed in this section.

2.4.1. Theoretical approach

Considering that the measured fluorescence anisotropy is based on the difference in rotational diffusion rates of the bound and free forms of the chiral molecule in the presence of a chiral selector, it stands that in a solution of mixed enantiomeric composition the anisotropy will vary between two extreme values for the each of the pure enantiomers. For a given chiral compound in the presence of a chiral selector, the measured anisotropy is then a function of the enantiomeric composition.

In order to elucidate the exact relationship between the measured anisotropy and the enantiomeric composition, a mathematical model was developed that is based on the additive nature of polarization measurements. For a solution containing a single enantiomer, A_R, and a selector, S, the association reaction is given by

$$A_R + S \leftrightarrow A_R S \tag{10}$$

where A_R represents the R-enantiomer of the analyte and S represents the selector. Under equilibrium conditions the measured anisotropy is a weighted average of the free and bound species (Eq. 11)

$$r_R = f_{b,R} r_{b,R} + (1 - f_{b,R}) r_{f,R} \tag{11}$$

The anisotropy can then be expressed in terms of the binding constant and it is typically appropriate to drop the second term of the equation because the anisotropy of the free species typically approaches zero, resulting in the Eqs. 12 and 13 for the R and S enantiomers, respectively.

$$r_R = \frac{K_R[S] r_{b,R}}{(1 + K_R[S])} \tag{12}$$

$$r_R = \frac{K_S[S]r_{b,S}}{(1 + K_S[S])} \tag{13}$$

Thus, the magnitude of the anisotropy from each enantiomer in solution is a function of the binding constant, the concentration of free selector, and the anisotropy of the bound species (i.e. diastereomer). A solution containing a chiral selector and a mixture of two enantiomers will, under typical conditions, exhibit an average anisotropy (r_{avg}) that is the sum of the anisotropy of the bound species (A_RS and A_SS) weighted by the molar fractions of each species, where ϕ_R represents the fraction of R-enantiomers in the mixture (Eq. 14).

$$r_{avg} = \phi_R r_R + (1 - \phi_R) r_S \tag{14}$$

Thus, in a solution of mixed enantiomers the anisotropy can be given by a combination of Eqs. 12–14.

$$r_{avg} = \frac{\phi K_R[S]r_{b,R}}{(1 + K_R[S])} + \frac{(1 - \phi)K_S[S]r_{b,S}}{(1 + K_S[S])} \tag{15}$$

Considering that K_R, K_S, $r_{b,R}$, $r_{b,S}$ are constant in a given system and that the concentration of free selector is approximately constant when the selector is in excess, a linear relationship is expected between the average anisotropy and the molar fraction of the enantiomers. This can more easily be observed by combining the constants to a single term (γ) for each enantiomer (Eqs. 16 and 17),

$$\gamma_R = \frac{K_R[S]r_{b,R}}{1 + K_R[S]} \tag{16}$$

$$\gamma_R = \frac{K_S[S]r_{b,S}}{1 + K_S[S]} \tag{17}$$

and expressing a form of the equation that is linear with respect to the enantiomeric composition (Eq. 18),

$$r_{avg} = \phi_R (\gamma_R - \gamma_S) + \gamma_S \tag{18}$$

where a plot of the measured anisotropy as a function of enantiomeric composition produces a linear response where the intercept is determined by γ_s and the slope is determined by the difference between γ_R and γ_S. Figure 11 shows the results of modeling the major parameters of Eq. 15.

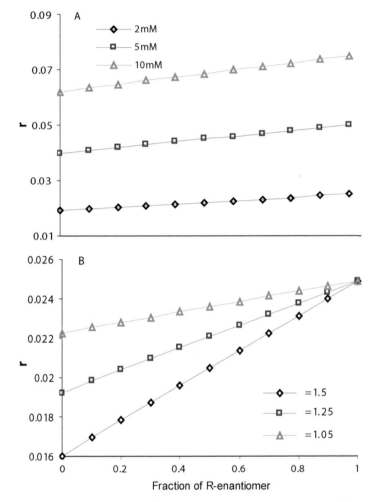

Fig. 11. Calculated anisotropy as a function of enantiomeric composition at (A) various selector concentrations ($K_R = 100\,\text{M}^{-1}$, $K_S = 80\,\text{M}^{-1}$, $r_{b,R} = 0.15$, $r_{b,S} = 0.14$, $[A_{total}] = 50\,\mu\text{M}$ ($\alpha = 1.25$) and (B) various selectivities ($K_R = 100\,\text{M}^{-1}$, $r_{b,R} = 0.15$, $r_{b,S} = 0.14$, $[S] = 2\,\text{mM}$, $[A_{total}] = 50\,\mu\text{M}$). Reproduced with permission from Ref. [38]. Copyright 2004, Am. Chem. Soc.

2.4.2. Experimental results

The modeling results shown in Fig. 11 and examination of Eqs. 15 & 18 indicate not only that the response should be linear, but that the slope is a function of the selectivity and the intercept is a function of the selector concentration. These predictions have been evaluated experimentally, the results of which are shown in Figs. 12 and 13. Figure 12 shows the experimental response for BNP in the presence of α- and β-CD. The linear response predicted from theory is clearly observed and the prediction that the slope is a function of the selectivity is also

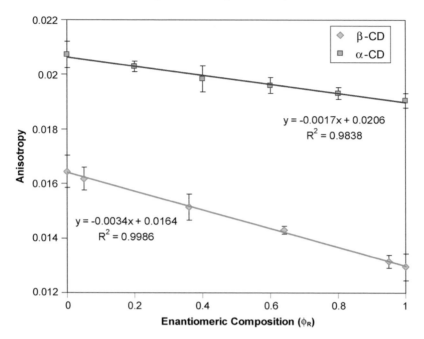

Fig. 12. Anisotropy of BNP as a function of enantiomeric composition in the presence of α- and β-CD. Reproduced with permission from Ref. [38]. Copyright 2004, Am. Chem. Soc.

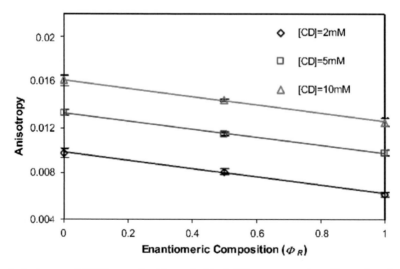

Fig. 13. Anisotropy of BNP as a function of chiral CD concentration. Reproduced with permission from Ref. [38]. Copyright 2004, Am. Chem. Soc.

Table 1
Determination of enantiomeric composition of BNP

% Composition (actual)	% Composition (measured)	Absolute error
95	92.3	−2.7
64	62.9	−1.1
36	33.7	−2.3
5	5.71	0.71

Source: Reproduced from Ref. [38].

confirmed, as β-CD has a significantly greater selectivity for BNP than does α-CD. Figure 13 shows the experimental data for evaluation of the prediction of concentration dependence, which shows the measured anisotropy at various CD concentrations. Importantly, the slope is invariant with CD concentration. To test the potential of the method in the quantitative determination of enantiomeric composition, a two point calibration was obtained (pure R- and pure S-enantiomer) and used to evaluate series of solutions of known enantiomeric composition. The determined enantiomeric compositions were in excellent agreement with the known values, having an average absolute error of ~1.7% (Table 1) [38]. It should be noted that this determination was carried out in standard 1 cm cuvettes using ~3 pg of material and could be pushed to even smaller quantities by use of micro cuvettes and on-capillary detection.

3. Future direction

The studies presented in this chapter clearly demonstrate the potential that exist for using fluorescence anisotropy to study chiral recognition in CD-based systems, which has ramifications from both fundamental and applied perspectives. For example, preliminary work in our laboratory has already shown the potential of using the system for evaluating and optimizing chiral separation system *a priori* [39]. Other areas worthy of pursuit lie in further investigation of the method for determination of enantiomeric excess and the determination of enantioselective thermodynamic parameters, primarily in terms of increasing the precision and sensitivity of the measurements.

References

[1] FDA, FDA's Policy Statement for the Development of New Stereoisomeric Drugs, US Food and Drug Administration, Center for Drug Evaluation and Research, Department of Health and Human Services. 1992.
[2] S.C. Stinson, Chiral Chemistry, Chemical and Engineering News, 2001. p 45.

[3] L. Pasteur, Ann. Chim. Phys. 24 (1848) 442.
[4] W.H. Pirkle, T.C. Pochapsky, Adv. Chromatogr. (NY) 27 (1987) 73.
[5] W.H. Pirkle, M.H. Hyun, B. Bank, J. Chromatogr. 316 (1984) 585.
[6] W.H. Pirkle, T.C. Pochapsky, Chem. Rev. 89 (1989) 347.
[7] R. Vespalec, P. Bocek, Chem. Rev. 100 (2000) 3715.
[8] K.B. Lipkowitz, R. Coner, M.A. Peterson, J. Am. Chem. Soc. 119 (1997) 11269.
[9] M. Rekharsky, Y. Inoue, J. Am. Chem. Soc. 122 (2000) 4418.
[10] W.A. Tao, D. Zhang, E.N. Nikolaev, R.G. Cooks, J. Am. Chem. Soc. 122 (2000) 10598.
[11] W.A. Tao, F.C. Gozzo, R.G. Cooks, Anal. Chem. 78 (2001) 1692.
[12] Z. Yao, T.S.M. Wan, K. Kwong, C. Che, Anal. Chem. 72 (2000) 5394.
[13] Z.P.W. Yao, T.S.M. Wan, K. Kwong, C. Che, Anal. Chem. 72 (2000) 5383.
[14] R. Corradini, G. Sartor, R. Marchelli, A.J. Dossena, J. Chem. Soc. Perkin Trans. (1992) 1979.
[15] X. Mei, C. Wolf, J. Am. Chem. Soc. 126 (2004) 14736.
[16] G.E. Tumambac, C. Wolf, Org. Lett. 7 (2005) 4045.
[17] T. Yorozu, K. Hayashi, M. Irie, J. Am. Chem. Soc. 103 (1981) 5480.
[18] D. Avnir, E. Wellner, M. Ottolenghi, J. Am. Chem. Soc. (1989) 2001.
[19] K. Kano, Y. Yoshiyasu, S.J. Hashimoto, J. Chem. Soc., Chem. Commun. (1989) 1278.
[20] T.D. James, K. Sandanayake, S. Shinkai, Nature 374 (1995) 345.
[21] P. Tundo, J.H. Fendler, J. Am. Chem. Soc. (1980) 1760.
[22] C. D. Tran, P. Tundo, J. Am. Chem. Soc. (1980) 2923.
[23] Y. Kubo, S. Maeda, S. Tokita, M. Kubo, Nature 382 (1996) 522.
[24] Y. Yan, M.L. Myrick, Anal. Chem. 71 (1999) 1958.
[25] A.B. Rawitch, G. Weber, J. Biol. Chem. 10 (1972) 680.
[26] J.R. Lakowicz, Principles of Fluorescence Spectroscopy, Plenum, New York, NY, 1983.
[27] A.R. Al Rabaa, F.F.M. Tfibel, P. Pernot, M. Fontaine-Aupart, J. Chem. Soc. Perkin Trans. 2 (1999) 341.
[28] M.E. McCarroll, F. Haddadian, I.M. Warner, J. Am. Chem. Soc. 123 (2001) 3173.
[29] F.H. Billiot, M.E. McCarroll, E.J. Billiot, I.M. Warner, Electrophoresis 25 (2004) 753.
[30] Y. Xu, M.E. McCarroll, J. Phys. Chem. B 109 (2005) 8144.
[31] M.P. Gasper, A. Berthod, U.B. Nair, D.W. Armstrong, Anal. Chem. 68 (1996) 2501.
[32] A.M. Stalcup, K. Gahm, Anal. Chem. 68 (1996) 1369.
[33] S. Shamsi, Anal. Chem. 73 (2001) 5103.
[34] J.E. Leffler, J. Org. Chem. 20 (1955) 1202.
[35] E. Grunwald, C. Steel, J. Am. Chem. Soc. 117 (1995) 5687.
[36] J. Lin, Q.S. Hu, M.H. Xu, L. Pu, J. Am. Chem. Soc. 124 (2002) 2088.
[37] L. Pu, Chem. Rev. 104 (2004) 1687.
[38] Y. Xu, M.E. McCarroll, J. Phys. Chem. A 108 (2004) 6929.
[39] I. W. Kimaru and M. E. McCarroll (submitted).

Cyclodextrin Materials Photochemistry, Photophysics and Photobiology
Abderrazzak Douhal (Editor)
© 2006 Elsevier B.V. All rights reserved
DOI 10.1016/S1872-1400(06)01004-1

Chapter 4

Photochemistry and Photophysics of Cyclodextrin Caged Drugs: Relevance to Their Stability and Efficiency

Maged El-Kemary[a,b], Abderrazzak Douhal[a]

[a]*Departamento de Química Física, Sección de Qumicas, Facultad de Ciencias del Medio Ambiente, Universidad de Castilla-La Mancha, Avda. Carlos III, S.N., 45071 Toledo, Spain*
[b]*Chemistry Department, Faculty of Science, Tanta University, 33516 Kafr ElSheikh, Egypt*

1. Introduction

Cyclodextrin (CD) encapsulation has recently emerged as a novel chemical technology complementary to the conventional approaches [1]. CDs are enzyme-modified starch derivatives with a truncated-cone, or donut-shaped molecular structure. The natural derivatives, synthesized by bacteria, are made by six, seven or eight glycoside units, linked together by α-1,4-glycoside bonds (Fig. 1).

CD is capable of forming inclusion complexes with a wide variety of organic and inorganic compounds mainly due to non-covalent interactions [2]. The chemistry of CD caged drugs is a hot topic in current chemical research, in which photophysics and photochemistry of the drug upon encapsulation by CD nanocavities show significant changes due to the local environment. Therefore, photophysical and photochemical studies are often an adequate approach to analyze the possible mechanisms through which phototoxic effects are pro-duced. CDs have high molecular recognition ability to accommodate guest molecules with suitable polarity and size because of their hydrophobic internal

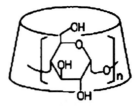

Fig. 1. Schematic representation of CD; $n = 6$, 7 or 8 in α-, β- and γ-CD, respectively.

and polar or hydrophilic exterior nanocavities [3]. This molecular recognition process of CD has attached much attention due to its wide potential applications in chemistry, pharmaceuticals, agriculture, foods, cosmetic and related technologies [1, 4–6]. An increased number of reviews have been dedicated to their industrial and pharmaceutical applications [1,7–11]. Furthermore, CD can constitute an efficient protector agent of photoreactive drugs. Photodecomposition results in unstable drug formulations and/or leads to undesirable side effects. A loss of potency of the drug and the rise of adverse effects may be attributable to the formation of photoproducts during storage or administration of the drug. Therefore, knowledge of the photostability of drug molecules has developed into an important area of research. There is a clear need to evaluate and provide information on handling, packing, labelling, and on adverse and therapeutic aspects of drug delivery systems. The dynamic aspects of some photosensitive drugs in the nanocavities can be used to gain information for new design of the caged drugs. CD is able to influence both the fate of the reaction intermediates and the deactivation pathways of the excited states [12]. These changes account for the observed decrease in the phototoxicity of the drugs.

Furthermore, the interest in CD–drug complexes has been related to the problem of the biological photosensitization by drugs. Indeed, serious phototoxic reactions (mainly diseases of epidermis and dermis), as well as photoallergic and photomutagenic effects, can be induced in patients sensitive to sunlight irradiation while treated with pharmacologically important drugs such as antibacterials, tranquillizers, antimicotics and non-steroidal anti-inflammatory drugs (NSAIDs). Such noxious effects, correlated to the drug photochemical reactivity significantly [13], are decreased in presence of CD with these drugs *in vitro* cellular systems [14,15].

The objective of this chapter is to discuss the effect of CD on the photochemical and photophysical properties of some drug molecules. We will focus on behaviour of the excited drugs in CD nanocavities. Attention has been paid to the study of the excited-state dynamics of drug molecules in the nanocavities [16,17]. We will also discuss different factors that can influence the encapsulation of the drug.

2. Why cyclodextrin caged drugs

The three-dimensional structure of CD endows them with properties that are useful for many applications. The principal advantages of natural CD as drug carriers are: (1) a well-defined chemical structure, yielding many potential sites for chemical modification, (2) the availability of CD of different cavity sizes, (3) low toxicity, low pharmacological activity, and (4) the protection of the included drug molecules from biodegradation. The formation of drug–CD inclusion complexes increases the stability of drugs against dehydration, hydrolysis, oxidation and photodecomposition, minimizes the biological damage photoinduced by drugs [18,19], improves the performances of intravenous formulation, prolongs the pulmonary absorption, sustains the release rate, enhances the peak concentration of drugs in blood, improves bioavailability and enhances the extent and rate of absorption in organs [19]. Moreover, CD can also be used to convert liquid drugs into microcrystalline powders, to prevent drug–drug or drug–additive interactions, to enhance the drug activity, to induce selective transfer and/or reduction of side effects and reduce or eliminate unpleasant taste, bad smell and irritation power, to reduce evaporation and stabilize flavours, reduce haemolysis and prevent admixture incompatibilities [1]. They have also been used as chiral selectors for capillary electrophoresis enantiomer separations and analytical purposes [20]. The main advantages of caging excited drugs are summarized in Scheme 1.

In the following subsections, we will discuss the driving forces and main factors that influence the encapsulation of drug molecules into the CD nanocavity. Knowledge of these factors is important in order to prepare drug–CD complexes with desirable properties.

2.1. The driving forces

Several studies show that the most probable mode of inclusion involves the insertion of the less polar part of the guest molecule into the nanocavity, while the more polar and often charged group of the guest is exposed to the bulk solvent, outside the wider opening of the cavity [21]. An exception to this general rule is the aromatic hydroxyl group, which can penetrate deeply into the CD nanocavity where it H-bonds to one of CD peripheral hydroxyl groups [22]. In an aqueous environment, CD form inclusion complexes with many lipophilic drugs through a process in which water molecules located inside the cavity are replaced by either the whole drug molecule or more frequently, by some lipophilic structure of the molecule. Since water molecules located inside the lipophilic CD cavity cannot satisfy their hydrogen-bonding potential, they are of higher enthalpy than bulk water molecules located in the aqueous environment [23].

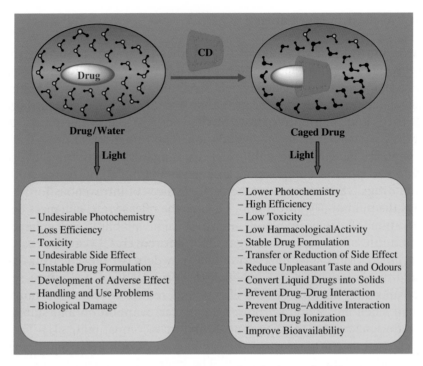

Scheme 1. The main advantages of caging an excited drug.

Owing to the chair conformation of the glucopyranose units, the shape of CD molecules is like a truncated cone rather than a perfect cylinder (Fig. 1). The hydroxyl functions are orientated to the cone exterior with the primary hydroxyl groups of the glucose residues at the narrow edge of the cone and the secondary hydroxyl groups at the wider edge. The cavity of the molecule has a pronounced hydrophobic character, and this is one of the most important characteristics. The main driving forces that have to do with the processes, at least in the case of β-CD and its derivatives appear to be a release of these enthalpy-rich water molecules from the CD nanocavity, which lowers the energy of the system. However, other forces, such as van der Waals interactions, hydrogen bonding, hydrophobic interactions, release of structural strains and changes in surface tension, may also be involved in the inclusion complex formation [7]. The association process, assisted by these weak interactions, produces novel ground and excited state properties. In aqueous solutions, drug molecules located in the central cavity are in a dynamic equilibrium with free drug molecules and the drug/CD complexes are constantly being formed and broken at rates close to those of diffusion-controlled limits [24]. Both CD and drug/CD complexes do possess some surface activity in aqueous solutions and are known to form loosely connected aggregates or micelles [25].

2.2. Factors influencing inclusion process

2.2.1. Size of nanocavity

The type of CD can influence the formation as well as the nature and stability of drug/CD complexes [26,27]. For complexation, the cavity size of CD should be suitable to accommodate a drug molecule of particular size [28,29]. The three natural α-, β- and γ-CD differ in their ring size and solubility (Table 1) [7]. CD with less than six units cannot be formed due to steric hindrances while the higher homologues with nine or more glucose units are very difficult to purify. However, recently Endo *et al.* established an isolation and purification method for several kinds of large ring CD and obtained a relatively large amount of δ-CD (cyclomaltonose) with nine glucose units [30]. The cavity size of α-CD is too small for many drugs and γ-CD is relatively expensive. δ-CD, in general has weaker complex formation ability than conventional CD. With drugs like digitoxin and spiranolactone, δ-CD showed greater complexation than α-CD but the effect of δ-CD was lesser than that of β- and γ-CD.

2.2.2. Change in ionic state (effect of pH)

For weak electrolytes, the strength of binding to a CD is dependent on the charged state of the drug, which is dependent on dissociation constant/s of the drug and the pH of the environment. For most molecules, the ionized or charged form of the molecule has poorer binding to CD compared to the non-ionized or neutral form of the drug especially when bound to a neutral CD [31]. Weak acids (A) and bases (B) can form complexes with various CD as shown in Schemes 2 and 3, where K_a and K_a' are the ionization equilibrium constants for the free and encapsulated forms, respectively, and K_1 and K'_1 are the complexation constants for the neutral and ionized forms.

Compared to neutral CD, the complexation can be larger when the CD and the drug carry opposite charges but may decrease when they carry the same charges [32]. For many acidic drugs forming anions, the cationic (2-hydroxy-3-(trimethyl ammonio) propyl)-β-CD acted as an excellent agent for solubilization [7]. In general, ionic forms of drugs are weaker complex-forming agents than their non-ionic forms [33], but in the case of mebendazole, the non-ionized form

Table 1
Some characteristics of α -, β -, γ -, and δ-CD Ref. [7,30]

Type of CD	Cavity diameter (Å)	Molecular weight (g)	Solubility (g/100 ml) in water
α-CD	4.7–5.3	972	14.5
β-CD	6.0–6.5	1135	1.85
γ-CD	7.5–8.3	1297	23.2
δ-CD	10.3–11.2	1459	8.19

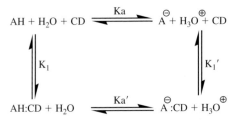

Scheme 2. Apparent equilibrium involved in the binding of an acidic drug (AH) and its anionic form (A⁻) to a CD cavity.

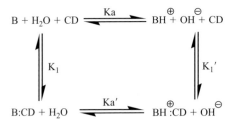

Scheme 3. Apparent equilibria involved in the binding of an anionic drug (B) and its cationic form (BH⁺) to CD.

was less included in hydroxypropyl-β-CD (HP-β-CD) than the cationic form [34].

2.2.3. Effect of temperature

Obviously, the variation of the temperature can affect complexation process. In most cases increasing the temperature decreases the magnitude of the apparent stability constant of the formed complex. The effect was explained in terms of possible reduction of internal interaction forces, such as van der Waals and hydrophobic forces when the temperature increases [33]. However, temperature changes may have negligible effect when the drug–CD interaction is predominantly entropy driven (i.e., resulting from the liberation of water molecules through inclusion complexation) [32]. The thermodynamic parameters of the binding, that are, the enthalpy and the entropy are obtained from the dependence of the binding constant on the temperature through the van't Hoff equation, $R\ln K = \Delta S^\circ - \Delta H^\circ / T$, provided these magnitudes remain constant within the considered temperature range.

The main factors that contribute to the complexation thermodynamics of CDs are (a) penetration of the hydrophobic part of the guest molecule into the cavity, (b) dehydration of the guest, (c) the release of water molecules from CD cavity to bulk water, (d) the conformational change of the host cage and of the guest, (e) and the stability of the formed complex [21].

3. Few examples

In this chapter, we will describe and discuss different examples illustrating different families of CD caged drugs, and we apologise if we forgot any other systems. Figures 2, 3, 6–8 show the molecular structures of the considered drugs.

3.1. Non-steroidal anti-inflammatory drugs

3.1.1. Photostability and efficiency

Photoexcitation of drug–CD systems has been mostly used for the assessment of the drug photostability [35]. The effects of β-CD on the stability and efficiency of some NSAIDs has been reported [36–39]. Monti *et al.* reported that UV irradiation of 2-arylpropionic acids bearing a benzophenone-like aromatic moiety and a substituent with a carboxyl group (ketoprofen (**1**), suprofen (**2**), tiaprofenic acid (**3**) and tolmetin (**4**)), (Fig. 2), in aqueous medium at neutral pH, undergo photodecarboxylation from the dissociated forms as the sole photoreaction [35]. Ketoprofen is photobleached with a quantum yield of 0.75 [36], and converted in an anaerobic condition to the corresponding ethyl derivative I (see Scheme 4). Moreover, in aerated solutions, a series of oxygenated derivatives were also formed [36]. In the presence of β-CD the authors found a decrease of quantum yield of photodecarboxylation from 0.75 to 0.42. Additional photoprocess, likely reductive, contributes to keep the global photodegradation quantum yield of the drug close to the high value of the homogeneous aqueous medium [40]. The formation of product II (Scheme 4) was confirmed by both UV absorption and NMR spectroscopy [40].

The second molecule within the NSAID's family, for which photostability and *in vitro* phototoxicity change upon interaction with β-CD, is diflunisal (**5**, Fig. 2). The results show that in neutral aqueous solutions, this molecule undergoes photodefluorination [41]. However, in presence of β-CD, and upon irradiation, the photodegradation is reduced by ~four-fold, the formation of the photoproduct in free solution is suppressed and a new photoproduct was observed [41].

Another study, which used nanosecond laser flash photolysis and steady-state photolysis experiments, showed that the decarboxylation of the tolmetin is markedly reduced upon inclusion [17]. The rate constants of the decay of the intermediate transients involved in its photodecomposition were slowed down due to the effect of the hydrophobic CD nanocavity. Similar work carried out on the NSAID ketoprofen [42] and suprofen [43] has shown that inclusion in the β-CD cavity leads to a significant decrease in decarboxylation efficiency and to the opening of new photoreactive channels. Moreover, a modification of the distribution of the photoproducts and a considerable reduction of the quantum yield of singlet oxygen sensitized production either by the starting compound or

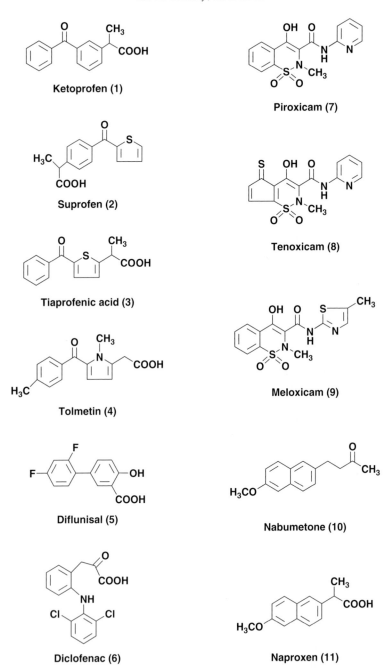

Fig. 2. Molecular structures of the NSAIDs discussed in the text.

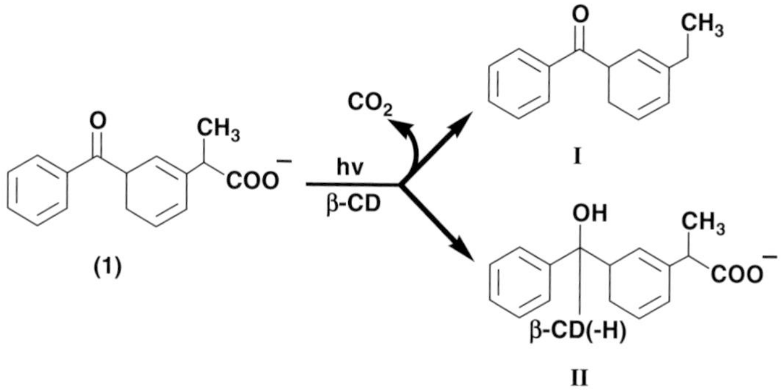

Scheme 4. Photoinduced conversion of ionized ketoprofen (**1**):β-CD entity to different structures in anaerobic (**I**) and aerated (**II**) solutions.

by its photoproducts were also observed [42,43]. The encapsulation of keto-profen by β-CD cavity does not significantly modify the efficiency of the triplet state population but leads to dramatic effects on the kinetics of the triplet state. Moreover, the triplet absorption of the caged drug having the maximum in-tensity at 525 nm is much longer lived [40]. In this regard, the most significant finding is that inclusion of tolmetin in the CD cavity does not seem to influence the efficiency of the electron transfer responsible for the deactivation of the triplet and triggering the decarboxylation. However, it does play a key role in reducing the efficiency of formation of the precursor triplet itself. Therefore, a significant reduction of the decomposition quantum yield was observed [43]. The modification of this drug photochemical and photophysical pathways pro-vides a rational for the CD-mediated protection against its phototoxicity *in vitro*.

3.1.2. Role of conformational aspects and mode of inclusion

The conformational factors and mode of inclusion of the drug may affect its emission properties upon encapsulation. A relatively limited number of studies were dedicated to the conformational aspects of the excited drugs encapsulated within the CD nanocavities and mainly concerned with NSAID's family. These studies attempted to link the structural and dynamical features to get insight into the mode of interaction of the drug with the cell components and infor-mation on the origin of the adverse biological photoeffects.

Oxicams are characterized by rich dynamic structural features. Such charac-teristics reflected in an emission behaviour highly sensitive to the environment. Of particular concern, the inclusion of piroxicam (**7**, Fig. 2) within CD nano-cavities at various pH values [44–47]. The goal was to clarify the reported differences in stoichiometry of the formed inclusion complexes. They found that the pH-dependent self-association phenomena, detected in both alkaline and acidic media, are the defining features for complex stoichiometry [44]. For example at acidic pH, the inclusion complexes of piroxicam zwitterions with β-CD, methyl-β-CD (M-β-CD) and HP-β-CD have 2:2 stoichiometry. At pH > 4 more stable inclusion complexes of 2:5 stoichiometry are observed [44] where at pH = 3.5 a 1:2 stoichiometry was found [45]. This molecule showed several prototrophic forms (Fig. 3) resulting from acid–base equilibria. These involve inter- or intramolecular H-bonds and tautomers formed by proton transfer either in the ground or in the excited state [46,47]. In water at pH = 4 the wavelength at the maximum of the absorption spectrum shifts from 360 nm to 330 nm in presence of β-CD, due to a change in the pK_a value for the deprotonation of the encapsulated hydroxyl group. A weak fluorescence with maximum intensity at 390 nm and emission lifetime shorter than 20 ps was attributed to the anionic form (Fig. 3). An extremely weak emission in the range of 450–550 nm, i.e., with a large Stokes shift, was attributed to the presence of a small amount of the phototautomer, formed by an excited state intramolecular proton transfer (ESIPT) from the hydroxyl group of the benzothiazine ring to the *ortho* carbonyl group. Emission in the wavelength region close to 415 nm was assigned to the zwitterionic form of the drug (Fig. 3) [46].

In the presence of β-CD, the emission intensity around 460 nm was enhanced in a well-defined band, longer lived (*ca.* 60 ps lifetime) and its quantum yield increased from < <0.0001 to ~0.001 [46]. This change was interpreted as being due to enhanced yield of ESIPT reaction. The intramolecular process was favoured by the formation of a piroxicam:β-CD complex. The authors attributed the observed emission component, with a lifetime of ~130 ps in presence of β-CD, to the inclusion complex of the open conformer (Fig. 3) [46].

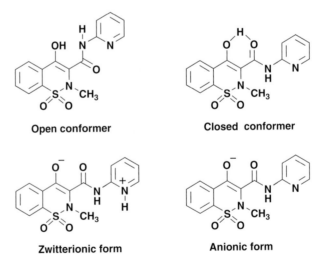

Fig. 3. Molecular structures of prototropic forms of piroxicam.

The authors concluded that the hydrophobic effect of β-CD nanocavity was thought to facilitate the tautomer emission through inclusion of the benzothiazine moiety of the drug in the "closed" geometry (Fig. 3). In this conformation, and owing to the intramolecular H-bond, intramolecular proton transfer is favoured within the cavity. In contrast, H-bonding with water may favour the emission of the "open" conformer in the homogeneous environment [35]. Addition of α-CD does not induce any variation of its emission [46]. This leads to conclude that α-CD cavity is too small to accommodate piroxicam. Studies of molecules showing intra- and intermolecular H-bond using steady-state, as well as fast and ultrafast spectroscopy has shown the conversion of water-bond aromatic dyes to guest having intramolecular H-bond upon encapsulation. Furthermore, the encapsulated structures undergo a femtosecond ESIPT reaction within the cavities [48–53], and the subsequent structure may undergo a twisting motion. This motion depends on the size of CD cavity. A relationship between the wavelengths of the maximum of emission band within the nature of CD cavity was observed. The results obtained in homogeneous media indicated the roles of twisting motion as well as intra- and intermolecular H-bond on the relaxation pathways of photoexcited NSAIDs [54].

A detailed photophysical study of the inclusion interaction of this drug and three CD (β-CD, hydroxypropyl-β-CD (HP-β-CD) and carboxymethyl-β-CD (CM-β-CD)) was carried out [55]. Analysis of the data revealed the following inclusion ability: β-CD > CM-β-CD > HP-β-CD. For β-CD complexes, a 1:2 (guest:host) complex was suggested to form [55]. The authors concluded that the inclusion behaviour depends on the size fit between guest and host. Based on NMR data, Scheme 5 shows the proposed conformation modes of the piroxicam in the nanocavities [55]. Table 2 gives the apparent binding constants for the related complexes in aqueous acidic media.

A recent attempt to clarify the reported differences in the stoichiometry of encapsulated piroxicam found that the pH-dependent self-association phenomena, detected in both alkaline and acidic media, are the defining features for complex stoichiometry [44]. It was also observed that the inclusion complexes of

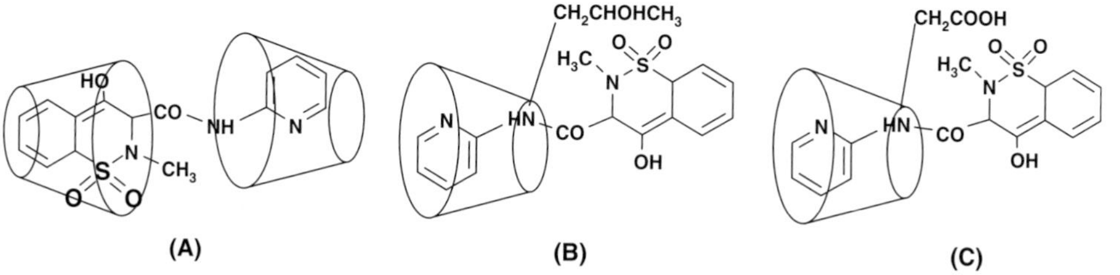

Scheme 5. Proposed conformation of the piroxicam:CD inclusion complexes in water. (A) β-CD, (B) CM-β-CD and (C) HP-β-CD. Reproduced from Ref. [55].

Table 2
Apparent equilibrium constant (K) between free and caged piroxicam in acidic aqueous solutions
at 298 K

CD	pH	Stoichiometry	$10^{-4} K$
β-CD	3.0	1:1	$134 \, M^{-1}$
β-CD	3.5	1:2	$2.0–2.8 \, M^{-2}$
HP-β-CD	3.5	1:1	$0.0122 \, M^{-1}$
CM-β-CD	3.5	1:1	$0.0299 \, M^{-1}$

Source: Data from Refs. [45,55,56].

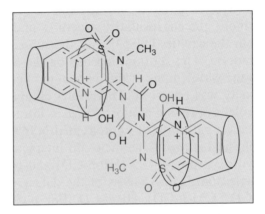

Scheme 6. Possible structure of the 2:2 inclusion complex of piroxicam with CD, in water solutions at $0.9 < pH < 7$. Piroxicam insertion can occur also from the primary end of the CD cavity.
Adapted from Ref. [44].

the piroxicam zwitterions with β-CD, CM-β-CD and HP-β-CD have 2:2 stoichiometry in acidic pH (Scheme 6), through inclusion of the benzyl ring moiety
[44]. When the pH increases, the different ionic states of the molecules created
new geometries, involving also the pyridyl ring in the complexation. Large
inclusion complexes (2:5 stoichiometry) were suggested at pH > 4 [44]. The different stoichiometry values and dimers formation were investigated by using the
two-dimensional ^{1}H NMR spectroscopy.

Further progress in understanding this system focused on the spectroscopic
and molecular modelling techniques to study the interaction of the oxicam group
(piroxicam (**7**), tenoxicam (**8**) and meloxicam (**9**)) with β-CD (Fig. 2) [56]. The
study concluded that in all cases the neutral forms of these drugs were incorporated into β-CD cavity with a binding stoichiometry of 1:1. Molecular modelling studies suggested that the minimum energy configuration gives favourable
energy interaction between the host and the guest when the conjugated rings of
the drugs are inside the hydrophobic cavity of β-CD and the third ring is exposed
to the solvent [56].

Tolmetin (**4**, Fig. 2) is another NSAID system in which photostability and *in vitro* phototoxicity triggered by a photodecarboxylation decrease in presence of β-CD [35]. Upon addition of β-CD to tolmetin, the fluorescence intensity at ~400 nm increases by ~25 times. In the absence of β-CD the fluorescence decays with a short lifetime (0.5 ns), and in its presence the lifetime is longer (2.7 ns). It was concluded that two complexes are formed, both having 1:1 stoichiometry but involving different inclusion modes. These are through either the tolyl or the pyrrole ring in to the CD nanocavity [35]. Lengthening of the fluorescence lifetime in the second complex is accompanied by a less efficient photoconversion to the triplet state. These facts are related to an increased photostability of the drug in presence of β-CD, because the triplet state is a precursor for adiabatic decarboxylation [17].

Diclofenac (**6**, Fig. 2) provides further NSAID example of how the CD nanocavity can influence the emission behaviour of the drug. In the presence of β-CD, inclusion of the aryl moiety holding the acetic acid group induces an increase in the fluorescence intensity of the drug, in both anionic and protonated forms [57]. However, the emission band does not show a significant spectral change. Analysis of the emission change in presence of β-CD suggested the formation of a 1:1 stoichiometry of diclofenac:β-CD complex, in which the acid–base properties of the carboxylic group of the drug are weakly affected (the excited pK_a* is 3.68 in the absence and 4.01 in the presence of CD) [57]. On the other hand, the association constants at room temperature for the neutral and anionic forms are similar (log $K = 3.11$ and $3.15\,M^{-1}$, respectively), suggesting the location of the carboxylic moiety outside and the aryl moiety inside the cavity [57].

At alkaline pHs, the fluorescence intensity of diclofenac drops due to an OH-induced quenching (Fig. 4). In the presence of β-CD this quenching process still takes place but occurs at higher pHs, in agreement with a protective action of the CD against attack by the base [57]. The acidic and anionic forms of the drug also interact with α-CD giving 1:1 complexes of similar stability and similar inclusion geometry as with β-CD. A significant fluorescence intensity enhancement and some spectral modification were observed upon addition of α-CD to a drug water solution (Fig. 5) [58]. A deep inclusion of the phenyl acetic moiety of the drug in the CD nanocavity decreases the non-radiative deactivation rate, thus increasing the emission intensities of the excited diclofenac.

Further interesting example of NSAIDs, whose fluorescence properties within CD are determined by the mode of inclusion, are nabumetone (**10**), a naphthalene derivative with a butanone side chain; and naproxen (**11**), a parent propionic acid derivative (Fig. 2). The dominant features of the inclusion of nabumetone inside β-CD are the formation of two conformations of the drug, but the folded structure is the main one, probably due to the presence of water in the proximity of the hydroxyl groups at the gates of CD [59]. This situation

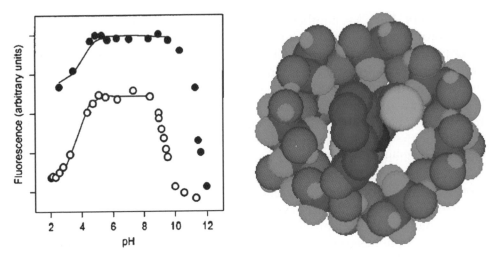

Fig. 4. Influence of pH on the fluorescence intensity of 5.5×10^{-6} M diclofenac (O) in the absence and (●) in the presence of 10^{-2} M β-CD, $\lambda_{exc} = 289$ nm, $\lambda_{em} = 362$ nm. Inset shows the expected inclusion of diclofenac inside β-CD. The left part is printed from Ref. [57] with permission by the Royal Society of Chemistry, 2005.

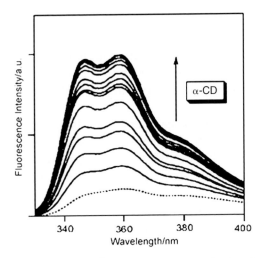

Fig. 5. Fluorescence spectra of 2.8×10^{-6} M diclofenac, (—) in the absence and (—) in the presence of increasing amounts of α-CD (from 1.6×10^{-4} to 7.2×10^{-3} M) in neutral aqueous media, $\lambda_{exc} = 289$ nm. Data from Ref. [58].

may explain the high sensitivity of the drug to the presence of water molecules, and it is likely that the polarity provided by water molecules favours the formation of a bipolar structure, and therefore, the intramolecular interaction. Theoretical calculations supported the formation of both *extended* and *half-folded* conformations. The latter has the carbonyl group pointing to the aromatic ring [59]. The extremely weak emission of nabumetone in water, with maximum at 355 nm, is explained in terms of non-radiative processes of the keto group leading to folded conformations [59]. These twisting motions may affect intramolecular charge transfer or radical-like reaction, involving the excited carbonyl and the naphthalene moiety [59]. Addition of α-CD to aqueous solutions of the drug induces a strong enhancement of the emission intensity and the observation of a structured spectrum. This effect also reflected by a change in the absorption spectrum, which is red-shifted by $\sim$9 nm. Naproxen (**11**) (Fig. 2) also has a fluorescence band peaked at 355 nm, which exhibits very little sensitivity to the α-CD environment. In presence of β-CD, the emission intensity of both drugs increases and changes in a similar way. Comparative examination of the values of ΔH° and ΔS° and ^{1}H NMR peaks shift upon complexation of both nabumetone and naproxen to α- and β-CD indicated that the butan-2-one chain of nabumetone is preferentially inserted in the α-CD cavity, whereas the methoxynaphthalene moiety of both drugs is accommodated in the larger β-CD cavity. A conformational effect by the α-CD cavity was thus proposed to determine the emission behaviour of the nabumetone complex [60]. Indeed, inclusion of the butanone chain in the α-CD cavity hinders the formation of the folded conformation responsible for its emission quenching [59,60].

3.1.3. CD-containing polymer

Of specific interest is the use of CD-containing polymers to gain more insight into the environmental factors, which determine the photostability of the encapsulated drugs. CD-containing polymers are recently used in drug delivery [1]. The photochemistry findings in most published studies of the ternary complex naproxen:CD:polymer were thoroughly characterized with steady-state measurements [61–63]. The interaction of the drug with the different components of the ternary complex varies with the nature of the polymer and the CD used. Studies of the effect of polyethylene glycol (PEG-35,000) on the inclusion between naproxen and β-CD demonstrated the formation of ternary complexes of naproxen:β-CD:polymer [62]. These studies showed that ternary complexes formation reduced the photoreactivity of the naproxen, whereas the binary one naproxen:β-CD does not do it [62]. The effect of the polyvinyl pyrrolidone (PVP) and/or HP-β-CD on the photostability of aqueous solutions of naproxen was also studied [63]. Kinetic analysis of the data revealed that the formation of ternary complexes reduced drug photodegradation. The presence of free naproxen clearly sensitizes the photoperoxidation of linoleic acid and the photosensitizing effect decreases where the PVP concentration increases [63]. The hydrogen bonds between the drug and CD in the binary complexes are cleaved due to the formation of ternary complexes. Therefore, in ternary complexes the formation of phototoxic radical intermediate decreases. These results confirm the idea that the phototoxicity of the drug, at least for naproxen, in the presence of CD may be due to the different behaviour of the complexed radical inter-mediates, rather than to an increase in drug photostability or to the nature of the photoproducts formed [63].

3.2. Anticancer drugs

CDs have also shown their potential as media for controlling photochemical and photophysical properties of some anticancer drugs [64–67]. The related studies are mainly based on the use of steady-state methods. Flutamide (**12**), Fig. 6, is a non-steroidal antiandrogen medicine used in advanced prostate cancer therapy. This drug is photosensitive and induces liver toxicity [68]. The photochemistry of the flutamide in homogeneous media provides an example of nitro to nitrite intramolecular photorearrangement triggered by the twisted conformation of the nitro group [64]. Incorporation of flutamide in the β-CD cavity leads to large effects on both the efficiency and the nature of the pho-tochemical deactivation pathways of the guest molecule [64]. A 20-fold increase in the flutamide photodecomposition quantum yield, and the formation of photoproducts originated by both reduction of the nitro group and cleavage of the amide bond were observed upon encapsulation by the CD nanocavity [68]. The study concludes that the photoreactivity of the guest molecule cannot be simply rationalized in terms of polarity of the host molecule. The occurrence of a structural modification, as suggested by the induced circular dichroism (ICD) spectrum, and by the nature of the photoproducts appears the main reason for the enhanced photoreactivity of caged flutamide [64].

Other authors have addressed the formation of inclusion complexes between flutamide and α-, β-, γ-CD, HP-β-CD and CM-β-CD, HP-β-CD and heptakis (2,3,6-tri-*O*-methyl)-β-CD (DM-β-CD). Only 1:1 stoichiometry was considered with the used CD [65]. The stability constant determined by UV absorption changes revealed the following order in the inclusion ability α-CD $\approx \gamma$-CD $< \beta$-CD $\approx$ CM-β-CD $<$ TM-β-CD $\approx$ HP-β-CD $<$ DM-β-CD. Clearly, the size of CD cavity plays an important role in the complex formation. Molecular modelling of the host–guest geometry of these complexes suggested that the β-CD complex is energetically the most favourable entity [65]. The difference between TM-β-CD and DM-β-CD complexes in the stability constant may be ascribed to the change in the shape of the cavity. The cavity of TM-β-CD is deeper than that of DM-β-CD because methyl groups are located at both ends of the cavity. This steric hindrance can influence the geometry of host–guest interactions and causes a shallower penetration of the aromatic part of the flutamide molecule into the cavity of TM-β-CD than in DM-β-CD [65]. Note also that the methyl groups play a hydrophilic role protecting the guest from H-bonding interactions with water molecules. From the point of view of equilibrium dynamics, the presence of methyl groups at the gates of CD decreases the rate constant of decomplexation and therefore increase the apparent equilibrium constant of the free and caged drugs. However, other energetic factors play a role in the stability of the complex, as noted above.

The study of the effect of HP-β-CD on the spectroscopic features of hesperetin (**13**), and its 7-rhamnoglucoside, hesperidin (**14**, Fig. 6) in water by means of Uv-visible absorption and fluorescence spectroscopy, showed that the stoichiometry of the complexes is 1:1 [66]. The authors attributed the higher degree of interaction showed by hesperetin to the higher hydrophobicity and smaller size of the aglycone molecule, which therefore exhibited a greater affinity to the CD and better fit into its cavity. The effect of encapsulation on the hesperetin and hesperidin activity was evaluated by means of different biological assays, related to the different mechanisms of *in vivo* action. In all cases, the efficiency was larger for the encapsulated drugs, with respect to the free ones [66]. These results are of great interest for their potential application in pharmaceutics.

The possibility of encapsulation between β-CD and six anticancer drugs (pipobroman, melphalan, civicin, 7-chlorocamptothecin, thiocolchicine and hexahydro-TMC69, structures not shown) belonging to different classes of action was characterized by molecular dynamics simulations [67]. The trajectories of the insertion angles, rotation of the non-polar parts of the drugs inside the macrocycle and other geometrical features give information on the dynamics of the complexes. In all cases the relative binding energies indicate the possibility of formation of inclusion complexes between β-CD and the anticancer drugs either in a 1:1 or in 2:2 stoichiometries. The values of the energetic components for the non-bonded potential energy showed that, under the conditions considered, all inclusion complexes should in principle be stable and, in the case of symmetrical molecules, no preferential mode of insertion was detected. However, exception for hexahydro-TMC69, for which the octyl-OH primary conformation was favoured with respect to the alternative octyl-OH secondary by approximately 18 kJ/mol. Furthermore, in the case of the two larger molecules considered, thiocolchicine and hexahydro-TMC69, the energetic gain due to the formation of 1:2 (guest:host) complexes was also calculated from simulation. The results indicate that, under the conditions considered (excess of β-CD), the 1:2 complex is favoured for both drugs. The theoretical results obtained so far are in good agreement with similar studies on β-CD inclusion compounds in vacuum. The dynamic behaviour of the same 1:1 and 1:2 (guest:host) complexes has been examined using a combination of molecular mechanics derived from molecular dynamic (MD) simulations in explicit solvent, and solvation free energy derived from a continuum solvation model. In general, it was found that van der Waals interactions and non-polar contributions to solvation always provide the basis for the favourable absolute free energy of binding.

The spectral characteristics of the inclusion complexes of polypyrrolic sensitizers and CD used in photodynamic therapy (PDT) of tumours have been reported [68–70]. The objectives of these studies were to describe the photo-physical properties of the caged sensitizers inside CD nanocavities in suitable environments for the cultivation of cancerous lines and to determine the optimal radioactive conditions for exploiting PDT of cancerous cells. The PDT of cancer uses a combination of photosensitizer drugs and light to give rise to reactive oxygen species in the tumour environment, leading to tumour death. The inclusion interaction between meso-tetrakis(4-hydroxylphenyl) porphyrin (THPP, **15**, Fig. 6) and β-CD and its derivatives has been recently reported [69]. The dominant feature of this study is the enhancement of the fluorescence intensity of THPP by 300 times after addition of HP-β-CD [69]. When comparing to β-CD and CM-β-CD abilities, the strongest are of HP-β-CD. The result was explained in terms of a strong hydrogen bond between HP-β-CD and THPP [69]. The authors concluded that the large increase in the fluorescence intensity may have a potential application in the detection of THPP in pharmacokinetics and bio-distribution studies. A large enhancement of fluorescence intensity was also observed upon addition of β-CD to TDBHPP (**16**, Fig. 6) [70]. In addition, the Uv–visible absorption spectra showed that the inclusion by the cavity of HP-β-CD causes a de-aggregation of THPP, which is in agreement with the effect of acidity of water solution.

Curcumin (**17**, Fig. 6) has a wide array of pharmaceutical activities, including antioxidant, anti-carcinogenic and anti-inflammatory properties. Recently, Baglole *et al.* reported the effect α-, β- and γ-CD and their hydroxypropylated derivatives in the solubility and fluorescence of the curcumin [71]. The results indicated the formation of 1:2 (guest:host) inclusion complexes with CDs (Scheme 7). For parent CDs, the strongest binding was observed for α-CD, the smallest cavity, whereas in the case of hydroxypropylated CD, the strongest binding was observed for HP-β-CD (Table 3). Significant enhancement of fluorescence occurs upon binding to α- and β-CD, and HP-β-CD. This emission increase might be used for potential applications in fluorescence-based detection of this pharmaceutical compound. However, in the case of γ-CD, a decrease in the emission intensity was observed. The result was explained in terms of a folded drug inclusion molecule into the large size of γ-CD cavity. In all cases, the binding to these CDs provides a significant increase in the aqueous solubility of curcumin, with the largest increase (a factor of *ca.* 5000) in the case of 10 mM HP-β-CD. Such increase in drug solubility in aqueous solutions might be useful for applications in pharmaceutics.

3.3. Antibiotic drugs

Few examples on the photophysics and photochemistry of encapsulated antibiotic drugs have been represented [72–75]. Ciprofloxacin (**17**, Fig. 7), a fluoroquinolone carboxylic acid, is an antimicrobial agent. Addition of HP-β-CD

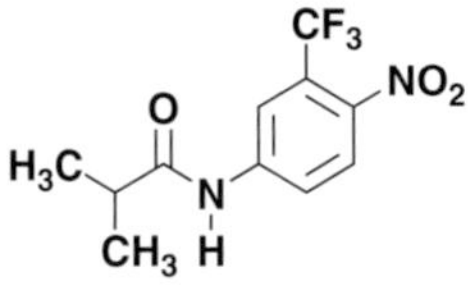

Flutamide (12) 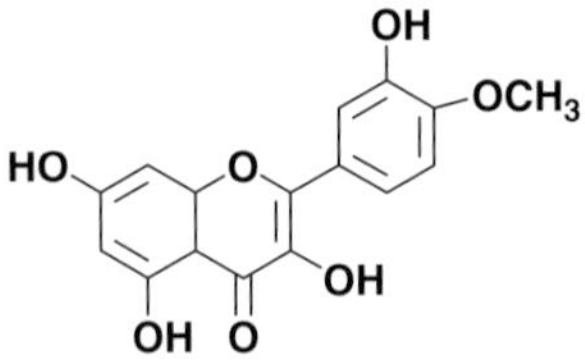 Hesperetin (13) 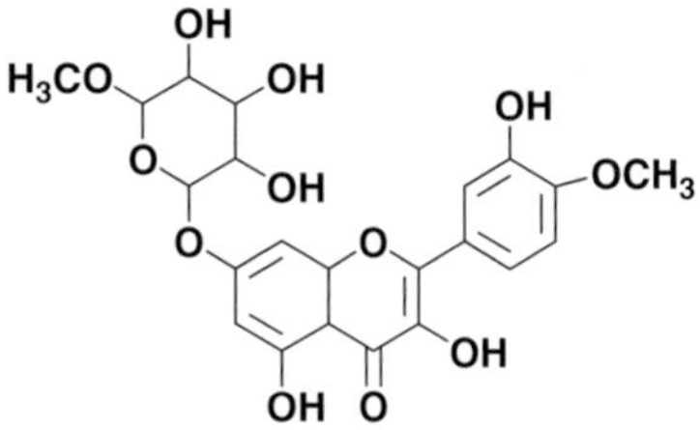 Hesperidin (14)

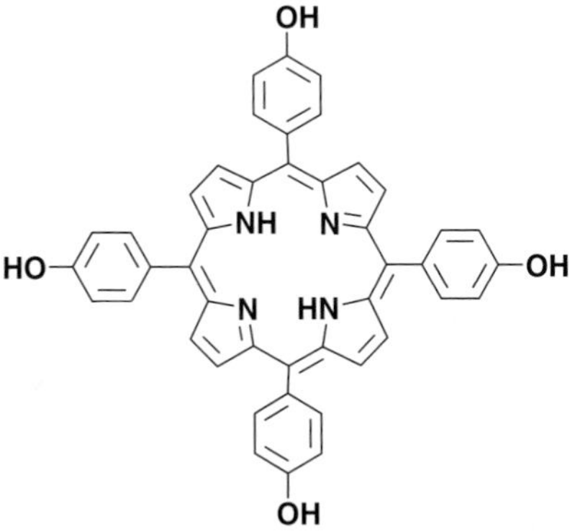

THPP (15) 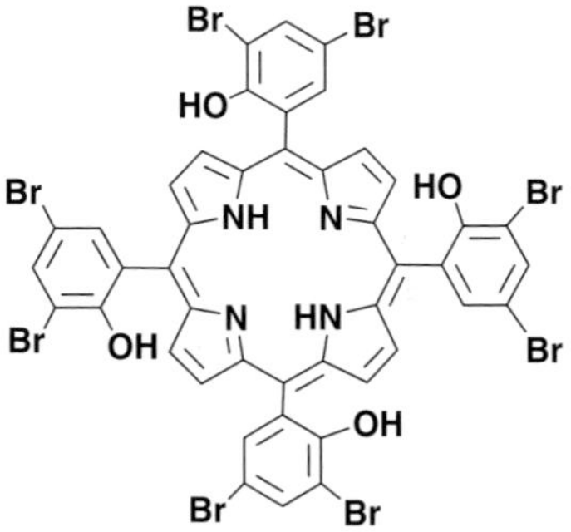 TDBHPP (16)

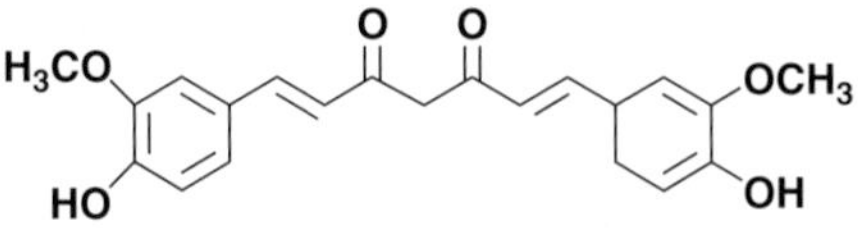

Curcumin (17)

Fig. 6. Chemical structures of anticancer drugs discussed in the text.

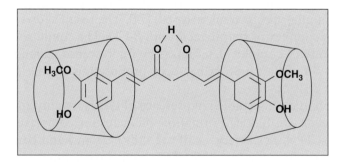

Scheme 7. The proposed structure of the 1:2 curcumin:CD inclusion complex following Ref. [71].

Table 3
Equilibrium constants values for 1×10^{-5} M curcumin:CD in 1% methanol/water at room temperature

CD	$K_1(\text{M}^{-1})$	$K_2(\text{M}^{-2})$
α-CD	3300 ± 1200	5.4 ± 1.3
HP-α-CD	12000 ± 3000	28 ± 2
β-CD	280 ± 120	6.6 ± 5.2
HP-β-CD	3400 ± 1800	120 ± 40
γ-CD	≈ 8000	≈ 140
HP-γ-CD	21000 ± 1200	8.4 ± 22

Note: K_1 and K_2 are for 1:1 and 1:2 complexes respectively.
Source: Data from ref. 71.

to ciprofloxacin solution results in a significant enhancement of the fluorescence intensity [72]. Analysis of the experimental data revealed the existence of 1:1 inclusion complex between ciprofloxacin and HP-β-CD and an equilibrium constant of 343 M^{-1} at room temperature. Based on two dimensions NMR spectra, the authors proposed a confined structure of the complex, where the binding force is behind its formation and stability is mainly due to dipole–dipole, hydrophobic and van der Waals interactions [72].

Sparfloxacin (**19**) and norfloxacin (**20**) (Fig. 7) are also fluoroquinolone antibiotic drugs. Their complexation with β-CD in water and in mixed organic/aqueous media has been studied by UV-visible absorption spectroscopy [73]. The authors concluded that only 1:1 complexes were formed and the presence of organic solvents causes the degradation of the complexes. The binding constants of the inclusion complexes of sparfloxacin and norfloxacin with β-CD in water are larger than those with mixed organic–water media at room temperature. The higher the hydrophobicity of the co-solvent, the weaker the complexation of sparfloxacin and norfloxacin by β-CD [73]. Chao et al. also studied the interaction of sparfloxacin with β-CD using several analytical techniques (^{1}H-NMR, ^{13}C-NMR, fluorescence spectroscopy, infrared spectroscopy, thermal analysis and scanning electron microscope) [74]. The result shows the formation of 1:1 inclusion complex of sparfloxacin with β-CD. The equilibrium constant at room temperature of complex determined by fluorescence technique is 50 M^{-1}. The configuration of the complex has been proposed using 2D ^{1}H NMR technique (Scheme 8).

Another antibiotic drug studied in presence of CD is natamycin (**21**, Fig. 7), a polyene macrolide antibiotic. Formation of inclusion complexes with β-CD, HP-β-CD or γ-CD causes a very small red shift (≈ 1 nm) and band broadening of its UV absorption spectrum [75]. This band broadening might be due to electron density change of the guest trapped inside CD, which partially shields

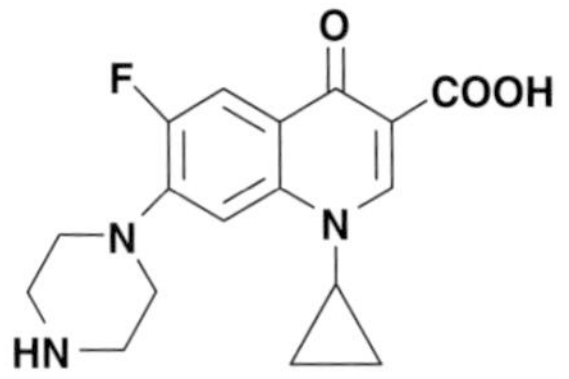

Ciprofloxacin (18) 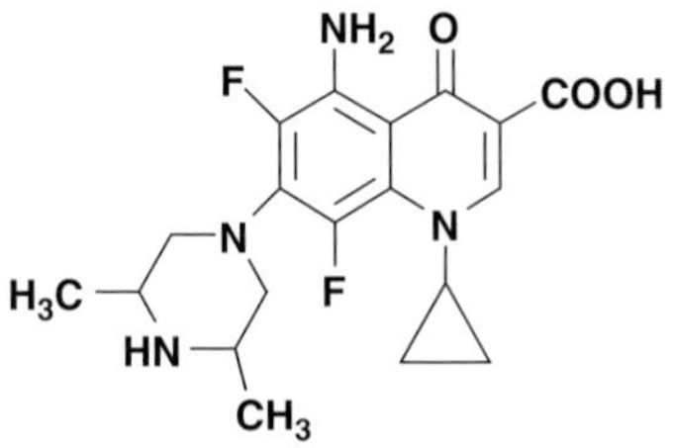   Sparfloxacin (19) 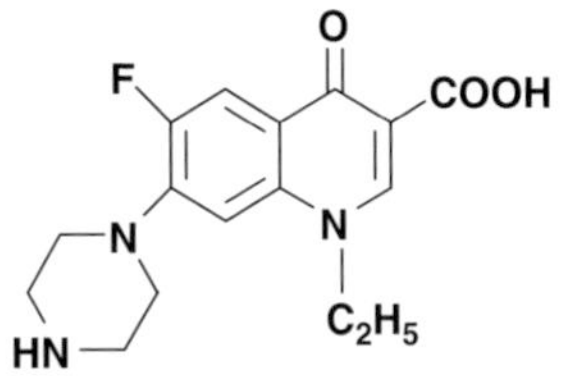   Norfloxacin (20)

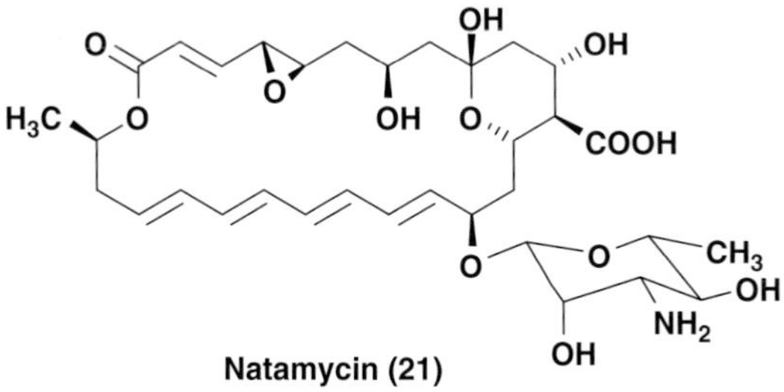

Natamycin (21)

Fig. 7. Molecular structures of the antibiotic drugs discussed in the text.

the electrons of the guest [76]. The small spectral shift was interpreted as an indication of inclusion complex formation. In the presence of α-CD, the UV absorption spectrum does not exhibit any shift, which suggests the absence of inclusion complex using a smaller cavity [75].

3.4. Anaesthetic drugs

The physicochemical properties of the inclusion complex formed between β-CD and an anaesthetic drug, procaine (**22**, Fig. 8) have been studied by means of spectroscopic (UV-visible, steady-state fluorescence and NMR) and thermodynamic (density and speed of sound) techniques [77]. The global picture of the results suggests that the drug penetrates the CD cavity by the wider gate. The NH$_2$ group of the guest penetrates first, bringing the aromatic ring of the drug inside the cavity, and the stoichiometry of the complex is 1:1. The author proposed a model to determine the binding constants from UV-visible spectra, based on wavelength shift instead of an absorbance change upon addition of CD to the drug solution. The association constant values, obtained from emission and absorption data, range from 400 to 200 M^{-1} on going from 285 to 313 K. The variations of standard enthalpy and entropy are $\Delta H^\circ = -19 \pm 5\,\mathrm{kJ\,mol^{-1}}$, and $\Delta S^\circ = -15 \pm 7\,\mathrm{J\,K^{-1}\,mol^{-1}}$, respectively. The author concluded that the encapsulation of procaine hydrochloride by β-CD is an exothermic and enthalpy governed process, with a balance between van der Waals interactions,

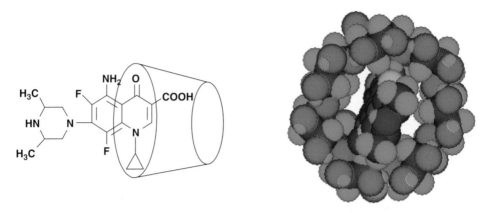

Scheme 8. Proposed conformation of sparfloxacin:β-CD inclusion complex, reproduced from Ref. [74].

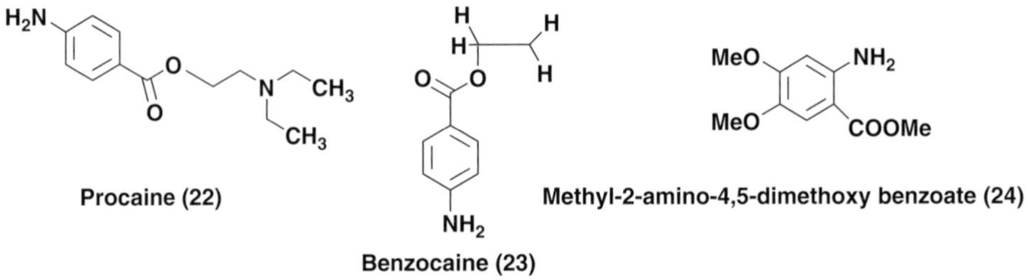

Procaine (22)

Benzocaine (23)

Methyl-2-amino-4,5-dimethoxy benzoate (24)

Fig. 8. Chemical structure of anaesthetic drugs discussed in the text.

hydrophobic effect, and solvent reorganization being mainly responsible for the overall stability of the complex. A thermodynamic study has also shown that in the reorganization of water molecules after the association of CD and the drug, four to five water molecules are expelled from host and four to five water molecules are removed from the hydration shell of procaine [77].

Benzocaine (**23**, Fig. 8) is a local anaesthetic drug whose low water solubility limits its application to topical formulations. Recently, the physicochemical characterization of inclusion complexes of this molecule inside β-CD cavity has been reported [78]. Differential scanning calorimetry and electron microscopy gave evidences of the formation and on the morphology of the complex. Analysis of the fluorescence emission spectra revealed the existence of benzocaine: β-CD complex of 1:1 stoichiometry. Phase solubility diagrams allowed the determination of the association constants between the drug and β-CD (549 M^{-1} at room temperature). They revealed that a three-fold increase in benzocaine solubility can be reached upon complexation with β-CD. Experimental data of 2D ^1H NMR spectroscopy provided information to analyze the molecular interactions within the complex and suggest an inclusion model for this drug into β-CD where the aromatic ring of the anaesthetic is located near the gate of the cavity. Moreover, toxicity studies showed that the complex was less toxic than the free drug alone, since it induces a decrease in the *in vitro* oxidation of human haemoglobin. These results suggested that the formed complex represents an effective novel formulation to enhance benzocaine solubility in water, making it promising for use outside of its traditional application, i.e., in infiltrative anaesthesia.

Tormo *et al.* [79] reported on fast relaxation of a caged molecule, methyl 2-amino-4,5-dimethoxy benzoate (**24**, Fig. 8), an anaesthetic analogous to procaine and tetracaine, in order to shine some light on the role played by the α-, β-, and γ-CD confinement effect on the photorelaxation of caged drugs. The stoichiometry of the formed complexes is 1:1 in all the used cavities, and the equilibrium constant using β-CD is $12 \pm 3\,M^{-1}$ at 293 K. Encapsulation by β-CD induces a large increase in emission quantum yield and lifetime values, suggesting a hydrophobic effect of the nanocage on the photophysical behaviour of the guest. The lifetimes of the caged guest are 0.8 ns and 3.7 ns, while those in water are 47 ps and 110 ps. The time-resolved emission spectra showed a red shift ($720\,cm^{-1}$) when using β-CD cavity: this spectral relaxation is larger than the observed one in water ($490\,cm^{-1}$), and the difference suggests a shortening of the NH_2 ... CO_2Me intramolecular H-bond of the caged guest at the S_1 state, when compared to the free molecule. The anisotropy experiment shows that the trapped guest can still move within the cage. The authors suggest that the reported results are relevant to those found in photochemistry and photophysics of encapsulated drugs and give new information for a better understanding of the photochemical events of medicines used in phototherapy.

4. Conclusion and outlook

In this chapter, we focused on some photophysical and photochemical properties of encapsulated drugs (NSAIDs, anticancer, antibiotics and anaesthetic) inside CD cavities. Several factors play an important role on the effect of CD complexation and stability of the caged drugs. In this regard, a discussion of various relaxation processes (intramolecular charge transfer, inter- and intra-molecular proton transfer, twisting motion) was also covered in this chapter. The effects of photoexcitation of caged drugs are used for understanding drug photostability. In this direction, increasing the lifetime of singlet excited state leads to an increased photostability of the drug inside CD cavity. In most of these cases, the triplet state of the drug is the precursor of formation of photoproducts. Moreover, in some cases, the changes in the drug photochemistry upon inclusion into CD might be related to the decrease in its phototoxicity. The modification of some NSAID's photochemical and photophysical pathways discussed in this chapter, provides a basis for the CD-mediated protection against drug phototoxicity *in vitro*. The overall results confirmed that the use of CD as a protecting cavity may represent a simple and useful strategy to increase drug photostability, and to minimize the biological damage caused by NSAIDs. The relationship between the fluorescence properties of some drugs in presence of CD and the conformational features of the molecule may help to clarify the mode of interaction of the drug with the cell components and will provide information on the origin of the adverse biological photoeffects.

Furthermore, from the point of view of biology the interest in drug:CD complexes is related to the development of *in vivo* photosensitization by drugs. Serious phototoxic reactions (mainly diseases of epidermis and dermis), as well as photoallergic and photomutagenic effects, can be induced in patients under sunlight irradiation while treated with pharmacologically important chemicals such as antibacterials, tranquillisers, antimicotics and NSAIDs. Such noxious effects, correlated to the drug photochemical reactivity, significantly decrease from phenothiazine and NSAIDs using CD (*in vitro* cellular systems). Therefore, the application of CD in the design of protective systems for the clinical administration of photosensitising drugs is a real promising issue.

Reduction of the drug photoproducts upon addition of polyvinylpyrrolidone (PVP) to a mixture of naproxen and HP-β-CD might suggest that polymer used for the increasing of the drug photostability in addition to CD caging. Moreover, encapsulation of the anaesthetic drugs as benzocaine and procaine inside the β-CD complex represents an effective novel formulation to enhance benzocaine solubility of drugs in water. The results reported for fast relaxation of a caged anaesthetic analogous to procaine and tetracaine may help for a better understanding of the photochemical events of medicines used in phototherapy.

Several laser-based (spectroscopy and microscopy) techniques are available to examine drugs–CD interactions at atomic and ultra-short timescales. The expected results will be of great importance not only for a better understanding of these systems, but also for their storage, handling and use.

Acknowledgments

This work was supported by the MEC and the JCCM through the projects: MAT2002-01829, CTQ2005-00114 and SAN-04-000-00. M.E. thanks the MEC (Spain) for supporting his sabbatical stay at the UCLM through project SAB2004-0086.

References

[1] M. Davis, M. Brewster, Nat. Rev. Drug Dis. 3 (2004) 1023.

[2] J. Szejtli, Cyclodextrins and Their Inclusion Complexes, Akadémiai Kiadò, Budapest, 1982.
[3] K. Connors, Chem. Rev. 97 (1997) 1325.
[4] D. Duchen (Ed.), Cyclodextrin and Their Industrial Uses, Editions de Santé, Paris, 1987, p. 448.
[5] J. Szejtli, Cyclodextrin Technology, Kluwer Academic Publishers, Dordrecht, 1988, pp. 450.
[6] K. Förmming, J. Szejtli, Cyclodextrins in Pharmacy, Kluwer Academic Publishers, Dordrecht, 1994, pp. 224.
[7] T. Loftsson, M. Brewester, J. Pharm. Sci. 85 (1996) 1017.
[8] K. Uekama, F. Hirayama, T. Irie, In drug targeting delivery. in: A. Boer, (Ed.), Drug Targeting Delivery, Vol. 3, p. 411, Harwood Publishers, Amsterdam, 1993.
[9] J. Babu, J. Pandit, Cyclodextrin Inclusion Complexes: Oral Applications, Eastern Pharmacist, pp. 37–42, 1995.
[10] K. Uekama, F. Hirayama, T. Irie, Chem. Rev. 98 (1998) 2045.
[11] T. Loftsson, M. Masson, Rev. Int. J. Pharm. 225 (2001) 15.
[12] P. Bortolus, S. Monti, Adv. Photochem. 21 (1996) 1.
[13] F. Bosca, M. Marin, M. Miranda, Photochem. Photobiol. 74 (2001) 637.
[14] T. Hoshino, K. Ishida, T. Irie, F. Hirayama, M. Yamasaki, J. Incl. Phenom. 6 (1988) 415.
[15] M. Partyka, B.H. Au, C.H. Evans, J. Photochem. Photobiol. A: Chem. 140 (2001) 67.
[16] M. Valero, S. Costa, M. Santos, J. Photochem. Photobiol. A: Chem. 132 (2000) 67.
[17] S. Sortino, J.C. Scaiano, G. De Guidi, S. Monti, Photochem. Photobiol. 70 (1999) 549.
[18] P. Caliceti, S. Salmaso, A. Semenzazo, T. CaroWglio, R. Fornasier, M. Fermeglia, M. Ferrone, Bioconjug. Chem. 14 (2003) 899.
[19] T. Hoshino, K. Ishida, T. Irie, F. Hirayama, M. Yamasaki, J. Incl. Phenom. 6 (1988) 415.
[20] J. Chao, J. Li, D. Meng, S. Huang, Spectrochim. Acta A. 59 (2003) 705.
[21] M. Rekharsky, Y. Inoue, Chem. Rev. 98 (1998) 1875.
[22] P. Ross, M. Rekharsky, J. Biophys. 71 (1996) 2144.
[23] R. Bergeron, Cycloamylose-substrate binding, in: J. Attwood, J. Davies, D. MacNicol (Eds), Inclusion Compounds, Academic Press, London, 1984, p. 391.
[24] V. Stella, R. Rajewski, Pharm. Res. 14 (1997) 556.
[25] L. Szente, J. Szejtli, G. Kis, J. Pharm. Sci. 87 (1998) 778.
[26] J. Castillo, J. Canales, J. Garcia, J. Lastres, F. Bolas, J. Torrado, Drug. Dev. Ind. Pharm. 25 (1999) 1241.
[27] P. Mura, M. Faucci, P. Parrini, S. Furlanetto, S. Pinzauti, Int. J. Pharm. 179 (1999) 117.
[28] M. Blanco, J. Moyano, J. Martinez, J. Gines, J. Pharm. Biomed. Anal. 18 (1998) 275.
[29] H. Akasaka, T. Endo, H. Nagase, H. Ueda, S. Kobayashi, Chem. Pharm. Bull. 48 (2000) 1986.
[30] T. Endo, H. Nagase, H. Ueda, S. Kobayashi, T. Nagai, Isolation, Chem. Pharm. Bull. 45 (1997) 532.
[31] K. Okimoto, R.A. Rajewski, K. Uekama, J.A. Jona, V.J. Stella, Pharm. Res. 13 (1996) 256.
[32] Y. Nagase, M. Hirata, K. Wada, H. Arima, F. Hirayama, T. Irie, M. Kikuchi, K. Uekama, Int. J. Pharm. 229 (2001) 163.
[33] M. Tros déllarduya, C. Martin, M. Goni, M. Oharriz, Drug Dev. Ind. Pharm. 23 (1998) 301.
[34] D. Diaz, M. Bernad, J. Mora, C. Uaons, Drug Dev. Ind. Pharm. 25 (1999) 111.
[35] S. Monti, S. Sortino, Chem. Soc. Rev. 31 (2002) 287.
[36] L. Costanzo, G. De Guidi, G. Condorelli, A. Cambria, M. Famá, Photochem. Photobiol. 50 (1989) 359.

[37] G. De Guidi, L. Chillemi, S. Costanzo, S. Giuffrida, S. Sortino, G. Condorelli, J. Photochem. Photobiol. B: Biol. 23 (1994) 125.
[38] F. Boscá, M. Miranda, G. Garganico, D. Mauleon, Photochem. Photobiol. 60 (1994) 96.
[39] F. Boscá, M. Miranda, J. Photochem. Photobiol. B: Biol. 43 (1998) 1.
[40] S. Monti, S. Sortino, G. De Guidi, G. Marconi, New. J. Chem. 22 (1998) 1013.
[41] S. Sortino, S. Giuffrida, S. Fazio, S. Monti, New J. Chem. 25 (2001) 707.
[42] S. Monti, S. Sortino, G. De Guidi, G. Marconi, New J. Chem. 22 (1998) 599.
[43] S. Sortino, G. De Guidi, G. Marconi, S. Monti, Photochem. Photobiol. 67 (1998) 603.
[44] S. Rozou, A. Voulgari, E. Antoniadou-Vyza, Eur. J. Pharm. Sci. 21 (2004) 661.
[45] G. Escandar, Analyst 124 (1999) 587.
[46] Y. Kim, D.W. Cho, S.G. Kang, M. Yoon, D. Kim, J. Lumin. 59 (1994) 209.
[47] J. Martins, M. Sena, R. Poppi, F. Pessine, Appl. Spectrosc. 53 (1999) 510.
[48] A. Douhal, F. Amat-Guerri, A.U. Acuña, Angew. Chem. Int. Ed. Engl. 36 (1997) 1514.
[49] A. Douhal, T. Fiebig, M. Chachisvilis, A.H. Zewail, J. Phys. Chem. A 102 (1998) 1657.
[50] I. García-Ochoa, M.A. Díez López, M.H. Viñas, L. Santos, E. Martínez Ataz, F. Amat-Guerri, A. Douhal, Chem. Eur. J. 5 (1999) 897.
[51] D.P. Zhong, A. Douhal, A.H. Zewail, Proc. Natl. Acad. Sci. (USA) 97 (2000) 14052–14055.
[52] A. Douhal, Chem. Rev. 104 (2004) 1955.
[53] A. Douhal, M. Sanz, L. Tormo, Proc. Natl. Acad. Sci. (USA) 102 (2005) 18807.
[54] M. El-Kemary, Chem. Phys. 295 (2003) 1.
[55] G. Xiliang, Y. Yu, Z. Guoyan, Z. Guomei, C. Jianbin, S. Shaomin, Spectrochimica Acta: A 59 (2003) 3379.
[56] R. Banerjee, H. Chakraborty, M. Sarkar, Biopolymers 75 (2004) 355.
[57] J.A. Arancibia, G. Escandar, Analyst 124 (1999) 1833.
[58] J.A. Arancibia, M. Boldrini, G. Escandar, Talanta 52 (2000) 261.
[59] M. Valero, S. Costa, J. Ascenso, M. Velásquez, L. Rodriguez, J. Incl. Phenom. Macro. Chem. 35 (1999) 663.
[60] M. Valero, S. Costa, M. Santos, J. Photochem. Photobiol. A: Chem. 132 (2000) 67.
[61] T. Loftsson, Pharmazie 53 (1998) 733.
[62] M. Valero, C. Carrillo, L. Rodríguez, Int. J. Pharm. 265 (2003) 141.
[63] M. Valero, B. Estéban, J. Photochem. Photobiol. B: Biol. 76 (2004) 95.
[64] S. Sortino, S. Giuffridal, G. De Guidi, R. Chillemi, S. Petralial, G. Marconi, G. Condorellil, S. Sciuto, Photochem. Photobiol. 73 (2001) 6.
[65] J. Taraszewska, K. Migut, M. Koźbiał, J. Phys. Org. Chem. 16 (2003) 121.
[66] S. Tommasini, M. Calabró, R. Stancanelli, P. Donato, C. Costa, S. Catania, V. Villari, P. Ficarra, R. Ficarra, J. Pharm. Biomed. Anal. 39 (2005) 572.
[67] M. Fermeglia, M. Ferrone, A. Lodi, S. Pricl, Carbohyd. Polym. 53 (2003) 15.
[68] D. Leory, A. Dompmartin, C. Szczurko, Photodermatol. Photoimmunol. Photomed. 12 (1996) 216.
[69] X. Guo, W. An, S. Shuang, F. Cheng, C. Dong, J. Photochem. Photobiol. A: Chem. 173 (2005) 258.
[70] Y. Zhang, W. Xiang, R. Yang, F. Liu, K. Li, J. Photochem. Photobiol. A: Chem. 173 (2005) 264.
[71] K. Baglole, P. Boland, B. Wagner, J. Photochem. Photobiol. A: Chem. 173 (2005) 230.
[72] J. Chao, D. Meng, J. Li, H. Xu, S. Huang, Spectrochim. Acta A 60 (2004) 729.
[73] N. Al-Rawashdeh, Asian J. Chem. 16 (2004) 483.
[74] J. Chao, H. Tong, S. Huang, D. Liu, Spectrochim. Acta A 60 (2004) 161.
[75] J. Koontz, J. Marcy, J. Agric. Food Chem. 51 (2003) 7106.

[76] J. Szejtli, Inclusion of Guest Molecules, Selectivity and Molecular Recognition by Cyclo-dextrins, in: J. Szejtli, T. Osa (Eds), Cyclodextrins; Comprehensive Supramolecular Chemistry, Vol. 3, Elsevier Science, Amsterdam, 1996, p. 189.
[77] C. Merino, E. Junquera, J. Jiménez-Barbero, E. Aicart, Langmuir 16 (2000) 1557.
[78] L. Pinto, L. Fraceto, M. Santana, T. Pertinhez, J. Junior, E. Paula, J. Pharm. Biomed. Anal. 39 (2005) 956.
[79] L. Tormo, J. Organero, A. Douhal, J. Phys. Chem. B. 109 (2005) 17848.

Cyclodextrin Materials Photochemistry, Photophysics and Photobiology
Abderrazzak Douhal (Editor)
© 2006 Elsevier B.V. All rights reserved
DOI 10.1016/S1872-1400(06)01005-3

Chapter 5

Excited-State Properties of Higher-Order Cyclodextrin Complexes

Sandra Monti, Pietro Bortolus

Istituto per la Sintesi Organica e la Fotoreattività, C.N.R. Area della Ricerca, I-40129 Bologna, Italy

1. Introduction

The most common complexation stoichiometry of cyclodextrins (CDs) with a guest is 1:1, i.e. one host CD includes in the cavity one guest molecule. In general, the photochemical properties of the guest dissolved in homogeneous aqueous solution are affected by the inclusion in the CD cavity. The extent of the modification depends, besides the chemical structure, on the depth of the inclusion in the cavity. A review of the influence of the CD complexation on the photochemical properties of a large variety of guests has been published in 1996 [1]. As "higher-order complexes" we refer to systems in which the host:guest stoichiometry is 1:2, 2:1, 2:2 and higher, or a third partner is involved in the associated structure. On account of a more complete encapsulation and/or a more tight interaction of the guest with the CD cavity, an alteration of the photophysical and photochemical properties larger than that observed in 1:1 complexes is expected. This chapter is not intended to be an exhaustive review of every reported case in which higher-order association induced excited-state behaviour modifications. Selected examples from literature published up to the first 6 months of 2005 have been considered. Focus has been centred on higher-order CD complexes, whose photophysical and/or photochemical properties have been mechanistically rationalized in the light of available structural information.

Comparison with the features of the related 1:1 associates has been performed when possible. This approach has been adopted to evidence the specific role of the intermolecular host–guest, host–host and guest–guest interactions in the peculiar excited-state behaviour of the higher-order CD associates.

2. Singlet excited state

2.1. Molecular emission

Cyclodextrin complexation very frequently brings about an enhancement of the guest fluorescence intensity due to steric hindrance to molecular degrees of freedom assisting non-radiative deactivation of the singlet excited state. The change of microenvironmental polarity is a further parameter that can influence the emission features of CD:guest systems. Cavity size generally controls such effects. However, complex stoichiometry also plays an important role, since in higher-order complexes the guest is generally more tightly interacting with the CD environment and more protected from water contact.

With α-CD and β-CD the excited singlet state properties of 1,4-dimethoxy-benzene (**1**) appear to be strongly modified [2]. The changes are particularly important with α-CD, where 1:1 ($K_{11} = 13\,\text{M}^{-1}$) and 2:1 ($K_{21} = 8000\,\text{M}^{-2}$) complexes are sequentially formed, with the last one largely predominating in solution. The lowest energy absorption band, $\lambda_{\max} = 286\,\text{nm}$, is red-shifted by few nanometres and becomes structured at relatively low α-CD concentration ($9 \times 10^{-3}\,\text{M}$); the total fluorescence quantum yield increases from 0.135 in water to 0.52 due to the increase of the radiative rate; the internal conversion appears to be completely suppressed (Table 1). The comparison with solvents like ethanol and cyclohexane indicates that the polarity of the medium plays a minor role, the excited-state properties being mainly controlled by the constraints imposed by CD inclusion. The structure of the complexes (Fig. 1), determined by an approach involving a combination of computational techniques based on

Table 1
Fluorescence quantum yields (Φ_f), lifetimes (τ_f), radiative rate constant (k_r), rate constants for intersystem crossing (k_{isc}) and for internal conversion (k_{ic}); intersystem crossing quantum yield (Φ_T), bimolecular rate constants for triplet quenching by O_2 (k_{O2}) and ground state (k_{GS}), for **1** in various media. Data from Ref. [2]

	Φ_f	τ_f/ns	$k_r/10^7\,\text{s}^{-1}$	Φ_T	$k_{isc}/10^7\,\text{s}^{-1}$	$1-(\Phi_f+\Phi_T)$	$k_{ic}/10^7\,\text{s}^{-1}$	$k_{O2}/10^7\,\text{M}^{-1}\text{s}^{-1}$	$k_{GS}/10^7\,\text{M}^{-1}\text{s}^{-1}$
H_2O	0.135	1.75	7.7	0.425	24.0	0.44	25.0	700.0	25
EtOH	0.20	2.67	7.5	0.75	28.0	0.05	1.9		
C_6H_{12}	0.21	3.03	6.9						
α-CD	0.52	2.60	20.0	0.60	23.0	0		32.0	0.19
β-CD	0.19	2.13	8.9	0.51	24.0	0.30	14.0	3.0	7.0

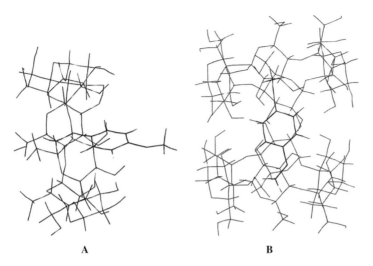

Fig. 1. Computed conformations of 1:1 (A) and 2:1 (B) complexes of α-CD with 1,4-dimethoxy-benzene. Reprinted with permission from Ref. [2]. Copyright (1996), American Chemical Society.

conformational calculations by a Dynamic Monte Carlo method, including solvation, and on induced circular dichroism (icd) calculations, was able to account for the observed phenomena. Indeed, in the 2:1 α-CD associate (structure B) the guest is completely isolated from water, whereas in the 1:1 structure (structure A) the aromatic ring is largely exposed to the bulk medium. A 1:1 complexation with β-CD, due to a deeper penetration of the guest in the cavity and higher association constant ($630\,M^{-1}$), produces a higher perturbation.

Derivatives containing two 1,4-dimethoxybenzene-like residues (e.g. **2**) exhibit a photophysical behaviour influenced by CD complexation via cavity size-controlled inclusion of bichromophoric moieties [3]. The main features of these systems is the presence of a distribution of ground-state conformations with two main equilibrium geometries, a monomer-like species, characterized by an emission at *ca.* 330 nm and a dimer-like species emitting at *ca.* 380 nm. The conformational equilibrium is controlled by the cavity size of the complexing CD. Indeed for **2** 1:1 association with α-, β- and dimethyl-β-CD leads to a quenching of the dimer-like emission and enhancement of the monomer-like one, whereas 1:1 association with γ-CD induces the opposite effect. This was explained with the inclusion of two aromatic rings in the large γ-CD macrocycle (Scheme 1, Case b), whereas only one of them can be hosted in the smaller α-, β- and dimethyl-β-CD cavities (Case a).

Host–host hydrogen bonding interactions involving hydroxylic groups is also a factor that can control the guest emission by affecting the complex flexibility. This has been shown in higher-order CD complexes of naphthalene derivatives [4,5]. The emission of naphthalene, **3**, ($[3] \geqslant 5 \times 10^{-5}\,M$) in presence of α-, β-, and

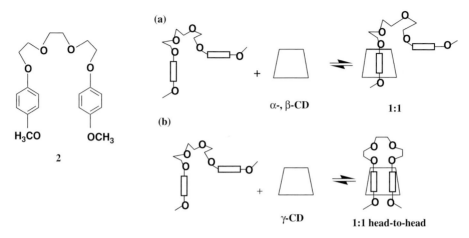

Scheme 1. Inclusion of 2 in CD, reproduced in part from Ref. [3] by permission of The Royal Society of Chemistry.

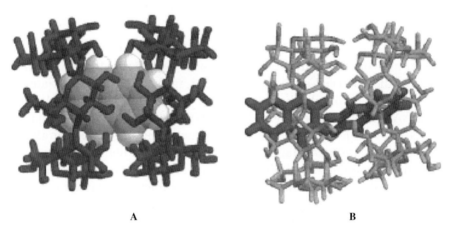

Fig. 2. Computed conformations of 2:1 (A) and 2:2 (B) complexes of α-CD with naphthalene. Reprinted in part with permission from Ref. [4]. Copyright (2000), American Chemical Society.

γ-CD is modified by the formation of higher-order complexes [4,6,7]. The stoichiometry of the predominant complex with α-CD is 2:1 host:guest. This arrangement has the effect of strongly enhancing the vibronic structure of the fluorescence spectrum and increasing both quantum yield and lifetime of **3** emission with respect to neat aqueous solution ($\Phi_f = 0.11$, $\tau_f = 39$ ns in water and $\Phi_f = 0.19$, $\tau_f = 86.2$ ns in 0.05 M α-CD). The structure of the complex afforded by Dynamic Monte Carlo calculations [4] shows that **3** is tightly encapsulated and well shielded from the aqueous environment (Fig. 2, structure A); the most stable conformations appear to be those where the secondary hydroxylic rims are facing each other and hydrogen bonding between host hydroxylic groups occurs.

Dimethyl- (**4**) and diethyl-2,6-naphthalene-carboxylate (**5**) also exhibited fluorescence modifications by formation of higher-order complexes with α- and β-CD [5,8]. The intensity greatly increased by addition of α-CD, whereas the addition of β-CD caused a small increase of **5** emission and had no effect on that of **4**. In all the cases, except for **4** in presence of β-CD, the fluorescence variations as a function of [CD] gave evidence for a 2:1 stoichiometry. Van'T Hoff plots for the formation of the complexes pointed to an enthalpy-driven complexation. Molecular mechanics calculations indicated that in the 2:1 β-CD-**5** complex the guest is well shielded from the solvent and stabilized by H-bonding between the faced β-CD, while in the 2:1 α-CD-**5** it is partially exposed to the aqueous medium.

$$4 \quad R = -CH_3$$
$$5 \quad R = -C_2H_5$$

The fluorescence of 3*H*-indoles is very sensitive to the environment so that they were proposed as potential probes for microstructures. In a series of papers [9–14], the CD effect on the fluorescence of various aryl-substituted derivatives was examined and the formation of 1:1 and 2:1 complexes was inferred either from the variations of Φ_f or from the analysis of the fluorescence decay as a function of [CD]. A 3:1 stoichiometry was assessed in some indoles containing long-chain substituents [15].

2.2. Excimer emission

The ability of CD to bring two guest molecules in close contact can provoke an excimer-like emission. Such effect was observed by addition of β- or γ-CD ([CD] $\geqslant 5 \times 10^{-3}$ M) to the **3** aqueous solution and was attributed to the interaction of two guest molecules within a 2:2 complex [4,6,7]. The formation of such a complex with β-CD was experimentally supported by the variations of absorbance and excimer fluorescence intensity as a function of [**3**] at fixed [β-CD] [6]. The experimental findings were confirmed by conformational calculations of the complex structure by a Dynamic Monte Carlo method (see structure B of Fig. 2) [4]. The most stable 2:2 structures show the important contribution of intramolecular hydrogen bonding between the secondary hydroxylic groups of the two CD moieties and an orthogonal position of the two **3** molecules, which appear rather free to move within the large cavity, by maintaining essentially the same relative orientation. The fluorescence lifetimes monitored at 340 (monomer) and 420 nm (excimer-like) were 60.9 and 90.7 ns, respectively; both are higher than the lifetime (39 ns) of free **3** in water, in agreement with the coexistence of 1:1 and 2:2 complexes in the adopted experimental conditions. Excimeric emission as a consequence of higher-order complexation was also exhibited by several methyl- and ethyl-**3** derivatives in the presence of CD [16–18]. The singlet excited, complexed naphthalene is protected from the external quenching by I^-: $k_q = 6 \times 10^9 \, M^{-1} s^{-1}$ for the free molecule, $3.9 \times 10^9 \, M^{-1} s^{-1}$ for the 1:1 complex, and $1.8 \times 10^9 \, M^{-1} s^{-1}$ for the 2:2 complex [6].

1-Cl-naphthalene (**6**) also gave excimer emission in 2:2 complexes in water with β-, γ-CD and a maltosyl-γ-CD [19,20]. Fluorescence polarization studies carried out in presence of D-glucose allowed to conclude that the guest pair and the two encapsulating β-CD rotate independently during the excimer lifetime according to a relatively weak interaction between the host and the guest [19].

Pyrene (**7**) fluorescence has a vibronic structure sensitive to the polarity of the medium. The value of the ratio (R) between the first and the third vibronic bands increases from ~ 0.6 in cyclohexane to ~ 1.9 in water [21], so that it has been extensively employed to characterize the polarity of microenvironments of organized systems such as micelles, vesicles and CD.

The 2:1 complexes were proposed to form with the three natural CD. In the complex with α-CD, R decreased from 1.85 to 1.7 while the fluorescence lifetime τ_f remained unaffected [22]. Two α-CD cap **7** from opposite sides; the low penetration of the guest in the host cavity leaves the central part of the molecule directly exposed to water and is responsible of the small R decrease and of the constancy of τ_f.

The sequential formation of 1:1 and 2:1 complexes β-CD:**7** is supported by several steady-state and time-resolved spectroscopic experiments. The effect of the complexation is a decrease in R [22–26] and an increase in the global fluorescence quantum yield [26], but it was difficult to disentangle the contribution of the two complexes to the above phenomena (overall association constant $6 \times 10^4 \, M^{-2}$ [25], $3.8 \times 10^4 \, M^{-2}$ [26]). On the other hand, the fluorescence decay of **7** is described by the sum of two exponentials. In air-equilibrated solutions, the lifetimes were ~ 130 and ~ 300 ns. The short-lived component was attributed to the free **7** and to the 1:1 complex. The long-lived component was attributed to the 2:1 complex [22,25]. It can be concluded that β-CD offers a better protection from water than α-CD.

The emission of a dilute **7** solution ([**7**] $\approx 3 \times 10^{-7} \, M$) in presence of γ-CD is characterized by a band with vibronic structure due to the monomer, and by a broad band peaked at ~ 475 nm due to an excimer-like species [27–32]. This emission was attributed to the formation of a 1:2 [27–30] or a 2:2 [28,32] complex. The discrepancies arise, in part, from the different experimental conditions adopted. The relative intensities of the monomer and excimer emission depend on [γ-CD]; the excimer intensity (I_{exc}) firstly increased with [γ-CD] and decreased when [γ-CD] $\geqslant 5 \times 10^{-3}$ M [32]. Moreover, the excimer emission disappeared in 0.4 N NaOH. The rationale is that the 1:1 complex accumulates with increasing [γ-CD] and associates to give a 2:2 complex in which the interaction of two molecules of **7** leads to excimer emission. An increase of [γ-CD] over 5×10^{-3} M favours the formation of the 2:1 CD:**7** complex, which emits as a monomer [31]. The ionization of the CD at high pH (CD p$K_a = 12.1$) hampers the self-association of 1:1 complexes inhibiting the interaction. This interpretation was confirmed by studies of complexation dynamics by using the stopped flow technique [32]; formation of the 2:2 complex was almost completed within 200 ms after the mixing of the solutions containing **7** and γ-CD; re-equilibration in several seconds pointed to the formation of the thermodynamically more stable 2:1 complex, evidenced by a decrease of the excimeric and increase of the monomeric emission. The rates of association/dissociation of the 1:1 complex to/from the kinetically favoured intermediate 2:2 structure were found to be $(6.3 \pm 1.3) \times 10^7 \, \text{s}^{-1}/(73 \pm 5) \, \text{s}^{-1}$, i.e. considerably slower than the typical rates relevant to guest association/dissociation to/from a single CD moiety (see in the triplet-state section the case of naphthylethanols [33]).

The fluorescence variations of diluted solutions of *p*-terphenyl (**8**, at concentration $< 10^{-6}$ M) [34] and of α-tertiophene (**9**) [35,36] solutions in presence of CD have some features common to those of pyrene. In presence of β-CD, the fluorescence of both **8** and **9** is similar to that observed in CH_3OH or CH_3OH/H_2O and is attributed to a locally excited (LE) singlet state. In presence of 10^{-2} M γ-CD, **8** displays, besides the structured emission due to the LE state, a structureless excimer-like emission centred at ~ 400 nm, attributed to the electronically excited 2:2 complex. In agreement with this assignment, the contribution of the structureless emission to the global fluorescence decreased by decreasing [**8**] or by adding further γ-CD. The long-wavelength emission also disappeared when the pH of the solution was raised to 13.4. The time-resolved emission spectra progressively shifted to the red with the increasing gate times and the decay at $\lambda > 400$ nm was described by a three-exponential function, whose pre-exponential factors changed with the analysis wavelength. The lifetime of the shortest-lived component ($\tau = 1.4$ ns) is that of **8** in presence of β-CD and was attributed to the emission of the LE state in singly occupied (1:1 or 2:1) complexes. The two longer lifetimes ($\tau = 7.6$ and 27 ns) were attributed to two excimers characterized by different mutual overlapping of the chromophores [34].

At low [γ-CD], the **9** emission spectrum has a maximum at 425 nm, attributed to a LE singlet state. By increasing [γ-CD] the maximum shifted to longer wavelengths (at [γ-CD] $= 10^{-2}$ M the maximum was at 506 nm). Upon further

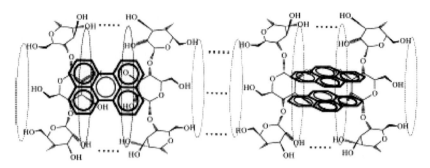

Fig. 3. Association geometry of the pyrene 2:2 and perylene 2:1 complexes with γ-CD, proposed in Ref. [37], reprinted with permission. Copyright (2004), American Chemical Society.

increase of [CD], the spectrum changed back to that obtained at low [γ-CD]. The same ipsochromic shift was observed when the pH of the solution was increased from 8.2 to 13.6. The fluorescence decay of **9** in 10^{-2} M γ-CD showed two components (4.96 and 0.35 ns) when the analysis was carried out at 400 and 500 nm, and only one (4.96 ns) when the analysis wavelength was 590 nm, where the LE singlet state does not emit. The 4.96 ns emission was attributed to an included excited dimer. These results are in agreement with the formation of a 1:1 complex that associates to give a 2:2 complex. The structure of the higher-order complex was disrupted to give back the 1:1 form either by increasing the pH over the pK_a of the CD or by increasing the concentration of γ-CD [36].

7 ([**7**] $= 2.6 \times 10^{-6}$ M) in 10^{-2} M γ-CD displays both the monomer and the excimer fluorescence when excited at 338 nm. Excitation at the same wavelength of a solution 0.1×10^{-6} M perylene (**10**) in 10^{-2} M γ-CD had a very low emission compared to that of 2.6×10^{-6} M **7**. When the two aromatics are simultaneously present in solution at the above concentrations, the **7** excimeric emission was quenched and a strong **10** fluorescence appeared [37]. The **10** emission was observed also when the solution was excited at 350 nm, a wavelength absorbed only by the 2:2 γ-CD:**7** inclusion complex. The authors proposed the occurrence of a singlet energy transfer from the excited **7**-dimer to **10**. Given the very low concentration of the two aromatics, they suggest an association of the complexed donor and acceptor able to bring close the two partners to form the supramolecular structure shown in Fig. 3. The presence of a 2 ns rise-time for the **10** emission when the solution was excited at 350 nm further supports the energy-transfer mechanism.

2.3. Proton transfer and proton dissociation

The excited-state intramolecular proton-transfer (ESIPT) process (Scheme 2) in 4-methyl-2,6-dicarbomethoxyphenol (**11**) was shown to be perturbed by CD complexation [38]. This molecule absorbs at 322 nm as an intramolecularly

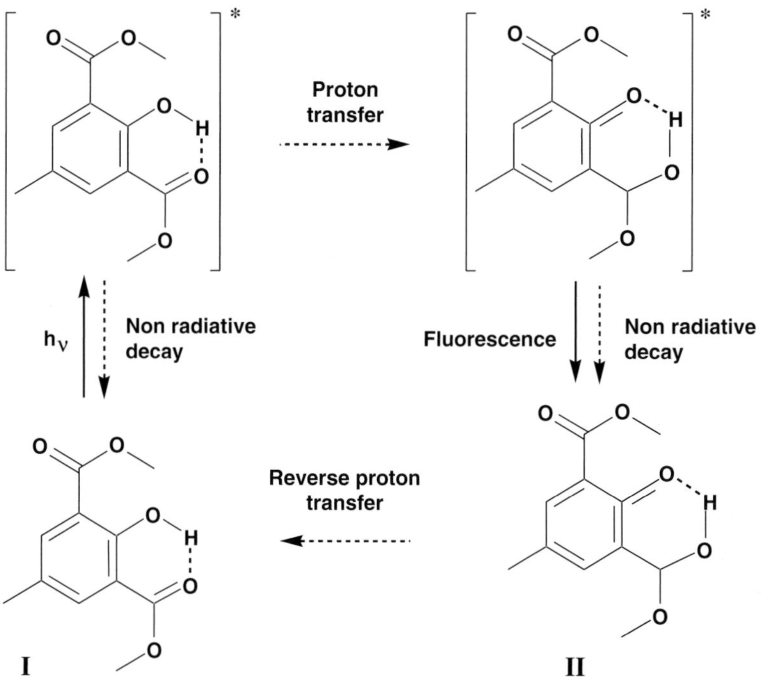

Scheme 2. ESIPT of **11**, adapted from Ref. [38].

hydrogen-bonded closed conformer (I) and emits with a very large Stokes shift, at 450–460 nm, as the phototautomer (II). In the presence of increasing concentration of CD, the phototautomeric emission of the molecule is enhanced. Complexes with 2:1 host:guest stoichiometry with α-CD and 1:1 stoichiometry with β-CD were proposed to form. These effects were rationalized as due to the intrinsic lower H-bonding ability and hydrophobicity of the CD cavity, minimizing the interference of intermolecular H-bonding interactions involving water molecules. On going from water to DMSO and DMF, a lower enhancement of the tautomeric emission was observed in spite of an increased stability of these complexes (Table 2). In such non-aqueous media, participation of H-bonding solvent molecules in host:guest:solvent ternary complexes of 1:1:2 stoichiometry with β-CD and 1:1:1 stoichiometry with α-CD was proposed to perturb ESIPT through external interaction (for a further paragraph on ternary complexes see below).

The proton dissociation process in CD included guests can be strongly perturbed by higher-order association. This is shown by the effect that CD exerts on the pK^* of naphthols [39–42] and hydroxybiphenyls [43,44]. The fluorescence emission of these molecules at pH = 6 is characterized by a blue-band with maximum in the region 330–350 nm, due to the emission of the undissociated

Table 2
Association constants and fluorescence enhancements of inclusion complexes of **11** with CD in water and non-aqueous media. Data from Ref. [38]

Host	Solvent	Stoichiometry	$K_{ass}(\mathrm{M}^{-1}$ or $\mathrm{M}^{-2})$	F/F_0
β-CD	Water	1:1	290.3	1.34
β-CD	DMSO	1:1	354.0	1.10
β-CD	Formamide	1:1	663.4	1.09
α-CD	Water	2:1	2.6×10^5	1.14
α-CD	DMSO	2:1	5.6×10^5	1.06
α-CD	DMF	2:1	36.7×10^5	1.05

Table 3
pK_a, excited-state pK_a^*, fluorescence quantum yield (Φ_f^n), proton dissociation yield (Φ^d) and rate constant (k_d), rate constant for fluorescence quenching by protons (k_q) for **12** in various environments. Data from Ref. [39]

	pK_a	pK_a^*	Φ_f^n	τ^n/ns	Φ^d	$k_d/10^7\,\mathrm{s}^{-1}$	$k_q/10^7\,\mathrm{M}^{-1}\mathrm{s}^{-1}$
$\mathrm{H_2O}$	9.52	3.05	0.2	7.5	0.38	7.6	2.4
β-CD	9.9	3.6	0.24	9.2	0.17	2.2	2.0
α-CD (1:1)	—	—	0.22	8.4	0.23	3.5	—
α-CD (2:1)	> 9.54	> 4	0.4	11.2	< 0.02	< 0.3	1.5

form (Ar–OH*), and by a weaker red-band peaked at 380–430 nm, corresponding to the emission of the deprotonated form (Ar–O$^-$*) generated by partial prototropic equilibration in the excited singlet state. The addition of CD has an opposite effect on the two bands: the blue-band increases in intensity, while the red-band tends to disappear.

α-CD forms with 2-naphthol (**12**) both 1:1 and 2:1 complexes, while β-CD gives only a 1:1 complex [39]. In Table 3 are collected the photophysical decay parameters of the excited species involved in the excited-state equilibrium and the H$^+$ quenching rate constant on **12***. The data in the table clearly show that the photophysical parameters of **12*** are much more influenced in the 2:1 complex than in the two 1:1 complexes. This fact suggested a very different environment of the molecule in the two types of complex. In the 1:1 species, **12*** experiences an environment similar to the aqueous medium, while in the 2:1 structure the water molecules are partially excluded from the cavity formed by the two α-CD which encapsulate the probe. This is confirmed by the structure of the complexes obtained by molecular mechanics calculations [39], which were found similar to those obtained for the interaction of naphthalene and CD.

1-Naphthol (**13**, $pK_a = 0.4$) in the excited state is an acid stronger than **12***, so that **13*** in water at pH = 7 emits exclusively from the dissociated form (ArO$^-$).

However, in the presence of a functionalized β-CD, the $N(N'$-formyl-L-phenyl-alanyl)-β-CD, the base emission was partially quenched and the growth of a fluorescence peaked at ~350 nm (corresponding to the emission of the neutral form in organic solvents) was observed [45]. This was attributed to the formation of a 2:1 complex between the functionalized CD and **13**, which partially hampers the prototropic equilibrium in the excited state.

The interaction of 3- and 4-OH-biphenyl (**14** and **15**) with α-CD leads to very stable 2:2 structures, formed by self-association of 1:1 complexes [43,44]. Values for log (K_{11}/M^{-1}) ~1.5–2 and log (K_{22}/M^{-2}) = 4.7–4.9 were obtained by fluorescence titration in neutral phosphate buffer at 295 K. Such a structural arrangement, evidenced in the icd spectra by a positive/negative splitting of the dichroic bands and by conformational calculations, produces a substantial enhancement of the blue-band accompanied by a depression (practically total for **15** and only partial for **14**) of the red-band (see Fig. 4 for **15**).

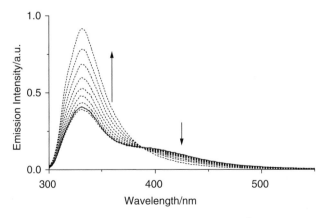

14 3-OH

15 4-OH

Fluorescence decay of **15** in presence of α-CD is well described by a three-exponential kinetic model. The lifetimes of the complexed species, longer than that of the free molecule ($\tau_{\text{free}} = 1.55$ ns), are 4.3 ns for the 1:1 complex and 10.8 ns for the 2:2 complex. The quantum yields exhibit a different trend: that of the free molecule is 0.075, that of the 1:1 complex is estimated to be lower by

Fig. 4. Variations of the fluorescence intensity from **15** 2.2×10^{-5} M in phosphate buffer 0.05 M pH 6.76 upon addition of α-CD from 4.4×10^{-5} to 2.0×10^{-2} M. Adapted from Ref. [43].

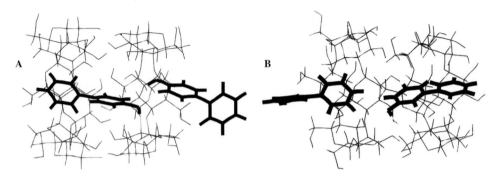

Fig. 5. Computed structure of 2:2 complexes with α-CD of (A) **15** from Ref. [43], reprinted in part by permission of Wiley; and (B) **14** from Ref. [44], reprinted in part by permission of the PCCP Owner Societies.

one order of magnitude and that of the 2:2 complex is higher by more than a factor five [43]. All this is consistent with the formation of 2:2 complexes with structures A and B (Fig. 5), where the two OH groups of the guest molecules are not (for **15**) or are only partially exposed (for **14**) to the aqueous solvent so that proton dissociation is totally or partially hampered. The "external" location of the guest with respect to the CD cavity in the 1:1 complexes does not produce appreciable perturbation of the proton dissociation process, and an overall decrease of the emission intensity is the only signature of the initial 1:1 association step detected at low α-CD concentrations ($<10^{-4}$ M) [43,44].

2.4. Intramolecular charge transfer

p-Dimethylaminobenzonitrile (**16**) is the prototype system for this photoprocess; its dual fluorescence in organic solvents of moderate polarity originates from two emitting states, whose nature and structure have been largely debated (for a review see Ref. [46]). It is accepted that the high-energy emission occurs from a LE state and that the low-energy emission arises from an intramolecular charge transfer state (ICT state), which undergoes structural reorganization in polar solvents. Two main parameters control the degree of charge transfer and structural change: the local polarity of the medium and the size of the free volume around the reorganizing groups.

16

Moreover, the emission capability of the ICT state is influenced by the environment *via* the energy gap to the ground state. For the above reasons, studies in CD cavities were undertaken since 1984 [47] with the aim to infer the polarity of the environment and the geometry of the inclusion complexes (for a overview of the relevant literature up to 1994 see Ref. [1]).

In **16**, the intensity of the ICT emission band is considerably enhanced and the λ_{max}, at 520 nm in water, is progressively shifted up to 440 nm, for concentrations of α-CD above 2×10^{-2} M; addition of β-CD or γ-CD makes both the LE and ICT emission to moderately enhance and the long-wavelength emission to shift to 500 nm; the emission maxima depend on the excitation wavelength; the excitation spectra for the LE and ICT emission are different; the time-resolved fluorescence decays are complex; the solutions exhibit aging effects [47–52] and strong influence of temperature and guest concentration on the emission spectral shapes were observed [53–55]. It was speculated on the formation of several different 1:1 inclusion complexes and/or of complexes with higher stoichiometries, 2:1 for α-CD and 2:2 for β-CD. Recently, the interaction of **16** with α- and β-CD was examined by an approach involving (i) a combination of computational techniques; (ii) quantitative analysis of multi-wavelength absorption and fluorescence data, obtained by varying temperature, CD and guest concentration; and (iii) laser-flash photolysis to access the triplet properties [54,55]. This study gave evidence for the formation of 1:1 and 2:1 complexes in the α-CD:**16** system and of 1:1 and 2:2 associates in the β-CD-DMABN system and provided the structures (Fig. 6), and an exhaustive characterization of the complexes. Table 4 and Fig. 7 report part of the information attained. In the 1:1 associates, the guest experiences a non-constrained, water-rich environment due to either a shallow penetration into the small α-CD cavity (see structure A) or the likely coinclusion of water molecules in the larger cavity of β-CD (see structure C), so that the emission of the 1:1 complexes is similar to that of the free molecule in water with prevailing ICT emission. In the 2:1 complex with α-CD (structure B) the guest does not experience any important steric hindrance to the excited-state geometry change, leading to the population of the ICT state. Moreover, the double CD cage protects the excited state from the contact with water, strongly enhancing the ICT emission by a reduced non-radiative deactivation. On the contrary, in the 2:2 complex with β-CD (see structure D) the head-to-head orientation of the guest molecules with the two dimethylamino groups facing each other does not allow population of the ICT state, so that the LE emission at 360 nm largely prevails. An additional emission at 420 nm, with a decay time constant of 14 ns and with the same excitation spectrum as the LE emission, is observed in this complex. This was assigned to an excimer-like emissive state, populated because of the close proximity of two guest molecules in the 2:2 dimer. An alternative structure with the two –CN groups facing each other was proposed for this complex by Pozuelo *et al.* [56] on the basis of molecular mechanics calculations.

The ICT emission in relation to the formation of higher-order complexes in CD complexes was explored in 4-dimethylaminoacetophenone (**17**) [57]. The

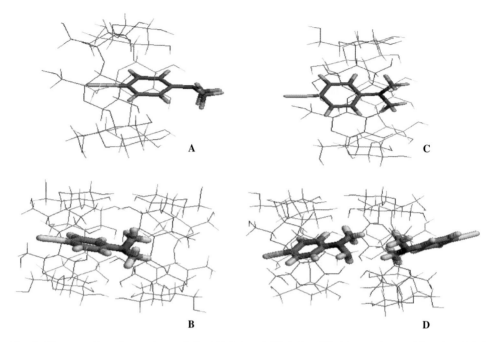

Fig. 6. Computed conformations of the (A) 1:1 and (B) 2:1 α-CD:**16** complexes; (C) 1:1 and (D) 2:2 β-CD:**16** complexes. From Ref. [55], reprinted in part by permission of the PCCP Owner Societies.

Table 4
Association constants (K_{ass}, M^{-1}, M^{-2} or M^{-3}) by UV absorption, total fluorescence quantum yield (Φ_f), lifetimes $\tau(\lambda_{em})$, intersystem crossing yields (Φ_{isc}), radiative (k_r) and non-radiative (k_{nr}) rate constants for **16** in various environments. Data from Refs. [54,55]

	$\log K_{ass}$ (UV)	Φ_f (LE + ICT)	Φ_{isc} (LE)	$\tau/ns(\lambda_{em}/nm)$	$k_r/10^6 s^{-1}$ (state)	$k_{nr}/10^9 s^{-1}$ (ICT)
Free **16**		7×10^{-4}	0.11	~0.5 (*500*)	1.4 (ICT)	~2
α-CD:**16** 1:1	4.46 ± 0.18	1.8×10^{-3}		~1.0 (*520*)	1.8 (ICT)	~1
α-CD:**16** 2:1	7.21 ± 0.19	9.8×10^{-3}		3.9 (*420*)	2.5 (ICT)	0.26
β-CD:**16** 1:1	2.18 ± 0.04	4.0×10^{-3}	0.22	~0.8–0.9 (*500* and *360*)	~5.0 (ICT)	~1.1
β-CD:**16** 2:2	9.08 ± 0.17	1.3×10^{-2}	0.44	1–2 (*360*) 13.6 (*420*)	0.6–13.0 (LE)	0.5–1

fluorescence spectrum of **17** in water only exhibited a LE band at 400 nm. In the presence of α-CD at concentrations higher than 1×10^{-2} M, a further emission band appeared at *ca.* 500 nm, accompanied by a lack of isosbestic points in the absorption spectra, and was attributed to the population of an emissive ICT state in a 2:1 host:guest barrel-type complex. Such a structure was consistent with disappearance of the red-shifted emission in alkaline solutions, where ionization of the hydroxyl groups of the CD rim induced electronic repulsion

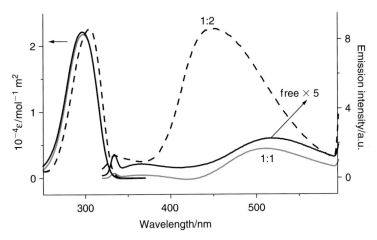

Fig. 7. UV-vis absorption (left) and emission (right) spectra of the free **16** and 1:1 and 2:1 α-CD:**16** complexes, reconstructed on the basis of the association constants in Table 4. Reprinted from Ref. [54] by permission of The Royal Society of Chemistry on behalf of the European Society for Photobiology and the European Photochemistry Association.

between the two CD moieties; moreover, it was in agreement with the biexponential kinetics observed for both the LE and ICT emission decays ($\tau_1 \sim 0.1$ ns, $\tau_2 \sim 3$ ns). Formation of the 1:1 complex at lower α-CD concentrations only afforded enhancement of the LE fluorescence band at 400 nm [57]. A red-shifted emission was found to appear on increasing the β-CD concentration and was attributed to an ICT process in a 1:1 complex on the basis of the presence of an isosbestic point in the absorption spectra of the system. On the contrary, 1:1 complexation with γ-CD enhanced the LE emission and did not induce any ICT emission, likely because water molecules, coincluded with the guest in the large cavity, increase the non-radiative rate of the ICT state [58].

17   **18**

Similar effects of the cavity size were observed on the emission features of CD complexes of methyl-4-(dimethylamino)benzoate (**18**) [59] and were similarly rationalized with formation of 1:1 plus 2:1 host:guest complexes with α-CD and 1:1 complexes with β- and γ-CD. In the 2:1 α-CD complex, the emission of the ICT state is at 450 nm, i.e. considerably shifted to the blue, a feature previously observed in the α-CD:**16** 2:1 complex and accounted for by the guest location in a double CD cage, where a lower polarity but no significant hindrance to structural rearrangements is found [54].

3. Triplet state

Triplet state peculiar features of inclusion complexes (spectral changes, lifetimes) proved to be useful for investigating both thermodynamic and kinetic aspects of binding and gave information on complex conformational dynamics. The rate for the intersystem crossing process is mainly controlled by intramolecular spin-orbit electronic interactions, so that it is not much influenced by the CD environment. However, the triplet quantum yield and decay kinetics of included guests are often strongly modified and, sometimes, a room temperature phosphorescence appears.

In presence of α-CD, the triplet decay of 1,4-dimethoxybenzene (**1**) is biphasic and strongly temperature dependent. From Arrhenius diagrams of triplet decay rates it was concluded that the 1:1 association step is characterized by a positive enthalpy ($38\,kJ\,mol^{-1}$) and a positive entropy ($150\,J\,mol\,K^{-1}$), whereas the association of the second CD molecule to give the 2:1 complex occurs with negative enthalpy ($-74\,kJ\,mol^{-1}$) and negative entropy ($-200\,J\,mol\,K^{-1}$). Thus, enthalpy–entropy compensation applies to both association processes in an opposite way: the first step is controlled by entropy through a large number of possible configurations, the second one is controlled by enthalpy and reflects the tight guest encapsulation and the well-defined geometry of the 2:1 complex (see structures in Fig. 1). The rate constants for bimolecular quenching of triplet **1** both by O_2 and by the ground state were found to be slowed down by more than one order of magnitude in presence of CD (see Table 1). The effect was particularly important with α-CD because of the formation of the 2:1 complex. In spite of this fact, the slope of the Arrhenius plot of the rate constant for O_2 quenching in this system was not significantly different from that of water. This fact pointed to a conformationally rigid structure of the 2:1 complex, which does not allow temporary opening needed for O_2 quenching to occur, in contrast to the cases of triplet 2:1 α-CD:naphthalene (**3**) and 2:2 α-CD:4-OH-biphenyl (**15**) complexes (see below) [4,43].

The triplet–triplet (T–T) absorption spectrum of **15** in water is characterized by a band extending from 330 to 440 nm with $\lambda_{max} = 377$ nm. At low α-CD concentrations (5×10^{-3} M) the spectrum was unchanged, but at higher host concentrations the λ_{max} was considerably red-shifted (by 9 nm at 5×10^{-2} M α-CD) and the band profile became asymmetric. These changes were rationalized by a partial hampering of the H-bonding to water due to the location of the hydroxyl donor groups of two **15** molecules within the cavity of the 2:2 structure (Fig. 5A). The absorption coefficient of the triplet 2:2 complex appeared to be *ca.* 3.2–4 times increased with respect to that of the monomer. This was interpreted as indicative of the presence of a strong interaction between the two guest molecules, presumably facilitated by a planarization of the aromatic system in the excited state, leading to an increase (approximately a doubling) in the molecular transition dipole moment [43]. This effect is rather unusual for dimeric CD species and was not observed in the 2:2 complex of the α-CD:3-OH-biphenyl (**14**) complex [44].

Concerning O_2 quenching, that of the triplet 2:2 α-CD:**15** complex occurs at a much slower rate (*ca.* $8 \times 10^8 \, M^{-1} s^{-1}$) than that for triplet monomer in water (*ca.* $7 \times 10^9 \, M^{-1} s^{-1}$) and with activation energy *ca.* two times higher ($60–65 \, kJ \, mol^{-1}$ compared with $30 \, kJ \, mol^{-1}$ in water). This effect is related to the conformational flexibility of the complex and indicates that an activated step, different from O_2 diffusion, is operative in the quenching process. A high activation energy for triplet quenching by O_2 was not observed in the 2:2 α-CD:3-OH-biphenyl complex [44]. The different spectroscopic and dynamic properties of the latter, compared to those of the 2:2 α-CD complex of **15**, may reflect a larger guest–guest separation and a more peripheral location of one of the two guest molecules, almost completely exposed to solvent (see for comparison structures A and B of Fig. 5).

α-CD complexation, predominant in the 2:1 stoichiometry, strongly affects the **3** triplet properties. The vibronic bands of the T–T absorption spectrum become narrower and λ_{max} shifts from 412 nm ($\varepsilon = 10500 \, M^{-1} cm^{-1}$) in water to 416 nm ($\varepsilon = 22500 \, M^{-1} cm^{-1}$) in α-CD. Moreover, quenching by O_2 is made less efficient. The activation energy for the latter process, $16 \, kJ \, mol^{-1}$ in water, increases to $66.5 \, kJ \, mol^{-1}$ in $5 \times 10^{-2} \, M$ α-CD, pointing to a conformational flexibility of the 2:1 complex similar to that observed in the hydroxybiphenyl derivatives [43]. Accordingly, the T–T absorption bands broadened and the fluorescence vibronic structure decreased with increasing temperature, indicating that the environment becomes more and more "water-like". Actually, the calculations showed the possibility of a wobbling motion of the two α-CD moieties in the 2:1 complex (see Fig. 2A) with temporary exposure of the guest to the bulk [4]. Due to higher flexibility of the relevant complexes and higher accessibility to water, β-CD modifies less than α-CD the triplet characteristics of **3**: the vibronic bands of the absorption spectrum are broader and λ_{max} shifts to 415 nm ($\varepsilon = 12000 \, M^{-1} cm^{-1}$); the activation energy for quenching by O_2 only exhibits a moderate increase to $26 \, kJ \, mol^{-1}$ [4].

The triplet quantum yield is increased in both 1:1 and 2:2 β-CD:**16** complexes but, as expected, the rate constant for triplet formation ($2.2 \times 10^8 \, s^{-1}$ in water) is not affected by complexation [54]. Information on ISC quantum yields and rate constants on α-CD:**16** complexes was not attained because of occurrence of aggregation phenomena. Temperature dependence of the rate constants for O_2 triplet quenching in aqueous solutions and in presence of α- and β-CD indicated protection with a lower degree of shielding with respect to **1** [2], **15** [44] and **3** [4]. Only a moderate increase of the activation energy with respect to water was observed (from 25 to 35 kJ mol^{-1}), which indicates that activated conformational changes of higher-order complexes do not play an important role with this guest [54].

α-CD includes 6-Br-2-naphthol (**19**) to give 1:1 and 2:1 complexes [60–62]. The complexation causes, besides the increase of the molecular fluorescence yield, the growth of a new emission in the region 480–700 nm, which was attributed to the phosphorescence of the 2:1 complex [60]. No phosphorescence was observed in water solutions of **19** even after prolonged nitrogen purging, while the phosphorescence appeared in aerated solutions containing 10^{-2} M α-CD, with lifetime 0.16 ms at room temperature. Therefore, the emission is not induced by CD protection of guest triplet from oxygen interaction, but is probably due to the suppression of radiationless deactivation processes of the probe excited state within the cavity. The study of the temperature effect on the phosphorescence lifetime [62] allowed the direct calculation of the exit rate constant of the triplet from the cavity: the value obtained, $(8.4 \pm 1.5) \times 10^3 \text{s}^{-1}$, is in a fairly good agreement with the value obtained for the exit rate of the triplet 2-naphthylethanol from the 2:2 complex with β-CD [33].

α-CD and 2-chloronaphthalene (**20**) form 1:1 and 2:1 complexes, which induce a decrease and a sharpening of the fluorescence with respect to the water solutions [63]. Interestingly, when α-CD was added to a **20** solution containing D-glucose, appearance of the phosphorescence emission was reported, together with a slight decrease of the fluorescence. In 3×10^{-2} M α-CD the room temperature phosphorescence quantum yield was found to be ~0.03, which should be compared with the value of 0.15 obtained for **20** in ethanol at 77 K. The phosphorescence variations as a function of [α-CD] indicated that only the 2:1 complex emits phosphorescence, also as a consequence of the high viscosity of D-glucose solutions, which restricts the CD movements. The addition of increasing concentrations of I$^-$ to the solution, increased the phosphorescence yield up to a maximum (observed at [I$^-$] = 0.03 M) and then quenched the emission: this behaviour was attributed to the formation of a ternary complex in which one or two I$^-$ ions are bonded to the 2:1 α-CD:**20** complex.

β-CD forms a 1:1 complex with 1-naphthylethanol (**21**) and 1:1 plus 2:2 complexes with the isomer 2-naphthylethanol (**22**). The results obtained by T–T absorption measurements gave a particularly interesting information. β-CD complexation shifts to the red by ~3 nm the T–T spectra of both molecules and increases the triplet lifetime, likely by protecting the chromophore from quenchers in solution. In water, the triplet of naphthylethanols was quenched by Mn^{2+} with rate constant $k_q = (2.6 \pm 0.4) \times 10^7 \text{M}^{-1} \text{s}^{-1}$. This value was reduced to $(4.3 \pm 0.4) \times 10^6 \text{M}^{-1} \text{s}^{-1}$ for the triplet 1:1 complex of **21**. The decay of **22** $(1.5 \times 10^{-4} \text{M})$ at [Mn^{2+}] $> 7 \times 10^{-5}$ M was fitted by the sum of two exponential terms: the fast one was attributed to quenching of the 1:1 complex and the corresponding k_q was $(2.5 \pm 1.7) \times 10^6 \, M^{-1} \, s^{-1}$. The slow component was attributed to a process involving the 2:2 complex and the corresponding k_q was estimated to be $(3.7 \pm 0.6) \times 10^4 \, M^{-1} \, s^{-1}$; this low value is due to the encapsulation of the probe by two β-CD. The triplet quenching by Mn^{2+} allowed to calculate the entry rate constant for the 1:1 complex of **22** $k_+ = (2.9 \pm 1.6) \times 10^8 \, M^{-1} \, s^{-1}$ (for **21**, $k_+ = (4.7 \pm 1.9) \times 10^8 \, s^{-1}$) and the corresponding dissociation rate constant $k_- = (1.8 \pm 0.7) \times 10^5 \, s^{-1}$ (for **21**, $k_- = (4.8 \pm 1.8) \times 10^5 \, s^{-1}$). For the 2:2 complex k_- was determined to be $\leqslant 2.4 \times 10^3 \, s^{-1}$. A slower dynamics for association processes involving more than one CD was also found in the case of pyrene 2:2 complexes [32].

21   22

The formation of the 2:2 complex for **22** leads to a long-lived triplet state. The contemporary detection of excimer emission [33] appears as a contradicting feature, since with two guest molecules in close proximity, self-quenching of triplet would be expected to be very efficient. Molecular mechanics calculations allowed to resolve this contradiction: in fact, two low-energy structures were found for the complexes with 2:2 stoichiometry. In one of them, the naphthyl groups are far away and this arrangement leads to a long-lived triplet excited state. In the other, the two naphthyl moieties are in close proximity, which allows the observation of the excimeric emission [64].

4. Chiral recognition of excited states

Owing to their pure enantiomeric nature, CDs are capable of chiral recognition of both racemic and single enantiomeric guests, evidenced by the different spectroscopic properties of the diastereomeric complexes [65]. Chiral discrimination mostly manifests on the thermodynamics of ground-state complexation; this fact promoted the use of CD as selectors in chromatographic methods for enantiomeric separations [66]. Chiral recognition on excited-states properties were rarely reported. In the two examples below, relevant to both 1:1 and higher-order complexation phenomena, both ground-state and excited-state discrimination effects were observed.

(1S)-(+)-**23**   (1R)-(-)-**23**

Chiral recognition of camphorquinone (**23**) enantiomers was exhibited by α-, β-CD and heptakis-(2,6-di-O-methyl)-β-CD [67]. α-CD proved to be the best chiral selector. The circular dichroism spectrum, peaking at 460–462 nm in pure water, was shifted to 472/486 nm for the (1R)-(−) and to 471/494 nm for the (1S)-(+) enantiomer in the presence of 2.1×10^{-2} M α-CD, whereas only minor changes were produced by the addition of the β-CD derivatives and of γ-CD. Fluorescence emission at 510 nm was enhanced by addition of CD, but the spectral distribution was modified only with α-CD; in the latter case, better evidence for a vibronic structure and appearance of the phosphorescence band at 560 nm were observed. These findings were rationalized with the formation of a 2:1 inclusion complex with equilibrium constants favoured by a factor of two for S(+) vs. R(−) enantiomer ($\Delta\Delta G \sim 1.9$ kJ mol^{-1}). Formation of 1:1 complexes exhibiting lower chiral discrimination with β-CD derivatives and no chiral discrimination with γ-CD was ascertained ($\Delta\Delta G \sim 0.2$–1.2 kJ mol^{-1}). These findings were in complete agreement with the structure (Fig. 8) of the most stable complexes, determined using conformational computations combined with theoretical interpretation of circular dichroism spectra (see Table 5).

T–T absorption indicated that the 2:1 α-CD complex is characterized by a very long-lived triplet state, not appreciably affected by the guest chirality (45 and 55 μs in the R and S form, respectively, vs. 5 μs in pure water); however, a remarkable chiral discrimination was evidenced for the triplet decay in presence of β-CD, with a 35 μs component observed with the R(−) and not with the S(+) enantiomer (see Table 3 in the chapter dedicated to triplet state and phosphorescence). These findings were rationalized by a differentiated binding constant of the enantiomeric guest triplet state in the 1:1 complexes, thus providing evidence for an unusual excited-state discrimination effect.

24   C H$_2$

Enantiomers of camphanate ester of 2,3-diazabicyclo[2.2.2]oct-2-ene (**24**) are discriminated by heptakis-(2,6-di-O-methyl)-β-CD for the fluorescence lifetimes,

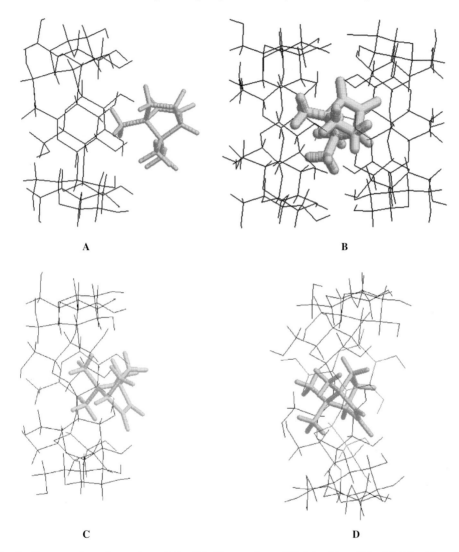

Fig. 8. Computed conformation of the CD:**23** complexes (A) 1:1 with α-CD; (B) 2:1 with α-CD; (C) 1:1 with β-CD; and (D) 1:1 with γ-CD. Reproduced from Ref. [67] with permission. Copyright (2002), American Chemical Society.

the most prominent effects being observed in the 2:1 complexes [68]. Indeed, the emission lifetimes for the two enantiomers were found to be $\tau(-) = 196$ ns, $\tau(+) = 220$ ns for the 1:1 complex and $\tau(-) = 190$ ns, $\tau(+) = 146$ ns in the 2:1 complex. Association of the second CD moiety to form the 2:1 complex induced preferential precipitation of one enantiomer, leading to an enantiomeric excess of 30% for the (-)-β-CD diastereomeric complex in the precipitate. In the association process, a statistically significant chiral recognition was evidenced on the

Table 5
Stoichiometry and energetic characterization of enantiomeric CD:23 complexes in water. ΔE_C gives the total complexation energy, defined as the difference between the sum of the potential and solvation energy of the complex and the potential and solvation energy of the isolated compounds. Data from Ref. [67]

Complex	Stoichiometry	$\Delta E_c/\text{kJ mol}^{-1}$
α:R(−)-23	1:1	−20.1
	2:1	−78.0
α:S(+)-23	1:1	−50.2
	2:1	−90.6
β:R(−)-23	1:1	−84.9
β:S(+)-23	1:1	−77.8
γ:R(−)-23	1:1	−69.8
γ:S(+)-23	1:1	−76.4

rate constants for entry of the **24** excited state in the cavity of the firstly bound CD molecule, *ca.* 20% faster for the (−)- than for the (+)-enantiomer ($k_{ass} = 1.28 \times 10^9\,\text{M}^{-1}\,\text{s}^{-1}$ vs. $1.12 \times 10^9\,\text{M}^{-1}\,\text{s}^{-1}$), despite the ground-state binding constants exhibit an opposite behaviour ($1190\,\text{M}^{-1}$ for the (−)- vs. $1440\,\text{M}^{-1}$ for the (+)-enantiomer). This observation implies that the exit rates also are characterized by the same trend and reveals that chiral discrimination effects in thermodynamics and kinetics of CD complexation may be not correlated with each other [69]. Unfortunately, the kinetics of formation of the 2:1 complex was too slow for being observed on the time scale of the experiment ($\sim 1\,\mu$s), which used the excited guest as a probe, thus preventing the study of the chiral discrimination in the association of the second CD moiety.

No evidence for chiral discrimination was obtained for the triplet behaviour of 1- and 2-naphthylethanol (**21** and **22**) complexed with β-CD; very low, if any, discrimination was obtained from fluorescence and NMR measurements [33,64].

5. Excited-state properties of nanotubes and ternary complexes

5.1. Nanotubes

A $10^{-2}\,\text{M}$ γ-CD solution containing the organic scintillator 2,5-diphenyl-oxazole (**25**) exhibited a long-wavelength emission ($\lambda_{\text{max}} \approx 425\,\text{nm}$) with lifetime $\tau = 16.6\,\text{ns}$ attributed to an excimeric species. The emission arises from extended aggregates (nanotubes), as suggested by the turbidity of the solutions, by laser scattering measurements, by the high polarization of the emission and by the temperature and pH effect on the fluorescence. The aggregates should be formed by the association of ≈ 60 complexes of 1:2 (γ-CD:**25**) stoichiometry in which the guest molecules partially overlap [70]. Above pH 13 and at temperature $> 333\,\mathrm{K}$ the **25** excimer emission decreased and the mono-mer fluorescence peaked at 375 nm grew in intensity, indicating disruption of the nanotube architecture [71]. Similar results were reported for other oxadiazole derivatives [72]. In the case of **26** it was found that both β- and γ-CD gave origin to the formation of nanotubes, which were believed to arise from the aggregation of 1:1 complexes with β-CD (estimated aggregation number ≈ 17) and of 2:1 complexes with γ-CD. α-CD only formed 2:1 complexes [73].

25   26

Another class of nanotubes-forming inclusion compounds is that of α,ω-di-phenylpolyenes. Fluorescence intensity and anisotropy variations of 1,3-diphe-nyl-1,3,5-hexatriene (**27**) in water/dimethylformamide (75/25) with added α- and β-CD indicated that only 1:1 complexes were formed, while a 2:1 complex was formed with permethylated γ-CD. The steep increase of the value of the flu-orescence anisotropy factor (r) with [γ-CD] indicated a strong association be-tween the CD and **27**. At [γ-CD] $= 2$–$3 \times 10^{-3}\,\mathrm{M}$ r reached its maximum value, 0.34, a value comparable to that of free **27** in EPA or glycerol at 77 K ($r = 0.38$). From this, the formation of nanotubes containing *ca.* 30 γ-CD interacting with each other by H-bonding was deduced. Accordingly, there was a large decrease of r when the pH of the solution was higher than 12.5. The only minor effect observed on r in the presence of permethylated γ-CD further supports the key role of H-bonding in the nanotube formation. The lack of fluorescence quench-ing by I^- on the (γ-CD:**27**)$_n$ system indicated that the probe is fully shielded from the aqueous medium [74]. The nanotube formation proceeds by a self-association chain reaction of 1:1 complexes and the length of the chain facilitates the for-mation of long tubes (aggregation number $\approx 3.5, 37, 50$ for two, three and four double bonds, respectively) [75]. The effect of $2 \times 10^{-2}\,\mathrm{M}$ γ-CD in water/ethylene glycol (60/40) on r and Φ_f for different chain lenghts is shown in Table 6.

5.2. Ternary complexes

In experiments of complexation by CD, the solutions are often prepared by diluting with water or a CD solution a concentrated alcoholic solution of the guest. This procedure can lead to the formation of ternary complexes

Table 6
Emission quantum yield (Φ_f), anisotropy factor (r) and average number of γ-CD units per nano-tube ($<j>$) in diphenylpolyenes with different number of double bonds N. From Ref. [75]

Number of double bonds	Concentration	Φ_f	r	$<j>$
$N = 2$	1.3×10^{-6} M	0.35	0.32	3.5
$N = 3$	8×10^{-8} M	0.56	0.34	37
$N = 4$	4×10^{-8} M	0.05	0.35	50

CD:guest:ROH. The change of the complexation stoichiometry influences the association constants and can produce variations of the spectral and photochemical properties of the system. The formation of a ternary complex between γ-CD, pyrene (**7**) and 1-butanol was reported by Kano already in 1982 [76,77].

In the following, we report in more detail on ternary complexes in which variations of the photophysical parameters of the aromatic guest were observed. The following equilibria describe these systems:

$$CD + X \rightleftharpoons CD : X \quad K_1 \tag{1}$$

$$CD + ROH \rightleftharpoons CD : ROH \quad K_2 \tag{2}$$

$$CD : X + ROH \rightleftharpoons CD : X : ROH \quad K_3 \tag{3}$$

$$CD : ROH + X \rightleftharpoons CD : X : ROH \quad K_4 \tag{4}$$

The addition of primary aliphatic (C_1–C_8) and alicyclic alcohols (cyclo-butanol, -pentanol, -hexanol) to a β-CD:**7** complex caused variation in the absorption and fluorescence spectra which were attributed to the association of one alcohol molecule [78]. The association constants for the formation of the ternary species were up to four orders of magnitude higher than K_1. The fluorescence lifetime increased from 192 ns, the value in the complex β-CD:**7**, to ~350 ns in the ternary complexes containing alicyclic alcohols; the lengthening of the lifetime was a function of the chain length when the concluded alcohol was an aliphatic one. The variation of the lifetimes paralleled the variation of the association constants. These results are in agreement with those of Nelson et al. [79], who reported the formation of ternary complexes of **7** with β-CD and γ-CD and primary, secondary and tertiary alcohols. In presence of 1% alcohol (by volume), the complexation constants were about two orders of magnitude higher than those obtained in water and those with β-CD were generally higher than the corresponding ones with γ-CD. The values of the association constants with γ-CD were dependent on the alcohol percentage [80,81]. The **7** fluorescence lifetime in all the ternary complexes was longer than that of the corresponding binary CD:**7** species.

An alternative stoichiometry 2:1:2 was proposed for the ternary complex β-CD:**7**:alcohol on the basis of fluorescence and NMR data [82]. The association constant relative to the equilibrium (5)

$$2\beta - CD + 7 + 2ROH \rightleftharpoons (\beta - CD)_2 : 7 : (ROH)_2 \qquad (5)$$

were estimated for a series of alcohols in 1% (v/v) concentration. The dependence of the association constants on the dimension and branching of the alcohols chain suggested that the alcohols cap the smaller mouth of the two CD complexed to **7**. Among the examined alcohols, cyclopentanol gave the largest association constant because of its proper geometry and volume.

Formation of ternary complexes of β-CD, **7** and amino acids were reported [25]. Eighteen amino acids were tested by UV-vis absorption, combined with steady-state fluorescence and only with tryptophan, leucine, phenylalanine, methionine and isoleucine the formation of ternary complexes was assessed. The calculated association constants at [amino acid] $= 10^{-2}$ M were lower than those obtained with alcohols at the same concentration, and this was attributed to a more rigid arrangement of the amino acid within the complex. The system with leucine at two β-CD concentrations (1×10^{-3} and 10^{-2} M) was characterized in detail also by multi-exponential analysis of the fluorescence decay. The obtained lifetimes in nanoseconds (in parenthesis the pre-exponential factor) were 116 ± 23 (0.44 ± 0.18), 170 ± 27 (0.39 ± 0.16) and 457 ± 33 (0.17 ± 0.04) in presence of [β-CD] 1×10^{-3} M and 191 (0.18) and 449 (0.82) at [β-CD] $= 10^{-2}$ M. The shortest lifetime in solutions containing 1×10^{-3} M [β-CD] was that of the free **7**, whereas the lifetimes 170 and 457 ns were attributed to the 1:1:1 and 2:1:2 complexes, respectively. At [β-CD] $= 10^{-2}$ M only the complexes were present, **7** was completely shielded from water so that the fluorescence lifetime and the ratio R of the vibronic bands were similar to those in hydrocarbons.

28   **29**

4H-1-benzopyran-4-thione (**28**) also forms ternary complexes with CD and alcohols [83]. This molecule proved to be a convenient probe of environment because of the photophysical properties of its S_2 state, which strongly depend on the medium. Formation of inclusion complexes with the three CD was indicated by the changes of the absorption spectrum and phosphorescence emission [83]. The complexation with β-CD was thoroughly examined in presence of various alcohols (1%, v/v) [84]. The fluorescence and phosphorescence quantum yields of **28**, very low in water ($\Phi_f = 8.5 \times 10^{-5}$, $\Phi_{ph} = 3.6 \times 10^{-4}$), were slightly increased by the addition of 10^{-2} M β-CD ($\Phi_f = 1.2 \times 10^{-4}$, $\Phi_{ph} = 6.5 \times 10^{-4}$), but were enhanced more than one order of magnitude in solution containing 1% 2-propanol ($\Phi_f = 1.6 \times 10^{-3}$, $\Phi_{ph} = 2.5 \times 10^{-2}$). The increase was found to be similar for other linear, branched and cyclic aliphatic alcohols and somewhat lower ($\Phi_f = 1 \times 10^{-3}$, $\Phi_{ph} = 1.3 \times 10^{-2}$) for 1% ethanol. On the basis of the changes in absorption and emission upon addition of β-CD and/or alcohols, a stoichiometry 2:1:2 was determined for the complexes. By comparing the properties of **28** in water and organic solvents of different polarity [85], the photophysical properties of the ternary complex appear to be unique. The high Φ_{ph} value in the presence of 2-propanol is remarkable, since thioketones are very reactive in H-abstraction processes, and this alcohol is a very good H-donor. The inefficient non-radiative deactivation of the triplet state of this thione was attributed to restrictions imposed on the relative orientation of **28** and alcohol by the β-CD cavity.

The photochemistry of xanthone (**29**) in the presence of CD [86] and added alcohols [86,87] was studied by absorption, fluorescence and laser-flash photolysis. The addition of CD to a **29** water solution caused a blue-shift of the fluorescence spectrum and a substantial decrease of its intensity; this allowed the determination of the ground-state association equilibrium constant ($K_1 = 36$, 1000 and 200 M^{-1} for α-, β- and γ-CD, respectively) [86]. The intensity decrease at equal [CD] was smaller in the presence of alcohols which suggested the formation of a ternary complex with equilibrium constant K_4. With β-CD and hydroxypropyl-β-CD (HP-β-CD) $K_4 < K_1$ while with γ-CD a slight increase of the complexation strength was observed. In the presence of HP-β-CD no K_4 dependence on the alcohol nature was observed while, in the presence of β-CD, K_4 increased in the order 1-butanol < cyclohexanol < cyclopentanol < 1-pentanol < tert-butanol < 2-butanol.

The **29** triplet has a π,π^* character in CD-containing solutions and the value of its λ_{max} is very sensitive to the polarity of the medium. In presence of β-CD, the λ_{max} shifted from 580 to 602 nm in few hundreds of nanoseconds after laser excitation; this change was attributed to the exit of the **29** triplet from the CD cavity into the aqueous phase. The triplet relocation occurred with rate constant $k_- = 1.1 \times 10^7 \, s^{-1}$; for α- and γ-CD the rate parameters were 1.9×10^7 and $6 \times 10^6 \, s^{-1}$, respectively [86]. In 3×10^{-2} M γ-CD solution containing also the above-listed alcohols (0.1–0.11×10^{-3} M), the exit rate constants decreased to 3–$6 \times 10^5 \, s^{-1}$. A smaller decrease was observed in 10^{-2} M β-CD-containing alcohol. The dissociation rate of the ternary complex CD: ^{3}**29***:ROH resulted to be independent of the strength of the ground-state complex [87].

6. Conclusions

The stoichiometry and the structural features of the higher-order complexes control the excited-state behaviour of the guest molecules. When more CD units encapsulate the guest, the steric constraints tend to hinder the molecular freedom slowing down the non-radiative deactivation of the guest excited singlet and, in some cases, bringing about an important increase in the radiative rate. The latter effect, unusual in other media, is related to distortions of the molecular symmetry brought about by the tight guest encapsulation. Host–host hydrogen bonding can promote such a perturbation. When more than one guest molecule is involved in the complex, excimer emission can originate owing to the strong guest–guest interactions in the restricted environment. Hydrophobicity of the CD environment can favour non-radiative processes like ESIPT because of substantial limitations imposed to competitive intermolecular H-bonding on the solvent. Proton dissociation can be also hampered because of (partial or total) water exclusion from the cavity. The pK^* of the excited state is generally increased. ICT is generally disfavoured by the low polarity of the environment and by the steric constraints imposed on the accompanying structural rearrangements. Isolation of the guest from the bulk solvent protects the excited state from interaction with external quenchers like oxygen and impurities and increases its lifetime. In the case of the triplet state, this fact can facilitate occurrence of room-temperature phosphorescence. Finally, in higher-order CD complexes the existence of severe steric constraints can allow chiral recognition of excited states, an infrequently documented event.

References

[1] P. Bortolus, S. Monti, Adv. Photochem. 21 (1996) 1.
[2] G. Grabner, S. Monti, G. Marconi, B. Mayer, C. Klein, G. Köhler, J. Phys. Chem. 100 (1996) 20068.
[3] P. Bortolus, S. Monti, M. Smoluch, H. Bouas-Laurent, J.P. Desvergne, J. Chem. Soc. Perkin Trans. 26 (2000) 1165.
[4] G. Grabner, K. Rechthaler, B. Mayer, G. Köhler, K. Rotkiewicz, J. Phys. Chem. A 104 (2000) 1365.
[5] I. Pastor, A. Di Marino, F. Mendicuti, J. Phys. Chem. B 106 (2002) 1995.
[6] S. Hamai, Bull. Chem. Soc. Jpn. 55 (1982) 2721.
[7] S. Sau, B. Solanki, R. Orprecio, J. Van Stam, C.H. Evans, J. Incl. Phenom. Macro. Chem. 48 (2004) 173.
[8] M. Cervero, F. Mendicuti, J. Phys. Chem. B 104 (2000) 1572.
[9] S. Nigam, G. Durocher, J. Phys. Chem. 100 (1996) 7135.
[10] X.H. Shen, M. Belletête, G. Durocher, J. Phys. Chem. B 101 (1997) 8212.
[11] S. Nigam, G. Durocher, J. Photochem. Photobiol. A: Chem. 103 (1997) 143.
[12] X.H. Shen, M. Belletête, G. Durocher, Langmuir 13 (1997) 5830.
[13] X.H. Shen, M. Belletête, G. Durocher, J. Phys. Chem. B 102 (1998) 1877.
[14] X.H. Shen, M. Belletête, G. Durocher, Chem. Phys. Lett. 301 (1999) 193.
[15] Y.L. Chen, T.K. Xu, X.H. Shen, H.C. Gao, J. Photochem. Photobiol. A: Chem. 173 (2005) 42.
[16] S. Hamai, Bull. Chem. Soc. Jpn. 69 (1996) 543.
[17] S. Hamai, A. Hatamiya, Bull. Chem. Soc. Jpn. 69 (1996) 2469.

[18] S. Hamai, J. Incl. Phenom. Mol. Rec. Chem. 27 (1997) 57.
[19] S. Hamai, J. Phys. Chem. B 103 (1999) 293.
[20] S. Hamai, J. Mat. Chem. 15 (2005) 2881.
[21] K. Kalyanasundaram, J.K. Thomas, J. Am. Chem. Soc. 99 (1977) 2039.
[22] W. Xu, J.N. Demas, B.A. DeGraff, M. Whaley, J. Phys. Chem. 97 (1993) 6546.
[23] A. Muñoz de la Peña, T.T. Ndou, J.B. Zung, I.M. Warner, J. Phys. Chem. 95 (1991) 3330.
[24] H. Yang, C. Bohne, J. Photochem. Photobiol. A: Chem. 86 (1995) 209.
[25] H. Yang, C. Bohne, J. Phys. Chem. 100 (1996) 14533.
[26] M. Hollas, M.A. Chung, J. Adams, J. Phys. Chem. B 102 (1998) 2947.
[27] T. Yorozu, M. Hoshino, M. Imamura, J. Phys. Chem. 86 (1982) 4426.
[28] N. Kobayashi, R. Saito, H. Hino, Y. Hino, A. Ueno, T. Osa, J. Chem. Soc. Perkin Trans. 2 (1983) 1031.
[29] K. Kano, H. Matsumoto, S. Hashimoto, M. Sisido, Y. Imanishi, J. Am. Chem. Soc. 107 (1985) 6117.
[30] K. Kano, H. Matsumoto, Y. Yoshimura, S. Hashimoto, J. Am. Chem. Soc. 110 (1988) 204.
[31] S. Hamai, J. Phys. Chem. 93 (1989) 6527.
[32] A.S.M. Dyck, U. Kisiel, C. Bohne, J. Phys. Chem. B 107 (2003) 11652.
[33] T.C. Barros, K. Stefaniak, J.F. Holzwarth, C. Bohne, J. Phys. Chem. A 102 (1998) 5639.
[34] G. Pistolis, Chem. Phys. Lett. 304 (1999) 371.
[35] C.H. Evans, S. De Feyter, L. Viaene, J. van Stam, F.C. De Schryver, J. Phys. Chem. 100 (1996) 2129.
[36] S. De Feyter, J. van Stam, F. Imans, L. Viaene, F.C. De Schryver, C.H. Evans, Chem. Phys. Lett. 277 (1997) 44.
[37] G. Pistolis, A. Malliaris, J. Phys. Chem. B 108 (2004) 2846.
[38] S. Mitra, R. Das, S. Mukherjee, J. Phys. Chem. B 102 (1998) 3730.
[39] H.R. Park, B. Mayer, P. Wolschann, G. Köhler, J. Phys. Chem. 98 (1994) 6158.
[40] T. Yorozu, M. Hoshino, M. Imamura, H. Shizuka, J. Phys. Chem. 86 (1982) 4422.
[41] D.F. Eaton, Tetrahedron 43 (1987) 1551.
[42] D. Wu, R.J. Hurtubise, Talanta 40 (1993) 901.
[43] P. Bortolus, G. Marconi, S. Monti, B. Mayer, G. Köhler, G. Grabner, Chem. Eur. J. 6 (2000) 1578.
[44] P. Bortolus, G. Marconi, S. Monti, G. Grabner, B. Mayer, Phys. Chem. Chem. Phys. 2 (2000) 2943.
[45] K. Takahashi, Chem. Commun. (1991) 929.
[46] Z.R. Grabowski, K. Rotkiewicz, W. Rettig, Chem. Rev. 103 (2003) 3899.
[47] G.S. Cox, P.J. Hauptman, N.J. Turro, Photochem. Photobiol. 39 (1984) 597.
[48] A. Nag, K. Bhattacharyya, Chem. Phys. Lett. 151 (1988) 474.
[49] A. Nag, R. Dutta, A. Chattopadhyay, K. Bhattacharyya, Chem. Phys. Lett. 157 (1989) 83.
[50] A. Nag, K. Bhattacharyya, J. Chem. Soc. Farad. Trans. 86 (1990) 53.
[51] K.A. Al-Hassan, U.K.A. Klein, A. Suwaiyan, Chem. Phys. Lett. 212 (1993) 581.
[52] K.A. Al-Hassan, A. Suwaiyan, U.K.A. Klein, Arab. J. Sci. Eng. 22 (1997) 45.
[53] A. Nakamura, S. Sato, K. Hamasaki, A. Ueno, F. Toda, J. Phys. Chem. 99 (1995) 10952.
[54] S. Monti, P. Bortolus, F. Manoli, G. Marconi, G. Grabner, G. Köhler, B. Mayer, W. Boszczyk, K. Rotkiewicz, Photochem. Photobiol. Sci. 2 (2003) 203.
[55] S. Monti, G. Marconi, F. Manoli, P. Bortolus, B. Mayer, G. Grabner, G. Köhler, W. Boszczyk, K. Rotkiewicz, Phys. Chem. Chem. Phys. 5 (2003) 1019.
[56] J. Pozuelo, A. Nakamura, F. Mendicuti, J. Incl. Phenom. Macro. Chem. 35 (1999) 467.
[57] Y. Matsushita, T. Hikida, Chem. Phys. Lett. 290 (1998) 349.
[58] Y. Matsushita, T. Suzuki, T. Ichimura, T. Hikida, Chem. Phys. 286 (2003) 399.

[59] Y. Matsushita, T. Suzuki, T. Ichimura, T. Hikida, J. Phys. Chem. A 108 (2004) 7490.

[60] S. Hamai, J. Phys. Chem. 99 (1995) 12109.

[61] R.E. Brewster, M.J. Kidd, M.D. Schuh, Chem. Commun. (2001) 1134.

[62] R.E. Brewster, B.F. Teresa, M.D. Schuh, J. Phys. Chem. A 107 (2003) 10521.

[63] S. Hamai, J. Phys. Chem. B 101 (1997) 1707.

[64] R.S. Murphy, T.C. Barros, B. Mayer, G. Marconi, C. Bohne, Langmuir 16 (2000) 8780.

[65] S. Li, W.C. Purdy, Anal. Chem. 64 (1992) 1405.

[66] S. Li, W.C. Purdy, Chem. Rev. 92 (1992) 1457.

[67] P. Bortolus, G. Marconi, S. Monti, B. Mayer, J. Phys. Chem. A 106 (2002) 1686.

[68] H. Bakirci, W.M. Nau, J. Org. Chem. 70 (2005) 4506.

[69] H. Bakirci, W.M. Nau, J. Photochem. Photobiol. A: Chem. 173 (2005) 340.

[70] R.A. Agbaria, A.F. Gill, J. Phys. Chem. 92 (1988) 1052.

[71] K.A. Agnew, T.D. McCarley, R.A. Agbaria, I.M. Warner, J. Photochem. Photobiol. A: Chem. 91 (1995) 205.

[72] R.A. Agbaria, A.F. Gill, J. Photochem. Photobiol. A: Chem. 78 (1994) 161.

[73] C.F. Zhang, X.H. Shen, H.C. Gao, Chem. Phys. Lett. 363 (2002) 515.

[74] G. Pistolis, A. Malliaris, J. Phys. Chem. 100 (1996) 15562.

[75] G. Pistolis, A. Malliaris, J. Phys. Chem. 102 (1998) 1095.

[76] K. Kano, I. Takenoshita, T. Ogawa, Chem. Lett. (1982) 321.

[77] K. Kano, I. Takenoshita, N. Ogawa, J. Phys. Chem. 86 (1982) 1833.

[78] S. Hamai, J. Phys. Chem. 93 (1989) 2074.

[79] G. Nelson, G. Patonay, I.M. Warner, Talanta 36 (1989) 199.

[80] G. Nelson, G. Patonay, I.M. Warner, Anal. Chem. 60 (1988) 274.

[81] J.B. Zung, M.A. Muñoz de la Peña, T.T. Ndou, I.M. Warner, J. Phys. Chem. 95 (1991) 6701.

[82] A. Muñoz de la Peña, T.T. Ndou, J.B. Zung, K.L. Greene, D.H. Live, I.M. Warner, J. Am. Chem. Soc. 113 (1991) 1572.

[83] M. Milewski, W. Augustyniak, A. Maciejewski, J. Phys. Chem. A 102 (1998) 7427.

[84] M. Milewski, M. Sikorski, A. Maciejewski, M. Mir, F. Wilkinson, J. Chem. Soc. Farad. Trans. 93 (1997) 3029.

[85] A. Maciejewski, R.P. Steer, Chem. Rev. 93 (1993) 67.

[86] M. Barra, C. Bohne, J.C. Scaiano, J. Am. Chem. Soc. 112 (1990) 8075.

[87] Y. Liao, C. Bohne, J. Phys. Chem. 100 (1996) 734.

Cyclodextrin Materials Photochemistry, Photophysics and Photobiology
Abderrazzak Douhal (Editor)
© 2006 Elsevier B.V. All rights reserved
DOI 10.1016/S1872-1400(06)01006-5

Chapter 6

Cyclodextrin Inclusion Complexes: Triplet State and Phosphorescence

Wei Jun Jin

*College of Chemistry, Beijing Normal University, Beijing 100875, P. R. China
School of Chemistry and Chemical Engineering, Shanxi University, Taiyuan 030006,
P. R. China*

Most of the sections of this book discuss the photophysical processes following absorption of electromagnetic radiation, therefore, a very concise review of the terms and concepts relevant to triplet states and phosphorescence is given at the beginning of this chapter. Then the quenching and protection of triplet states and phosphorescence and heavy atom effects are introduced briefly. The typical systems and methodologies on inclusion complex, and typical applications of CD induced-room temperature phosphorescence (CD-RTP) are then discussed. Finally, some important conclusions are made.

1. Brief introduction to triplet states and phosphorescence

As a term in quantum chemistry, multiplicity shows the total spin state of a molecule and can be defined as $M = 2S + 1$, where $S = \sum s_i$, with s_i being the spin quantum number of the electrons, and is equal to $+\frac{1}{2}$ or $-\frac{1}{2}$. When all electrons have an anti-paralleled counterpart ($\uparrow\downarrow$), the net spin is zero and $M = 1$, giving a singlet state, denoted by S; when one electron flips, there are two electrons that have a parallel counterpart ($\uparrow\uparrow$ or $\downarrow\downarrow$) giving a triplet state, denoted by T. Commonly, a stable single radical molecule gives a doublet state. In excited singlet states, the electron in the excited orbital is paired (of opposite spin) to the second electron in the ground-state orbital. In Jablonski diagram of molecular photophysical processes, the ground state is depicted as S_0, the lowest and higher excited singlet states are depicted as S_1, S_2, ... S_n, and the lowest and higher excited triplet states are depicted as T_1 and T_2, ... T_n, respectively. T indicates a threefold degeneracy of the state due to the three possible alignments of two unpaired spins.

The singlet state remains a single state while triplet state splits into three quantized states with magnetic sub-level split of $< 0.3\,cm^{-1}$ in the presence of an external magnetic field, i.e., the former is diamagnetic and the latter is paramagnetic. The triplet states are always of lower energy than the corresponding singlet states, this is a consequence of Hund's rule. Additionally, the triplet state has longer lifetime than singlet states because of spin-forbidden transition between them, going from sub-ms to *ca.* 10 s.

According to quantum mechanics, the transition between S and T states is spin-forbidden, so the lowest triplet states T_1 is populated almost exclusively as a result of singlet excitation produced by ultraviolet or visible radiation, the intersystem crossing from S_1 to T_1. But the status of spin-forbidden transitions can be changed by so-called heavy atom effect (HAE, see below).

The population of T_1 by absorption from S_0–T_1 can be realized for some compounds under rare case in the presence of heavy atom substitution and at low temperature [1,2]. Another pathway to populate triplet states is T–T energy transfer from excited triplet donor (D*) to triplet state of ground acceptor (A) as expressed by Eq. (1), in which spin-conversation rule must be obeyed, giving a sensitized triplet state [3]. Triplet states also can be populated by T–T energy migration [4]. These are very important ways for molecules with very low S_1–T_1 efficiency, and either energy transfer or migration is important probe for biological or material analysis.

$$D^*(T_1) + A(S_0) = D(S_0) + A^*(T_1) \tag{1}$$

The photophysical processes associated closely with triplet states T_1 include the delayed fluorescence, phosphorescence, nonradiative transition from T_1 to S_0, and T–T absorption or absorption to magnetic sub-levels of triplet state, as depicted in Fig. 1.

Phosphorescence, and some times the delayed fluorescence, is the main topic of this chapter. Phosphorescence of organic molecules can be defined as the radiative transition originating from the lowest excited triplet state, T_1, to the singlet ground state, S_0. In contrast to fluorescence, singlet S_1 to S_0, phosphorescence is a spin-forbidden process as mentioned above. Therefore, it is long lived as same order of magnitude as triplet state and more easily quenched especially by dioxygen molecules with triplet ground states. Phosphorescence is induced often by moderate heavy atom perturbation.

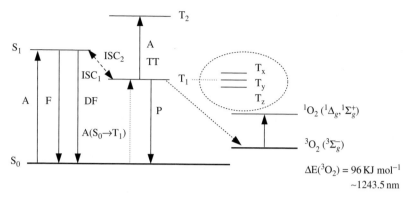

Fig. 1. Simplified Jablonski diagrams on photophysical processes relevant to triplet state and phosphorescence and oxygen quenching pathway.

Phosphorescence quantum yield is usually lower, which can be expressed as [5]

$$\phi_p = \vartheta_{ISC} \times \vartheta_P = \frac{k_{ISC}}{k_{ISC} + k_f + k_{nf} + \Sigma k_{q,f}[Q]} \times \frac{k_p}{k_p + k_{np} + \Sigma k_{q,p}[Q]} \quad (2)$$

where k_{ISC} is the intersystem crossing rate constant, k_f and k_p are the rate constants of fluorescence and phosphorescence, respectively, k_{nf} and k_{np} the rate constants of all unimolecular non-radiative decay, and $\Sigma k_{q,f}[Q]$ and $\Sigma k_{q,p}[Q]$ the sums of all effective (bimolecular) quenching rate constants of fluorescence and phosphorescence, respectively. The concentration term in $k_{q,p}[Q]$ indicates that the quenching efficiency depends on the concentration of quencher.

Triplet lifetime can be expressed as

$$\tau_T = (k_P + k_{nP} + \Sigma k_{q,P}[Q])^{-1} \quad (3)$$

2. Quenching of triplet states and phosphorescence

Quenching of triplet states usually occurs by triggering ISC to ground states. Oxygen molecule is one of the most efficient triplet state quenchers, and quenching pathways may be involved in oxidation of excited singlet $^1O_2$, the formation of ground complexes of O_2 molecule with luminophors and spin–orbit coupling, etc. But the triplet state energy transfer seems to be the common one, i.e. oxygen molecules first form the charge transfer complexes or encounter complexes with triplet state molecules [6–8], and then undergo energy transfer to produce a simultaneous S_1–T_1 transition in the perturbed molecule and a T–S transition in the oxygen molecule, as depicted in Fig. 1.

The effect of oxygen on triplet state should be concentration-dependent [9], therefore the efficience of the oxygen effect can be expressed as $k_{ST}^{O_2}[O_2]$, where

$k_{ST}^{O_2}$ is a bimolecular rate constant for oxygen perturbation and $[O_2]$ is the concentration of oxygen. Taking the typical value of $k_{ST}^{O_2}$ *ca.* 10^{10}–10^9 $1\,mol^{-1}\,s^{-1}$ and solubility of oxygen in water (at $298\,K$ and 1 atm.) *ca.* $10^{-3}\,mol/l$[10], $k_{ST}^{O_2}[O_2]$ is *ca.* 10^7–$10^6\,s^{-1}$. It means that the effect will be noticeable if $k_{ST}\sim10^7\,s^{-1}$ or less. As the prolongation of the conclusion, it may be considered as one of criterions to design lifetime- or kinetic-based optochemical sensors for oxygen.

T–T annihilation is a considerable quenching factor(s) under higher concentration. But it can be ignored because of the compartmentation of CD cavity or surfactant micelles if host concentration is much higher than the guest concentration in CD or micellar aqueous solutions.

Other quenchers include paramagnetic compounds, dioxide nitrogen and nitrite, iodide I^-, Cu(II) and Co(III) ions, etc. Nitrite and Cu(II) or Co(III) ions can used as anionic or cationic probe to probe the charged microenvironmental properties.

3. Stabilization of populated triplet states and enhancement of phosphorescence

The long-lived triplet state and phosphorescence are very easy to be quenched by dynamic collision or oxygen energy transfer as mentioned above. From Eq. (1) it can be seen that phosphorescence quantum yield can be improved by two ways. In the first method, k_{ISC} is increased by spin–orbit coupling, i.e. HAE, which describes the influence of heavy atom substitution or species on spin-forbidden transition [11]. The heavy atom species mixes pure singlet and triplet states to produce the states with a mixed character in spin multiplicity, and thus makes the spin-forbidden transition from S to T possible. The HAEs lead generally to the following four results: the enhancement of S_0–T_n absorption or S_1–T_n and T_1–S_0 processes; decrease in fluorescence lifetime and simultaneous increase in triplet quantum yield (or efficiency); decrease in phosphorescence lifetime and increase in phosphorescence quantum yield; decrease in both phosphorescence lifetime and quantum yield under over-perturbation. The HAEs on S–T intersystem crossing tend to be minimal for states which already possess substantial spin–orbit coupling (n-π^* states) or very fast fluorescence rates ($\sim10^9\,s^{-1}$). The heavy atom species are called heavy atomic perturbers as they produce the HAE, i.e. increase k_{ST} (k_P or k_{TS}) to a value that alters the singlet lifetime τ_S or triplet lifetime τ_T.

The HAE can be classified as intrinsic and extrinsic: in the former case the heavy atom acts as a chemically bonded moiety within phosphor and in the latter the heavy atom acts as a coexisted non-covalently species in the vicinity of phosphor. HAE appears selective in some cases and therefore the selective heavy atomic perturbed phosphorimetry can be established in analytical chemistry [12,13].

In the second way, non-radiative rate constants (k_{np}, $\Sigma\ k_{q,p}[Q]$) and (k_{nf}, Σ $k_{q,f}[Q]$ are decreased). At room temperature, many organized media at molecule scale can stabilize the populated triplet states and further enhance observed phosphorescence signals. Such media include deoxycholate dimmer aggregate [14–16], surfactant micelles [17] and cucurbit[n]urils [18]. CD as one kind of the media is used frequently to minimize most efficiently the nonradiative and quenching processes to realize high ϕ_T or ϕ_P based on formation of inclusion complexes.

Besides the two ways mentioned above, the change in the vibrational properties of C–H bonds, for example, isotopic effect caused by the substitution of deuterium for hydrogen [19] and the confined C–H vibrations of phosphore in the cavity of CD existing in suspension crystals [20], should be an efficient way to suppress the radiativeless processes.

4. Bi-component inclusion complexes

For the phosphores with intrinsic heavy atom substituent, their binary inclusion complexes with CD can produce strong phosphorescence at room temperature under efficient deoxygenation. The author's group has designed several phosphores for recognition of cationic ions in the presence of CD, as shown in Fig. 2 [21]. Inclusion and kinetic constants of the complexes are listed in Table 1.

Phosphorescence lifetime **1** and **5** are comparable. Lifetime of **4** is obviously shorter than that of **2** and **3** because of stronger HAE of iodo-atom. The quantum yield is as follows: completely protonated form: 0.069(**1**), 0.049(**2**), 0.073(**3**), 0.061(**4**), 0.067(**5**); completely free-protonated form: 0.00077(**1**), 0.003(**2**), 0.0014(**3**), 0.005(**4**), 0.00086(**5**). The pK_a (T_1) values derived here are substituent-dependent. Comparatively, **2–4** behave specially having higher pK_a (T_1), wider pH response range and also the emission was not completely "switched off" at high pH up to 11. The results can be rationalized as the retardation of PET in the apolar location with the rather hydrophobic substituents (Cl, Br, I). Also probably because of the hindrance from bulky hydrogen bond network around –OH or the crown ether ring in compounds **1** and **5**, the basicity of the amine group was lowered and hence the acidity of the conjugated acid was increased. These imply that the substituent in both type and position works well in the modulation of sensing properties. In addition, although an increase in phosphorescence intensity was expected upon protonation, RTP intensity decreased again at very low pH. Probably, the protonation enhanced the exiting rate of naphthyl from CD cavity because of the increase in water solubility.

In contrast to the above example acting as on–off switch, the inclusion complex of β-CD with bromonaphthyl iminodiacetic acid (BNIA) as shown in Fig. 3

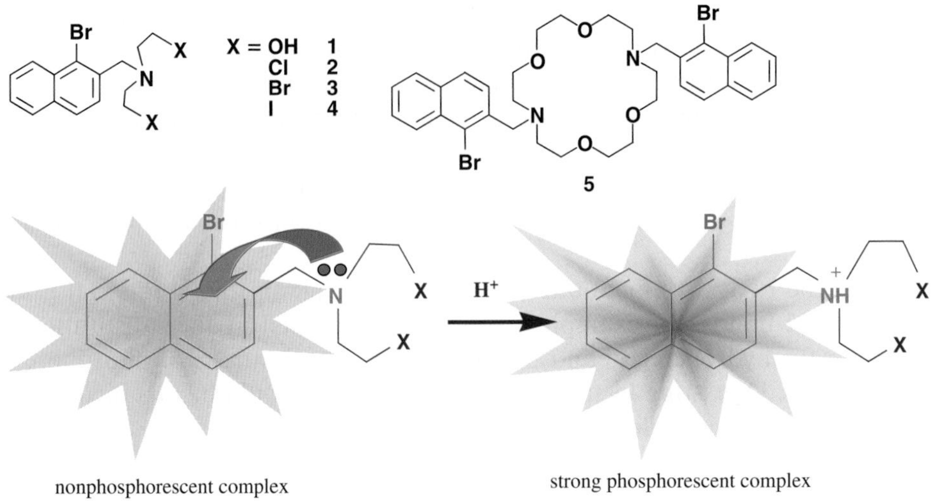

Fig. 2. Structures of five pH sensors (upper) and PET principle of sensors: off–on (bottom). **1**: *N*-(1-bromo-2-naphthylmethyl) diethanolamine; **2**: *N*-(1-bromo-2-naphthylmethyl)bis(2-chloro-ethyl)amine; **3**: *N*-(1-bromo-2-naphthylmethyl)bis(2-bromoethyl)amine; **4**: *N*-(1-bromo-2-naph-thylmethyl)bis(2-iodoethyl)amine; **5**: *N,N′*-bis(1-bromo-2-naphthylmethyl)diaza-18-crown-6.

Table 1
Comparison of inclusion consants and kinetic parameters of sensor **1–5**

Compound	**1**	**2**	**3**	**4**	**5**
K (l/mol)	230 ± 14	128 ± 33	113 ± 29	65 ± 17	67 ± 16
pK_a (T)	4.9	6	5.5	5.8	5
pH response range	4.3–6.5	3.5–10.5	3.4–10.5	3.4–10.5	3.4–6.1
$\Phi_P^{acid}/\Phi_P^{base}$ [a]	89.2	16.4	51.3	12.2	77.8
τ_P (ms)	0.42	0.36	0.34	0.28	0.47
k_P $(10^2\,s^{-1})$ [b]	1.66	1.37	2.18	2.16	1.19
k_D $(10^3\,s^{-1})$ [b]	2.21	2.64	2.72	3.36	2.01
k_{PET} $(10^5\,s^{-1})$ [b]	2.1	0.42	1.4	0.4	1.9

Note: (1) Ground-state $pK_a(S_0)$: trimethylamine, 9.80; triethylamine, 10.72. From Ref. [22], [β-CD] = 8 mM.
(2) K of parent 1-bromo-2-methylnaphthalene (BMN): 706 ± 141 l/mol.
[a]"Acid" and "base" refer to the protonated and deprotonated states, respectively.
[b]Obtained according to the following relations in acidic and basic solution: $\Phi_{p,acid} = k_p/(k_p + k_D)$, $\Phi_{p,base} = k_p/(k_p + k_D + k_{PET})$, $\tau_p = 1/(k_p + k_D)$, where k_p is the radiative rate constant, k_D the general non-radiative rate constant and k_{PET} to the photoinduced electron transfer (PET) rate constant [23].

behaves in off–on form, i.e. BNIA acts as a reverse PET phosphoroionophore with response to heavy metal ions [24]. The RTP intensity depends on the acidity, and the maximal value is obtained at pH 5.3. The quenching effect is observed upon the addition of heavy metal ions due to photoinduced electron

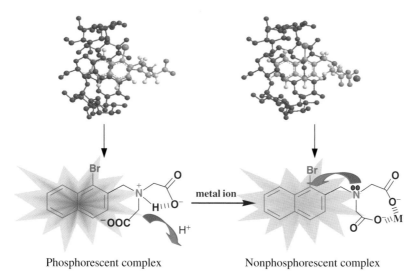

Phosphorescent complex   Nonphosphorescent complex

Fig. 3. Working principle of sensor: proton-transfer coupling reverse PET, on–off.

Table 2
Stern–Volmer quenching constants for BNIA in the presence of different metal ions

Metal ions	Cu^{2+}	Ni^{2+}	Pb^{2+}	Co^{2+}	Cd^{2+}	Zn^{2+}	Mn^{2+}	Cr^{3+}
K_{sv} ratios[a]	1	0.82	0.69	0.18	0.017	0.062	0.024	0.013
k_q ($10^9 \, l\,mol^{-1}\,s^{-1}$)	3.78	3.1	2.61	0.68	0.064	0.23	0.091	0.049

[a]$K_{sv, \, (Mn+)}/K_{sv, \, (Cu2+)}$, $K_{sv}(Cu^{2+}) = 1.51 \, l\,\mu mol^{-1}$

transfer process. The quenching ability is estimated by the Stern–Volmer quenching constants. Given the lifetime $\tau_0 \sim 0.40$ ms, bimolecular quenching rate constants obtained are listed in Table 2. For Cu^{2+}, Ni^{2+} and Pb^{2+}, k_q is comparable with diffusion rate constant, $\sim 10^{10} \, l\,mol^{-1}\,s^{-1}$. So the sensor can be used for screening Cu^{2+}, Ni^{2+}, Pb^{2+}, and the detection limits of Cu^{2+}, Pb^{2+}, Ni^{2+} are 0.017, 0.024 and 0.018 $\mu mol\,l^{-1}$ respectively ($S/N = 3$).

Bortolus *et al.* [25] investigated in detail the behavior of inclusion complexes of α-, β- and γ-CD with camphorquinone (CQ) enantiomers as shown in Fig. 4, and their triplet lifetimes are given in Table 3.

Gooijer's group reported the discrimination of CQ enantiomers aiming at analytical application [26]. Two kinds of inclusion complexes between CQ an-antiomers and α-CD in deoxygenated aqueous solutions are shown to exhibit the strongest room temperature phosphorescence. Interestingly, these signals differed significantly for the two enantiomers of CQ; the phosphorescence life-time of (+)-CQ was about four times longer than that of (−)-CQ, being 352 ± 16 and $89 \pm 6 \, \mu s$, respectively. This enantiomeric selectivity is attributed to

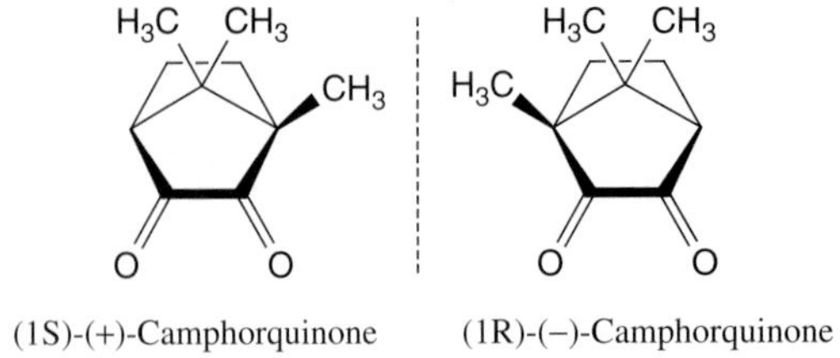

(1S)-(+)-Camphorquinone   (1R)-(−)-Camphorquinone

Fig. 4. Structure of the enantiomers of CQ.

Table 3
Triplet lifetimes of camphorquinone enantiomers in the presence of α-, β- and γ-CD in water at 295 K

	R-(−)-camphorquinone triplet lifetime		S-(+)-camphorquinone triplet lifetime	
	$\tau_1/(\mu s)$	$\tau_2/(\mu s)$	$\tau_1/(\mu s)$	$\tau_2/(\mu s)$
H$_2$O	6.7±1		7.0±1	
10^{-2} M α-CD	5±1	45±5	4±1	55±5
10^{-2} M β-CD	5±2	35±5	11±1	
10^{-2} M γ-CD	8±1		7±1	

a difference in dissociation rates (competing with the radiative emission process) for the diastereoisomeric inclusion complexes dealt with, which have a 2:1 stoichiometry (α-CD/CQ/α-CD). Time-resolved RTP detection using different delay times enables the determination of the two enantiomers in a mixture without involving a separation technique. The minimum detectable fraction of (+)-CQ in a 2 mM sample was 13%.

Phosphorescent inclusion complex can be used to probe heme accessibility in heme proteins [27]. The phosphorescence lifetime of 6-bromo-2-naphthyl sulfate (BNS) is several hundred microseconds and is self-quenched. Quenching of BNS phosphorescence does not occur for the non-heme protein lysozyme and apomyoglobin but occurs by a dynamic mechanism with a quenching constant of $1-2 \times 10^9 \, M^{-1} s^{-1}$ for cytochrome c and myoglobin and with a quenching constant of $6.2 \times 10^9 \, M^{-1} s^{-1}$ for protoporphyrin IX. The phosphorescence of an inclusion complex of 1-bromonaphthalene and β-CD is not quenched by heme-containing proteins. The temperature and viscosity dependencies of the rate with which BNS phosphorescence is quenched by microperoxidase-11 are consistent with unit quenching efficiency. These results indicate that quenching of BNS phosphorescence occurs only upon contact with the quencher, and the quenching constant can be used to assess the degree of accessibility of the heme group.

In bicomponent system consisting of CD and luminophor, ternary or more grade inclusion complex is possible [28]. For example, Brewster et al. [29]

constructed a temperature sensing system for 274.76–332.86 K based on the lifetime measurement of 1:2 ternary complex of 6-bromo2-naphthnol and α-CD. Turro *et al.* [30] inferred that the synthesized phosphorescence probes, *p*-benzoyl[5- (4-bromo-1-naphthoyl)-1-pentyl]benzoate (BNK$_5$BP), 5-(4-bromo-1-naphthoyl)-1-pentanol (BNK$_5$A) and 1-bromo-5-(4-bromo-1-naphthoyl)pentane (BNK$_5$B), may form 1:1 and 1:2 complex of the phosphor and the γ-CD based on the observation of the fast and slow decay, for example, 506 μs and 3.9 ms for BNK$_5$BP, in the solution containing γ-CD. The oxygen could completely quench the fast decay component, while the lifetime of the slow decay was not obviously influenced in the presence of or absence of oxygen.

The halogenated CD acts as both host and a heavy atom perturber towards non-halogenated guests. Femia and Cline Love [31] synthesized heptakis(6-bromo-β-CD) by replacing β-CD's primary hydroxyls with bromine which can form inclusion complex with several aromatic hydrocarbons in the solution containing *N,N*-dimethylformamide and water. This system induced strong RTP; however, the aromatic hydrocarbon outside the cavity of CD cannot phosphoresce. Hamai and Mononobe [32], and Hamai and Kudou [33] investigated RTP of iodine-substituted CD inclusion complex. Compared with parent β-CD, 6-iodo-6-deoxy-β-CD as a heavy atom perturber is more than 1.2 times as effective in enhancing the RTP of 2-chloronaphthalene. However, 6-iodo-6-deoxy-α-CD can form 2:1 inclusion complex with 6-bromo-2-naphthol and can reduce the RTP intensity by 18% relative to that for the 2:1 inclusion complex composed of parent α-CD. This may be attributed to the overperturbation, leading to increase in k_{ISC} too quickly.

5. Tri-component inclusion complexes

The synergetic interaction or effect exists widely in various scientific and technical fields. In terms of CD-induced room temperature phosphorescence it can be defined as the principal factor(s) and secondary factor(s) affecting the characteristic properties, e.g., spectrum, quantum yield, lifetime and anisotropy, of the system concerned match automatically with each other to reach an optimization state, and cause remarkable change in the properties. The third or both third and fourth components added to the system interested form ternary or higher inclusion complex together with the CD and phosphore, and they synergetically enhance the stability of triplet states of inclusion complex, leading to increase the phosphorescent intensity or lifetime.

In tri-component inclusion complex, the third component may display the large difference in stabilizing the triplet states and enhancing the phosphorescence of inclusion complex in terms of interaction mechanisms. Generally, the third component acts simultaneously as a heavy atom perturber to increase k_{ISC}

to certain degree by spin–orbit coupling because of the closest accessibility between perturber and luminophor confined in CD cavity and as a space regulator to intensify the rigidity of CD cavity by regulating the arrangement of phosphore and confining its motion.

Scypinski and Cline Love [34] first proposed to use the ternary inclusion complex in which 1,2-dibromoethane was used as HA perturber to enhance phosphorescence of polyaromatic hydrocarbons. The author's group reported different systems for interested analytes: β-CD/1-naphthaleneacetic acid/1,2-dibromopropane [35,36], β-CD/PAHs/epibromohydrin [37,38], and β-CD/7-methylquinoline/ bromocyclohexane [39].

Zhang et al. [35] suggested that three factors should be considered in choosing heavy atom perturber, namely (1) ensuring moderate spin–orbit coupling between phophor and perturber for increasing ϕ_P, (2) sufficient apolarity for entrance inside CD cavity; (3) moderate molecule geometric volume to match the residual space with luminophor together for maximum rigidization; and (4) chirality matching for discrimination of enantiomers noticed recently by Gooijer's group [40].

Wei et al. [41] investigated the rule of HAE of halide alkanes and concluded that (1) the heavy atom enhancement effect (HAEE) of halogens halide alkane is increased in the order $Br > Cl > I$; (2) among the bromoalkanes examined, dibromoalkanes have the strongest HAEE, followed by the monobromoalkanes. For tri- and tetrabromoalkanes, very weak HAE was observed. However, too many heavy atoms in the vicinity of phosphorescent molecule would increase nonradiation transition of the triplet state and an overperturbation such as in iodine with a S–O coupling constant of 66.1 kJ mol^{-1} was not beneficial for inducing RTP either; (3) in length, HAE is weakest as chain of organic compound equals to or exceeds six carbon atoms. This implies that the distance between centers of phosphore and heavy atom should be within ca. 5.35 Å.

CD-RTP can be easily observed under nitrogen stream, but chemical deoxygenation is also a good pathway to obtain RTP [42,43]. The quenching properties using NO_2^-, small quenching rate constants as shown in Table 4, indicate the efficient protection of CD to triplet states of phosphor.

The halide alkanes when used as HA perturbers make the system easily turbid under excessive amount. This makes the studying of inclusion equilibrium more difficult. While the halide alcohol perturbers, for example, 2-bromoethanol and 2,3-dibromopropanol, participating in the formation of inclusion complex as water-soluble third component is more favorable. First, Hamai [44] and sequentially Spanish researchers [45–47] reported the relevant results.

Sometimes, the third component acts as both space regulator and emulsification reagent for forming suspended crystal particles. The solubility of inclusion complex and CD per se was decreased because of dispersion of excessive

Table 4
Phosphorescence lifetime and quenching rate constants by NO_2^- – using SO_3^{2-} – as oxygen scavenger

β-CD/phosphores/heavy atom perturbers	Lifetime/s	$k_q/10^3$ ($l\ mol^{-1}\ s^{-1}$)
β-CD/acenaphthene/1,2-bromoethane	0.1	5.6
β-CD/acenaphthene/2-bromoethanol	0.1	8.5
β-CD/fluorene/1,2-dibromoethane	0.09	17
β-CD/fluorene/2-bromoethanol	0.3	64
β-CD/phenathlene/1,2-dibromoethane	0.2	5.6
β-CD/phenathlene/2-bromoethanol	0.3	3.4
β-CD/1,8-benzoquinoline/1,2-dibromoethane	0.4	16
β-CD/1,8-benzoquinoline/2-bromoethanol	0.4	2

amount of halide alkanes. Previously fluorescence enhancement by space regulation and association of cycloalkanes with CD were reported [48,49]. Jin *et al.* [50] used cyclohexane to induce RTP of some polyaromatic hydrocarbons and nitrogen heterocycles without an heavy atom substitute by efficient space regulation and crystal protection. Xie *et al.* [51] reported the inclusion constants of bromonaphthalene with β-CD/cyclohexane, β-CD/cyclohexanol and β-CD/methylcyclohexanol as 1.25×10^4, 0.10×10^4 and 0.16×10^4 1/mol, respectively. So large difference in inclusion constants show that cyclohexane indeed is a good space regulator. More importantly, Nazarov *et al.* [52] used adamantane (Ada) as the most efficient space regulator up to date to induce RTP of naphthalene and naphthalene-d8 by forming inclusion complex. Naphthalene-d8-β-CD-Ada has a lifetime of 10.3 s. The quenching rate constant $k_q^{O_2} = 112 l\ mol^{-1} s^{-1}$ for Naph-d8-β-CD-Ada system and $\sim 1000 l\ mol^{-1} s^{-1}$ for Naph-d8-β-CD-CH indicate that the dissolved dioxygen molecule could not contact phosphor easily in the presence of CH or Ada, respectively; moreover, it implies that adamantane is more efficient than cyclohexane in regulating the environment of CD cavity.

Zhang *et al.* [53] indicated that six-membered carbocyclic compounds could act as a molecular switch block of room temperature phosphorescence in non-deoxygenated β-CD solution. This may be a good phenomenon for possible application in future molecule computer design paralleled to the temperature sensor based on inclusion complex reported by Brewster *et al.* [29].

Ter-butyl chloride was found to have excellent ability to induce RTP of polycyclic aromatic compounds [54]. Mu *et al.* [55] investigated the effect of commonly used organic solvent of 1-bromonaphthalene, 1-bromo-2-methynaphthalene, 1-bromo-4-methylnaphthalene and potassium 6-bromo-2-naphthylsulfate in β-CD aqueous solution on RTP in the presence of dissolved oxygen. The enhancement effect of the different solvent were in the order cyclohexane > dichloromethane~1,2-dichloroethane~chloroform > tetrahydrofuran~ethylacetate~acetonitrile. The enhancement effect of tetrahydrofuran to all probes used and the enhancement effect of ethylacetate and acetonitrile to 1-bromonaphthalene were surprising.

Suspended crystal particles are important factors for further enhancing phosphorescence and leading to strong RTP in the presence of dissolved oxygen [28,50,52,56,57].

Wu *et al.* [58,59] investigated the synergetic effect of β-CD/1-bromo-4-(bromoacetyl)- naphthalene (BBAN)/Brij30 ternary complex on inducing non-deoxygenation RTP by ^{1}H-NMR and solvatochromic probe. Chemical shift changes in H-5 of β-CD were larger than those in H-3, which indicated that the phosphor was included in the hydrophobic cavity at the narrower end. H-6 also witnessed relatively large changes in its chemical shifts. So the exposed part of BBAN might be "locked" by the seven groups of methylol in the short section of the truncated cone molecule. ^{1}H-NMR spectra also indicated that the alkyl chain of Brij30 was partially inserted into the β-CD cavity so as to enhance the rigidity of the cavity that BBAN dwelled in, whereas the polar moiety was outside of the CD molecule at the narrower end. Additionally, ROESY (rotational nuclear Overhauser effect spectroscopy) validated the reciprocity between Brij30 and β-CD. To further elucidate the so-called synergetic effect (see above), fluorescence measurements of solvatochromic probe 4-(dicyano-methylene)-2-methyl-6-[4-(dimethylamino)stryryl]-4H-pyram(DCM) were also made in the same microenvironment as that of BBAN. The significant shifts to shorter wavelength of emission spectra implied that the BBAN resides in a more rigid environment with lower dielectric constant in the presence of both β-CD and Brij30 than exist when β-CD and Brij30 were present independently.

Interestingly, it could be recognized in some cases whether the third component was playing one of roles discussed above. Recognition of alcohol or other small organic molecules has been reported based on the change of triplet or phosphorescence characteristics of inclusion complexes. Typically, Gooijer's group [40] reported that the third component menthol (MT), as shown in Fig. 5, could induce strong room temperature phosphorescence (RTP) of 1-bromo-naphthalene (1-BrN) in aqueous β-CD suspensions, even under non-deoxygenated conditions. Interestingly, (−)-MT and (+)-MT enantiomers give rise to

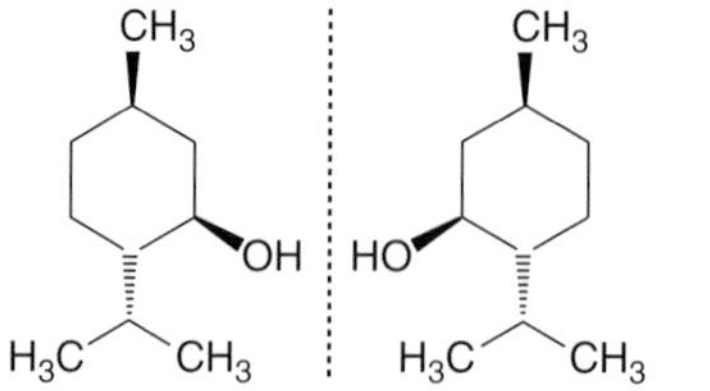

(1R,2S,5R)-(−)-Menthol   (1S,2R,5S)-(+)-Menthol

Fig. 5. Structures of the enantiomers of MT.

different phosphorescence intensities, the difference being $19 \pm 3\%$. It is argued that the signal can be mainly ascribed to the formation of ternary complexes β-CD/1-BrN/MT, which show different RTP lifetimes, i.e. 4.28 ± 0.06 and 3.71 ± 0.06 ms for $(-)$-MT and $(+)$-MT, respectively. Most probably, the stereochemical structure of $(-)$-MT provides a better protection of 1-BrN against quenching by oxygen than $(+)$-MT. This interpretation is in line with the observation that under deoxygenated conditions the phosphorescence intensity difference for the two complexes becomes very small, i.e. only about 4%.

The difference in lifetime values under aerated conditions makes possible the direct determination of the MT stereochemistry. For mixtures, in view of the 0.06 ms uncertainty in the lifetime, enantiomeric purity can be determined down to 10%. Furthermore, in the case of MT the concentration of the least-abundant enantiomer should be at least 3×10^{-4} mol/l, since otherwise complex dissociation would obscure the difference in lifetime. Sometimes it becomes incompatible when both the intensity and lifetime values become higher.

6. Tetra-component inclusion complexes and the fourth component

Because of limitation of CD cavity size, the formation of more than the quarternary inclusion complex may be difficult. Of course, if hydrogen-binding components by primary or secondary hydroxyl of CD rims are counted the inclusion complex of more than quarternary may be possible. However, the nomenclature herein especially counts the components included within the cavity with CD. In fact, it is very difficult to discriminate the third or fourth components. They may vary with the goals investigated. For example, β-CD/α-naphthanol/1,2-bromoethane or β-CD/β-naphthanol/1,2-bromoethane system displays strong phosphorescence, and the various alcohol molecules affect obviously the lifetime observed, resulting in the recognition of alcohol molecules, as shown in Table 5 [60]. The alcohol molecules here are named as the fourth component.

Du *et al.* [61] reported that the fluorescence from naphthalene dimer could be easily observed in β-CD/Naph or β-CD/Naph/α-BrN. When n-butanol was added as the fourth component to ternary inclusion complex, the dimer emission almost disappears accompanying remarkable enhancement in the RTP from α-BrN in quarternary inclusion complex β-CD/Naph/α-BrN/n-butanol.

7. Efficient triplet energy transfer in cyclodextrin cavity

In 1977, Tabushi *et al.* [62] proved that the specific and highly efficient energy transfer can occur between included component (acceptor) and benzophenone

Table 5
Effect of various fourth components on lifetime of inclusion complexes

β-CD/α-naphthanol/1,2-bromoethane					β-CD/β-naphthanol/1,2-bromoethane				
No.	Fourth component	Volume (μl/10 ml)	$\lambda_{ex}/\lambda_{em}$ (nm)	τ_p (ms)	No.	Fourth component	Volume (μl/10 ml)	$\lambda_{ex}/\lambda_{em}$ (nm)	τ_p (ms)
1	–	–	294/500	52.1	1	–	–	272/495	40.5
2	Methanol	40	294/494	85.9	2	Methanol	40	272/494	65.3
3	Ethanol	80	294/500	80.9	3	Ethanol	80	272/493	62.7
4	n-Propanol	40	294/500	104	4	n-Propanol	40	294/493	77.7
5	2-Propanol	80	294/498	84.1	5	2-Propanol	20	272/495	69.5
6	1-Butanol	40	294/500	51.7	6	1-Butanol	40	272/492	53.9
7	Glycol	80	294/500	108	7	Glycol	40	272/495	102
8	Acetone	40	294/500	53.9	8	Acetone	40	272/490	96.5
9	Butanone	60	294/500	43.4	9	Butanone	40	272/494	83.5
10	Acetonitrile	60	290/500	89.9	10	Acetonitrile	20	272/495	79.8

Table 6
Limit of detection of PAHs based on sensitized-phosphorescence

Acceptor	Donors	$\lambda_{ex}/\lambda_{de}$ (nm)	LOD (10^{-8} mol/l)
Biacetyl	α-BrN	285/511	6.8
	β-BrN	277/511	3.6
	Phenatherene	275/511	3.9
	Chryene	270/511	3.2

Note: Limit of detection, based on $3\sigma_{n-1}$; [β-CD] 9×10^{-3} mol/l, [Biacetyl] 1×10^{-4} mol/l; λ_{ex} and λ_{de}, excitation and detection wavelength, respectively.

(donor) derivative CD host molecule, as benzophenone covers the entrances of CD. In 1984, DeLuccia and Cline Love [63] systematically reported sensitized phosphorescence of biacetyl by PAHs and nitrogen heterocycles aiming at analytical application. The cavity can efficiently organize the energy transfer process between triplet states, resulting in the increase of energy transfer efficiency and enhancement of phosphorescence of biacetyl acceptor. If energy donor and acceptor are included simultaneously in CD cavity and match well in both geometry and size, the energy transfer efficiency should be highest [64].

In author's group, Gao *et al.* [65] used PAHs to sensitize biacetyl phosphorescence emission in CD medium and obtained quite good analytical properties as listed in Table 6.

8. Conclusions

The situation of triplet states and phosphorescence have largely improved because of the shield of CD from collisions by solute impurities, e.g., oxygen, in the inclusion complexes. Therefore, the CD inclusion complexes have become very useful in material fields, pharmaceutical preparation, environmental science and so on. Size-, shape- and chirality-dependent selectivity of CD to analytes make inclusion complexes very useful in analytical and bioanalytical chemistry.

Acknowledgments

The topics involved in this chapter were supported by the Natural Science Foundation of China and the Natural Science Foundation of Shanxi Province.

References

[1] S. Uppili, V. Marti, A. Nikolaus, S. Jockusch, W. Adam, P.S. Engel, N.J. Turro, V. Ramamurthy, J. Am. Chem. Soc. 122 (2000) 1102.

[2] A.A. Ruth, T. Fernholz, R.P. Brint, M.W.D. Mansfiel, J. Mol. Spectrosc. 214 (2002) 80.
[3] N. J. Turro, Modern Molecular Photochemistry, The Benjamin/Cummings Publishing Co., Inc., Menlo Park, CA, 1978, pp. 328–357.
[4] X.J. Duan, Z. Zhao, J.P. Ye, H.M. Ma, A.D. Xia, G.Q. Yang, C.C. Wang, Angew. Chem. Int. Ed. 43 (2004) 4216.
[5] J. Kuijt, F. Ariese, U.A. Th. Brinkman, C. Gooijer, Anal. Chim. Acta. 488 (2003) 135.
[6] C. Crewer, H.D. Brauer, J. Phys. Chem. 97 (1993) 5001.
[7] C. Grewer, H.D. Brauer, J. Phys. Chem. 98 (1994) 4230.
[8] I. Okura, Photosensitization of Porphyrins and Phthalocyanines, Gordon and Breach Science Publishers, Tokyo, 2000.
[9] N. J. Turro, Modern Molecular Photochemistry, The Benjamin/Cummings Publishing Co., Inc., Menlo Park, CA, 1978.
[10] J. A. Dean, Lange's, Handbook of Chemistry, 15th Ed., Beijing World Publishing-McGraw-Hill Book Co., Beijing, 1999.
[11] N. J. Turro, Modern Molecular Photochemistry, The Benjamin/Cummings Publishing Co., Inc., Menlo Park, CA, 1978.
[12] T. Vo-Dinh, J.R. Hooyman, Anal. Chem. 51 (1979) 1915.
[13] S.E. Chen, Y.J. Huang, X. Wang, W.J. Jin, J. Shanxi University(Nat. Sci. Ed.) 25 (1999) 150.
[14] H.R. Zhang, Y.S. Wei, W.J. Jin, C.S. Liu, Anal. Chim. Acta. 484 (2003) 111.
[15] Y. Wang, W.J. Jin, J.B. Chao, L.Q. Qin, Supramol. Chem. 15 (2003) 459.
[16] Y. Wang, J. J. Wu, Y. Feng Wang, L. P. Qin, W. J. Jin, Chem. Comm. (2005) 1090.
[17] W.J. Jin, C.S. Liu, Microchem. J. 48 (1993) 94.
[18] S.M. Liu, C. Ruspic, P. Mukhopadhyay, S. Chakrabarti, P.Y. Zavalij, L. Isaacs, J. Am. Chem. Soc. 127 (2005) 15100.
[19] S.M. Ramasamy, R.J. Hurtubise, Talanta 47 (1998) 971;
     J.S. Wang, R.J. Hurtubise, Anal. Chim. Acta. 332 (1996) 299.
[20] It can be reasonably deduced from suspension crystal system.
[21] Y. Wang, Z. Zhang, L.X. Mu, H.S. Mao, Y.F. Wang, W.J. Jin, Luminescence 20 (2005) 339.
[22] J. A. Dean, Lange's Handbook of Chemistry, 15th Ed., Beijing World Publishing-McGraw-Hill Book Co., Beijing, 1999.
[23] B. Birks, Photophysics of Aromatic Molecules, Wiley, New York, 1970.
[24] Y. Wang, Z. Zhang, Y.F. Wang, L.P. Qin, W.J. Jin, Anal. Lett. 38 (2005) 601.
[25] P. Bortolus, G. Marconi, S. Monti, B. Mayer, J. Phys. Chem. A. 106 (2002) 1686.
[26] C. García-Ruizl, M.J. Scholtes, F. Ariese, C. Gooijer, Talanta 66 (2005) 641.
[27] S.W. Bayles, S. Beckham, P.R. Leidig, A. Montrem, M.L. Taylor, T.M. Wight, Y. Wu, M.D. Schuh, Photochem. Photobiol. 54 (1991) 175.
[28] Y.L. Peng, Y.T. Wang, Y. Wang, W.J. Jin, Photochem. Photobiol. A: Chem. 173 (2005) 301.
[29] R.E. Brewster, M.J. Kidd, M.D. Schuh, Chem. Comm. 12 (2001) 1134.
[30] N.J. Turro, G.S. Cox, X. Li, Photochem. Photobiol. 37 (1983) 149.
[31] R.A. Femia, L.J. Cline Love, J. Phys. Chem. 89 (1985) 1897.
[32] S. Hamai, N. Mononobe, Photochem. Photobiol. A. 91 (1995) 217.
[33] S. Hamai, J. Kudou, Photochem. Photobiol. A 113 (1998) 135.
[34] S. Scypinski, L.J. Cline Love, Anal. Chem. 56 (1984) 322.
[35] S.S. Zhnag, C.S. Liu, Y.L. Bu, Chin. J. Anal. Chem. 16 (1988) 682.
[36] S.S. Zhnag, C.S. Liu, Y.L. Bu, Chin. J. Anal. Chem. 16 (1988) 494.
[37] S.M. Shuang, C.S. Liu, K.C. Feng, S.S. Zhang, Chin. J. Anal. Chem. 19 (1991) 1265.
[38] X.H. Bai, Y.S. Wei, C.S. Liu, Chin. J. Anal. Chem. 26 (1998) 243.
[39] Y.S. Wei, C.S. Liu, S.S. Zhang, Chin. J. Anal. Chem. 19 (1991) 533.
[40] C. García-Ruiz, X.S. Hu, F. Ariese, C. Gooijer, Talanta 66 (2005) 634.

[41] Y.S. Wei, W.J. Jin, R.H. Zhu, G.W. Xing, C.S. Liu, S.S. Zhang, B.L. Zhou, Spectrochim. Acta A 52 (1996) 683.

[42] C.G. Gao, Y.S. Wei, W.J. Jin, C.S. Liu, Chin. J. Anal. Chem. 24 (1996) 1015.

[43] C.G. Gao, Y.S. Wei, W.J. Jin, C.S. Liu, J. Anal. Sci. 14 (1998) 321.

[44] S. Hamai, J. Am. Chem. Soc. 111 (1989) 3954.

[45] A. Munoz de la Peña, I. Duran-Meras, F. Salinas, I.M. Warnar, T.T. Ndou, Anal. Chim. Acta. 255 (1991) 351.

[46] G.M. Escandar, A. Munoz de la Peña, Anal. Chim. Acta. 370 (1998) 199.

[47] A. Segura.Carretero, C. Cruces Blanco, Appl. Spectrosc. 52 (1998) 420.

[48] A. Ueno, K. Takahashi, Y. Hino, T. Osa, Chem. Comm. (1981) 194.

[49] T. Osajima, T. Deguchi, I. Sanemasa, Bull. Chem. Soc. Jpn. 64 (1991) 2705.

[50] W.J. Jin, Y.S. Wei, A.W. Xu, C.S. Liu, Spectrochim. Acta A 50 (1994) 1769.

[51] J.W. Xie, J.G. Xu, G.Z. Chen, C.S. Liu, Sci. China (Series B) 39 (1996) 416.

[52] V.B. Nazarov, V.G. Avakyan, M.V. Alfimov, T.G. Vershinnikova, Russ. Chem. Bull. (Int. Ed.) 52 (2003) 916.

[53] H.R. Zhang, Y.S. Wei, W.J. Jin, C.S. Liu, Anal. Chim. Acta 484 (2003) 111.

[54] S.Z. Zhang, Y.S. Wei, J.W. Xie, C.S. Liu, Chin. J. Anal. Chem. 28 (2000) 678.

[55] L.X. Mu, Y. Wang, Z. Zhang, W.J. Jin, Anal. Lett. 37 (2004) 1.

[56] G.M. Escandar, M.A. Boldrini, Talanta 53 (2001) 851.

[57] Y.L. Peng, W.J. Jin, F. Feng, Spectrochim. Acta A 61 (2005) 3038.

[58] J.J. Wu, Y. Wang, J.B. Chao, L.N. Wang, W.J. Jin, J. Phys. Chem. B 108 (2004) 8915.

[59] Z.Q. Gao, X.F. Shen, W.J. Jin, J. Shanxi University (Nat. Sci. Ed.) 26 (2003) 156.

[60] H. R. Zhang, MS Thesis, Shanxi University, Taiyuan, May 1999.

[61] X.Z. Du, Y. Zhang, X.Z. Huang, Y.Q. Li, Y.B. Jiang, G.Z. Chen, Spectrochim. Acta Part A 52 (1996) 1541.

[62] I. Tabushi, K. Fujlta, L.C. Yuan, Tetrahedron Lett. 29 (1977) 2503.

[63] F.J. DeLuccia, L.J. Cline Love, Anal. Chem. 56 (1984) 2811.

[64] F.J. DeLuccia, L.J. Cline Love, Talanta 32 (1985) 665.

[65] C.G. Gao, J.W. Xie, C.S. Liu, J.G. Xu, Chin. J. Anal. Chem. 26 (1998) 1424.

Cyclodextrin Materials Photochemistry, Photophysics and Photobiology
Abderrazzak Douhal (Editor)
DOI 10.1016/S1872-1400(06)01007-7

Chapter 7

Quantum Mechanics and Molecular Mechanics Studies of Host–Guest Stabilization and Reactivity in Cyclodextrin Nanocavities

Miquel Moreno

Departament de Química, Universitat Autònoma de Barcelona, 08193 Bellaterra (Barcelona), Spain

1. Introduction

Cyclodextrins (CD) range among the best known and most thoroughly studied molecules that possess a nanocavity able to accept different (smaller) molecules. In fact, CD were synthesized as early as at the beginning of 1970s. Since then, as the chemical industry has produced a large mass of CD, the interest in the chemistry inside CD has kept growing continuously and today CD are of fundamental interest in a variety of fields going from the basic to applied sciences. Among the thousands of contributions in this field that can be found in virtually all the science publications, the use of the CD as host molecules in supramolecular chemistry is of paramount interest as it is thought that the quantification of the driving forces involved in the molecular recognition of CD is fundamental for supramolecular chemistry as a whole [1].

As told elsewhere in this book, CD are macrocyclic oligomers of α–D-glucose, shaped like truncated cones with primary and secondary hydroxyl groups crowning the narrower rim and wider rim, respectively [2]. This particular shape makes the external part of the CD cavity quite polar (making CD soluble in water), whereas the interior of the cavity is clearly hydrophobic so that a host may easily enter the cavity. Different nanocavity sizes are obtained depending on the number of glucose units that form the macrocycle: α-CD is the smaller one and has 6 glucose units. β-CD, with 7 units, is the most widely used while γ-CD (8 units) and larger ones remain far less studied.

Given the wide interest that has recently arisen in the world of nanotechnology the possibility to modify the chemistry of small (guest) systems by using host molecules' nanocavities of the proper size, the theoretical understanding and evaluation of the host–guest interactions has become a fundamental issue as this knowledge may direct the obtainment of new nanocavities adapted to a particular family of compounds and/or to modify their physical properties and their reactivity. This change of reactivity upon encapsulation has also drawn the attention of groups working on molecular reactors and machines as cyclodextrin inclusion complexes can be used as miniature reaction vessels that control the outcome of a chemical reaction at a molecular level [3]. From a theoretical point of view, the study of the host–guest interactions is quite a tricky question as intermolecular forces are quite tenuous compared with the covalent bonds that are easily dealt within quantum electronic chemistry. Because the large size of the CD (147 atoms for the "medium"-sized β-CD) the use of the more sophisticated quantum electronic methods has been prohibitive until quite recently when the continuous growth in the power of calculation of computer has risen to a point where these calculations begin to be feasible. This does not mean that no theoretical studies were performed in the last decades. The so–called molecular mechanics (MM) techniques, that make use of empirical parameterizations of the intramolecular and intermolecular forces, have been massively used to analyze and predict the structure and behavior of a wide range of molecules inside CD cavities. Depending on the quality of the parameterization, MM methods can lead to very accurate predictions of the geometry of a particular host–guest system. However, they are less well suited to analyze the relative energies of the different conformations (results are usually to be considered qualitative) but the most severe drawback of MM methods is that they are totally unable to study a chemical reaction that implies the breaking and formation of chemical bonds. This makes the study of reactivity inside CD a task that can only be undertaken if some sort of quantum electronic method can be used.

In this chapter, we undertake a review of the previous work done to theoretically analyze and understand the host–CD interactions and to study the chemical (and photochemical) reactivity inside CD cavities. The emphasis will be to the use of quantum electronic methods to study host–guest CD complexes and the chemistry and photochemistry of different systems once encapsulated inside a CD cavity. As explained above, the use of the more sophisticated quantum methods to the macromolecular systems (such as the host–guest complexes analyzed here) is still in its infancy as the large-molecular size impeded these calculations until quite recently. For this reason, this work will also try to predict the future lines of progress that are expected in the near future. Given also the quite recent experimental progress that allow the study of chemical processes at the molecular level and using real-time sensors (femto-chemistry), the feedback between theoretical and experimental works can be quite efficient so that a fast progress in the field of supramolecular chemistry is expected in the near future.

2. Computational methods

This section will be divided in two subsections, the first one devoted to the theoretical evaluation of the energy of host–guest molecules (the primary goal of quantum electronic and MM methods). The second subsection will consider the different procedures available to obtain the conformations of minimum energy of the complex.

2.1. Obtaining the energy

In the real world, any system consists of a collection of atoms (nuclei and electrons). The whole energy of a system can only be accurately obtained through the resolution of the Schrödinger equation. It is quite customary to accept the so-called Born-Oppenheimer approximation based on the different mass scale of nuclei and electrons. Under this approximation the Schrödinger equation is divided in two parts:

$$\hat{H}_e \Psi_i^e(\mathbf{r}; \mathbf{R}) = E_i^e(\mathbf{R}) \Psi_i^e(\mathbf{r}; \mathbf{R}) \tag{1}$$

$$\left(\hat{T}_N + U_i(\mathbf{R})\right) \Gamma_N(\mathbf{R}) = E_i \Gamma_N(\mathbf{R}) \tag{2}$$

where

$$U_i(\mathbf{R}) = E_i^e(\mathbf{R}) + V_{NN} \tag{3}$$

and

$$\Psi(\mathbf{r}, \mathbf{R}) = \Psi_i^e(\mathbf{r}; \mathbf{R}).\Gamma_N(\mathbf{R}) \tag{4}$$

$U_i(\mathbf{R})$ is the potential energy hypersurface. Eq. 1 is called the electronic Schrödinger equation whereas Eq. 2 is the nuclear Schrödinger equation. In order to fully analyze the system, the nuclear Schrödinger equation should be solved. This implies the calculation of $U_i(\mathbf{R})$ at every configuration of the nuclei. Let us now consider the theoretical methods devised to obtain the potential energy surface. In what follows, a complete description of the theoretical methods is not presented but merely an overview of the different techniques that can be employed to deal with the complexation and reactivity inside CD.

2.1.1. *Quantum chemistry methods*

To exactly solve Eq. 1 the iterative Hartree–Fock (HF) procedure is the most widely used. To obtain exact results, a complete (i.e. infinite) basis set and a full Configuration Interactions (CI) method would be needed. This is unattainable so a finite basis set is used. Methods that evaluate all the integrals needed to solve Eq. 1 are called *ab initio* methods. The HF method does not consider the instantaneous interaction within pairs of electrons but only an average force. Not considering this instantaneous interaction (correlation energy) may lead to severe errors. Different methods exist to include the correlation energy, the already cited CI procedure (not at the usually unattainable full level) being the most evident way to overcome this problem. These methods, called post-HF, are quite costly in computational time so that they are restricted to small systems, well beyond the supramolecular level studied here.

Even the lowest HF level using a mimimum basis set (one basis set for each atomic orbital) may be a too costly procedure if the molecular system is really big (on the order of hundreds or even thousands of atoms). A further simplification involves the so-called zero differential overlap (ZDO) and the parameterization of a large number of the integrals involved in solving the electronic Schrödinger equation. These simplifications lead to the semiempirical methods. There are a lot of different semiempirical procedures that differ in the way the integrals are parameterized. It is not necessary to enumerate here a full list of the different semiempirical methods, as some of the oldest ones are no longer used. Among the ones that have been considered in the studies of CD and their complexes the AM1 (Austin Model 1) and the PM3 (Parametric Method 3) procedures are the most widely used. Quite recently, the PM5 method (an improvement of the PM3 one) has appeared but there are yet very few reports about how its performance is in obtaining energies and geometries of complex molecular systems [4].

In the last 15 years, the so-called density functional theory (DFT) techniques have emerged as a very powerful alternative to the HF-based methods. It is now well established that DFT methods can obtain quite accurate results with less computational effort than the sophisticated post-HF procedures. The DFT methods are based in a series of theorems by Kohn, which show that once the electronic density of a system of N electrons $\rho(r)$ is known the energy of the system can be obtained through the expression:

$$E[\rho] = T[\rho] + E_{ee}[\rho] + V_{ext}[\rho] \qquad (5)$$

where T is the kinetic energy of the electrons, E_{ee} is the electron–electron repulsion potential and V_{ext} is the nuclei–electron interaction.

Within the DFT formalism, the difficult point is the evaluation of the kinetic energy term. This is done by assuming that electrons are not interacting between them. Of course this approximation is not exact and the final expression of the energy has to be corrected. Depending on the formalism used different DFT methods have been developed. They are usually grouped in three families: the ones based on the local density approximation (LDA), the ones that make use of some gradient correction (GGA, generalized gradient approximation) and the hybrid methods that obtain the exact exchange energy from a HF calculation. Hybrid methods, such as the B3LYP one, [5,6] are the most widely used in chemistry up to now.

As HF and DFT procedures are based on a variational principle, they can only obtain the lowest energy of the molecular system. To obtain the energies of excited electronic states (and so be able to study photochemical processes) it is necessary to go to a CI calculation. The simple procedure is the CI-singles (CIS) that just considers monoelectronic excitations [7]. A more precise technique is the complete active space (CAS) method that performs a full CI over a selected (active) space of orbitals [8]. CAS methods are very powerful in the theoretical analysis of electronic spectra but are difficult to apply to reactivity as it is difficult to ascertain that the active space remains unchanged along all the reaction paths. Within the DFT formalism it is also possible to study excited electronic states using the time-dependent (TDDFT) formalism [9,10].

2.1.2. Molecular mechanics (MM) methods

A very drastic simplification to the above-mentioned procedure to obtain the potential energy hypersurface $U_i(R)$ is to consider the nuclei as point masses that evolve under the Newton mechanics within a conservative potential field created by the electrons. Under this classical approach the electrons do not explicitly appear and the only requirement is to have an expression for the force field. This is the theoretical basis of the MM methods. There are many different empirical force fields. They differ in the way the analytical function of the potential energy is defined and what kind of experimental data are used to fit the different parameters. They can be used for evaluating energies for systems of virtually any size so that supramolecular systems can be customarily obtained. Of course MM methods have also some severe limitations: the total energy has only a relative meaning as it cannot be compared with other systems that have different number of atoms or a different structure, and the electronic effects are not considered in the MM scheme.

2.1.3. Hybrid methods: The ONIOM procedure

The accurate description of a chemical system involves the use of the more sophisticated *ab initio* or DFT procedures. These methods cannot be used to deal with large systems as the ones found in the supramolecular system usually involves more than a hundred atoms. For these systems severe approximations (the ones that lead to the empirical MM methods for instance) have to be used but then the quality of the obtained energies is, at its best, of just qualitative value. A way to overcome this problem is to realize that, even with a very large system, the molecular region of interest is usually quite small. Then it is possible to divide the whole system in two or more parts. The central part that includes the atoms that are more relevant is to be dealt with a high-level method whereas the surrounding area is analyzed using a less-sophisticated procedure. Even if the basic idea is quite simple, there are several problems that have to be solved. In the following we will only refer to the ONIOM method as it is the only hybrid method used, up to now, to study the formation of host–guest complexes and the reactivity inside the cavity for CD.

The ONIOM method (our own *N*-layered integrated molecular orbital and MM) [11] allows the treatment of a large system using different methods of calculations for the different layers. The first layer is dealt at the highest level (presumably a quantum chemistry or molecular orbital method) whereas the second and third layers are studied at progressively less accurate levels. It is possible to use for the lower levels a more modest quantum chemistry model but also a MM force field (as implied in the name of the method). The application of the ONIOM to the CD host–guest complexes usually involves only two layers (high and low). The guest is dealt at the high level whereas the CD is included in the low-level layer. Two different molecular systems are then considered: the model one (in our case the isolated guest) and the real one (the full host + guest system). Calculations at the high level are performed only for the small model system whereas the low level of calculation is carried out for both the model and the real system. These three energies are then combined to obtain an approximation (E_{ONIOM}) to the energy of the real system at the high level as:

$$E_{ONIOM} = E_{LOW}(Real) + E_{HIGH}(Model) - E_{LOW}(Model) \tag{6}$$

It is to be noted that in a more general case the frontier between the low and high layers may imply the cleavage of covalent bonds. This makes the high-level layer unbalanced in terms of valence. This problem is circumvented by adding the so-called link atoms (usually hydrogen atoms) in the model layer so that the valence of the high-level layer is fully preserved.

2.2. Exploring the potential energy surface

The previous subsection has dealt with the different theoretical procedures that can be used to solve the electronic Schrödinger equation (Eq. 1) giving the potential energy at a given nuclear configuration $U_i(\mathbf{R})$. $\mathbf{R}$ is a vectorial magnitude that has the dimension of the total degrees of freedom of the molecular system ($3N$, N being the total number of atoms) so that the full evaluation of the potential energy hypersurface is virtually impossible for systems with more than four atoms (considering just 10 calculations per degree of freedom 10^{3N} energy calculations should be done). However, it is not really necessary to know the full potential energy surface but only its value at selected points. In chemistry, the more relevant geometries are the ones that correspond to minima in the potential energy surface, as these are the structures expected for a molecular system in thermal equilibrium (that is the one that can live enough time to be detected or even stored for a long time). In large systems, such as the host–guest complexes, where relatively weak intermolecular forces act to keep together the supramolecular system, there may exist a lot of nuclear configurations (conformations) of minimum energy. In that case the lowest energy conformation is expected to be the more likely one though, if there are small energy differences between different conformations, an equilibrium can be established between them. Things are more complicated if a chemical reaction is to be studied. From the viewpoint of the potential energy surface, a chemical reaction consists of a motion of the nuclei that evolve from one minimum (the reactants configuration) to another (the products configuration). This motion implies the crossing of an energy barrier. The highest energy point in the minimum energy path that goes from reactants to products is called the transition state (TS) of the reaction. In theoretical chemistry a lot of procedures have been devised to directly obtain the mimima and the transition states (TSs) of a given molecular system without having to calculate the full $U_i(\mathbf{R})$. These geometry optimization procedures are usually implemented in the quantum chemistry packages customarily used for theoretical calculations. These methods need the calculation of the gradient (first derivatives of the potential energy) and the Hessian (second derivatives of the potential energy). However, the latter has only to be exactly calculated if a TS is searched as minima are less sensitive to the shape of the potential energy surfaces and can be usually located using an approximate Hessian matrix.

The application of the direct localization algorithms to find minima and TS in supramolecular chemistry has some drawbacks. If a sophisticated molecular orbital quantum method is to be used, the evaluation of the gradient, not to speak of the Hessian, may be too costly from a computational point of view. More important is to realize that in complex molecular systems, such as the host–guest complexes, many minima and, consequently, many TS linking them, may exist. The localization of all the minima and TS may be a long and useless task, as not all the equilibrium geometries will be accessed for the actual supramolecular system. Usually the only relevant minima are the ones that have the lowest energy or, depending on the characteristics of the experiment, the ones that can be accessed from the initial reactants.

In order to obtain the lowest energy conformations of the host–guest complexes, a pure statistic approach can be used. These methods are based on the Monte Carlo (MC) algorithm that makes use of random numbers. Roughly speaking, MC methods perform a random walk through the potential energy hypersurface always looking for the lowest energy configurations. Of course this search has to be guided and a more sophisticated version of the statistical

methods are the genetic algorithms where only the best (lowest energy) geometries are used to generate new configurations. A less random approach is the
Molecular Dynamics (MD) method where the classical Newton equations of
motion are used to explore the potential energy surface. Again the goal of the
MD methods is to find minimum energy conformations along the obtained
classical trajectory. A relatively high kinetic energy has to be initially provided
to allow the system to surpass any energy barrier so that the system does not
remain in a very particular zone of the potential energy surface.

3. Applications

Given the large size of the whole molecular system, the majority of theoretical
works of CD complexes have been exclusively done using some kind of force
field (MM methods). While MM methods may perform well in the field of
structural chemistry (i.e. finding the more stable conformations of the complex),
they are unable to analyze any real chemical process that involves the breaking/
formation of new bonds. Photochemistry is also out of reach for MM methods
so that, as said in the introduction, this review will mainly focus on the (quite
few) theoretical studies performed within quantum chemistry-based methods
(molecular orbital, MO).

3.1. The molecular shape of the cyclodextrin nanocavities

The first step to be taken in computational chemistry in order to study chemistry inside CD is to analyze the shape of the CD cavity itself. That is, to exactly
know how the atoms are disposed in an isolated cyclodextrin molecule. From X-
ray and neutron diffraction studies it is well known that CD have the shape of a
truncated cone, as said in the introduction. In any case these diffraction experiments give the crystalline structure of CD which turns to be hydrated with
several water molecules (see refs. [12,13] and references therein).

Theoretical calculations have been carried out either to disclose the structure
of isolated or solvated CD or to explore the flexibility of the molecular shape of
CD when not in a rigid media. The older studies were confined to MM methods
but later on the MO methods became feasible and they have been applied to
decipher the CD geometry [14–18]. They have been carried out mainly at a
semiempirical level. Bodor *et al.* [14] used the AM1 method to study the isolated
α- and β-CD and to analyze several properties of the cavity size. They also
analyzed the changes produced upon partial methylation of the natural CD.
Nazarov and coworkers [15] studied β-CD within the PM3 method and found
that the more stable structure belonged to the C_7 symmetry group. Very recently
the group of Profs. H.F. dos Santos and W.B. de Almeida has carried
out several calculations on CD by using the more reliable *ab initio* and DFT

techniques. In a first work [17] they compared the structures of α, β and γ-CD obtained with different methods (MM, AM1, PM3 and HF) with the X-ray structures. In order to make the comparison more reliable, the solvation waters were also included in the calculation. It was seen that MM gave a better agreement to the "experiment" whereas AM1 performed quite poorly. PM3 and HF both gave a reasonable agreement with X-ray geometries. In a parallel work, [16] the same group studied the α-CD hexahydrate (with six solvation water molecules) with the energy evaluated at the DFT level but the geometry optimized at the semiempirical PM3 level. The hydration process was studied and seen to compare well with known thermodynamic data.

Finally, it has to be said that theoretical studies have also reached the dimers of CD. These studies are of interest, as it is known from stoichiometry relationships that some host molecules may need two CD to be encapsulated. Again most of the work is in the field of MM methods but the study of Prof. Nazarov's group on the dimer of β-CD was done at the semiempirical PM3 level. [15] At this level, the head-to-tail conformation where the wider (head) rim of one CD faces the narrower (tail) rim of the other is the most stable conformation. Very recently, Prof. dos Santos and his group has considered the α-CD hexahydrate dimer using the same DFT level of their previous work (BLYP energies with PM3 geometries) finding a preferred head-to-head conformation [18].

3.2. Structure and stability of CD–guest complexes

There are virtually hundreds of works that deal with the structure and stability of CD–guest complexes using some force field. MM methods are usually parameterized to take into account intermolecular forces so that they are well suited to analyze this problem. Therefore, the predicted structures for the more stable host–guest conformations are usually quite good but the energetics are far less accurate, MM results being, in most of the cases, just of qualitative value. Just to cite the more active groups in this field, the works of Profs. C Jaime, [19] F. Mendicuti, [20] H. Dodziuk, [21], A.C.S. Lino [22] E.A. Castro, [23,24] B. Mayer and G. Marconi [25] can be highlighted (more references are to be found in the cited works).

Not so many works have made use of MO methods to study the CD–guest minimum energy conformations [26–40]. It is not the purpose of this work a full analysis of the results for the different complexes but to present the capabilities of the MO methods in this field and analyze the limitations and future trends. Whereas the first works were almost restricted to the use of semiempirical methods, recently the more reliable *ab initio* or DFT methods have begun to be applied to such large molecular systems. A recent comparative study has shown that the more popular semiempirical methods AM1 and PM3 may give unphysical very short H···H distances between the host and guest molecules.

These short distances (below the sum of the van der Waals radii) do not appear in MM and *ab initio* calculations and are caused by an error in the evaluation of the core repulsion function that induces an unphysical interaction between host and guest [37]. It seems that the recent PM5 method, the successor of the PM3, does not present this drawback though, unfortunately, there are still very few reports on PM5 performance [4].

An overview of the MO works also reveals that they are mainly devoted to analyze the structure of the minimum energy conformations and the relative stability with respect to the separated molecules. A problem that appears now is the inability of the MO methods to perform a full search along the potential energy hypersurface (PES) (as the MM methods routinely do) as energy is not obtained from an analytical function but an iterative self-consistent procedure is needed. Given the large number of minimum-energy conformations that may be present in the host–guest complexes, a common procedure is to perform a previous search on the PES through a (cheap) MM method followed by MO optimizations starting from the more stable conformations obtained in the MM search. In some cases the difference in energies between different conformations is so small that a thermal equilibrium is predicted at regular temperatures.

A prominent point, that was the focus of many pioneering works of CD–guest complexes, is the analysis of the driving forces leading to the host–guest complex formation. These forces (interaction energies) can be classified in different ways, the more used terms being the non-bonded attractive (van der Waals) and repulsive (steric effects) forces and the electrostatic interactions. A special consideration has to be given to the hydrogen bonds that may exist between the host and the guest. In the real systems the motion of the surrounding water molecule also has to be considered when the host–guest complex is formed. This leads to the discussion of the complexation in terms of hydrophobic/hydrophilic forces and water redistribution energies. Other energy terms, more difficult to analyze from a theoretical point of view, are the changes in conformational energies of the cyclodextrin and its environment. A large number of MM studies have dealt with this problem and an interesting review was published some years ago [41]. However, at the MO level very few works have studied in detail the nature of the stabilization of the host–guest complex. This comes in part from the inability of MO methods to univocally separate the full energy in the terms familiar to chemists, and also from the difficulty (because of the present computer capabilities) to introduce, at the molecular level, the surrounding environment (such as the water molecules inside and outside the CD cavity). The more explicit study in this field is found for the β-CD–carvone complex [27] where it is concluded that the differences in the stability of the possible conformations of two enantiomeric isomers of the host are related to the number of "unconventional" C-H$\cdots$O $=$ C hydrogen bonds (distances of less than 3.0 Å) between the carbonyl oxygen of carvone and a hydrogen of the CD.

Usually the system studied is a simple 1:1 (host:guest) complex but sometimes dimers 2:2 [28,29] or 1:2 complexes [32] have been considered. The system studied in references [28] and [33] consists of two guests (cyclohexane or adamantane + naphthalene) inside β-CD. Some MO calculations have also considered the role of water as the solvent commonly used in the CD chemistry [38,40]. The solvent can be introduced self-consistently as a continuum that wraps up the molecular system or discretely by introducing some water molecules in the calculation.

Finally, the recent works done in the group of Prof. Belosludov [34–36,39] have to be highlighted as they are opening new perspectives in the field of host–guest complexes. These works analyze the encapsulation of polithiophene (PT) by several α- or β-CD molecules. This study is of interest for the semiconductor industry as PT is a conducting polymer and the study of PT–CD complexes may represent a significant advance toward the obtainment of electronic devices on a molecular scale. The theoretical model studied consisted of seven units of the thiophene monomer and 2 CD molecules. The dopant effect was also considered introducing some sodium atoms in the molecular model. Results using the hybrid ONIOM method that considers the PT at a DFT level and the CD at a MM level show that β-CD encapsulates PT without appreciably modifying the conducting properties of the polymer. Conversely, α-CD greatly distorts the polymer chain but when a molecular nanotube of crosslinking α-CD units is considered (two α-CD linked with two hydroxypropylene bridges in the theoretical model) the distortion of PT disappears and there is no charge transfer between CD and PT. This indicates that these complexes can be used as "insulated" molecular nanotubes.

3.3. The inclusion process

The study of the inclusion process, that is the way the host enters into the CD cavity, represents a step toward a more chemical view of the host–guest interaction problem that the mere structural study considered in the previous subsection. In fact much of the work referenced in the previous subsection could also be related to the inclusion process in that the stabilization of the complex is usually evaluated as the difference in energy between the complex and the isolated host + guest. In any case, such a thermodynamic study does not provide any clue about the actual molecular motion that leads to the formation of the complex. Attempts to answer this question are the works to be considered here.

Again the vast majority of theoretical studies are restricted to force field energy calculations. As previously said, pure MM calculations are not the main subject of this review though the quite systematic studies done by the groups of Profs. Jaime [42] and Mendicuti [43] can be highlighted.

In the quantum electronic side, there are very few works to comment. This is because, as previously said, the energy evaluation in the MO methods can be

quite time consuming for a supramolecular system. An evaluation of the molecular dynamics, even for the inclusion process that does not require the
cleavage or formation of covalent bonds, would usually need a huge number of
energy and gradient evaluations, out of the present computer capabilities. In
spite of that, some attempts have been already done to analyze the dynamics of
the inclusion process at a quantum chemistry level and a series of works by the
group of R. Lu and Q. Guo in China are notable in this field [44–46]. The
methodology used in this series is always the same: MO energy and gradient
evaluations are done at the semiempirical PM3 level. Consider the β-CD-styrene
and α-methyl styrene complexes as an example. There are two possible orientations of the guest: head-down, with the double bond of styrene pointing toward
the primary hydroxyls of the CD, and head-up with the double bond pointing
toward the secondary hydroxyls of the CD (Fig. 1). The formation of the
complex was simulated by displacing the guest along the Z-axis defined as in
Fig. 1. These calculations provide the energy profiles presented in Fig. 2 indicating a clear stabilization when the styrene is inside the CD cavity. The
energy minima along these profiles were used as starting structures for a full
geometrical optimization. The energy of the most stable structures was also

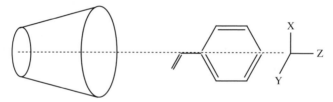

Fig. 1. Inclusion process of the styrene molecule inside a cyclodextrin. The reaction coordinate is
defined as the Z-axis that connects the centers of mass of both molecules.

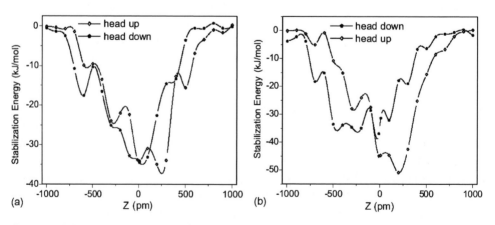

Fig. 2. Stabilization energy for the emulation of the inclusion process of styrenes into the CD
cavity along the Z-axis (see Fig. 1). (a) β-CD-styrene; (b) β-CD-α-methyl-styrene. From Ref. [45].

evaluated by a single point energy calculation at the HF/6-31G level. Figure 3 shows two views of the more stable minima so localized for both complexes. The energy variation involved in the inclusion emulation indicated that the complexes prefer to adopt inclusion geometry with the guest well inside the host cavity in order to increase the van der Waals attraction, H-bond interaction and dipole–dipole interaction between the host and the guest. Comparison of the structures and complexation energies for both complexes reveal that β-CD–α-methyl styrene complex is more stable than β-CD–styrene. This can be understood as α-methyl styrene having more interaction sites with β-CD than styrene. This could be anticipated by a careful comparison of the two geometries of Fig. 3 as α-methyl styrene fits "better" inside the β-CD cavity than styrene. The contrary happens with the smaller-sized α-CD. All these results are in agreement with experimental data.

A similar study was performed by N. Blidi Boukamel *et al.* for the (4-tert-butylphenyl)(3-sulfonatophenyl)(phenyl)phosphine/β-CD inclusion complex [47]. Plausible inclusion process pathways were studied through an energy coordinate calculation and the most stable structures of the 1:1 complex sought through a global potential energy scan.

3.4. Chemistry and photochemistry inside cyclodextrin cavities

In contrast with the previous subsections there are a very reduced number of works devoted to study chemical reactions inside CD. It is to be remarked that it is understood here that a chemical reaction involves the cleavage/formation of covalent chemical bonds. Of course this fact is to be attributed to the impossibility to deal with such a process within MM methods and the difficulty to apply pure MO methods to such big systems. The recent development of hybrid methods has opened the door of reactivity inside supramolecular systems to MO methods but these methodologies are still in their infancy so that previous calibration calculations are usually needed.

In apparent contradiction with the previous paragraph it has to be said that there have been some attempts to analyze reactivity within MM methods. Luzhkov and Åqvist presented, in two consecutive papers [48,49], a procedure based on the GROMOS force field to study the breaking/forming of new bonds. To account for the difference in the absolute energy between reactants and products (that cannot be obtained from pure MM calculations) they used experimental values. The MM energies for both reactants and products states were then used to fit an empirical valence bond potential (EVB). Once the potential energy is obtained, the free energy of activation $\Delta G^{\neq}$ was calculated through the free-energy perturbation method, which is based on MD simulations of several reactive trajectories that are statistically averaged to obtain a free-energy profile.

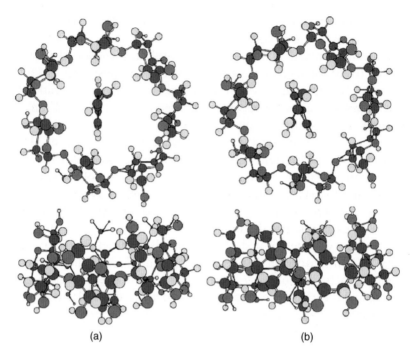

Fig. 3. Structures of the energy minimum obtained by the PM3 calculations for β-CD complexes. Seen from (top) β-CD cavity and (bottom) from the β-CD wall. (a) β-CD-styrene; (b) β-CD-α-methyl-styrene. From Ref. [45].

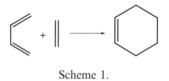

Scheme 1.

The maximum of G was then the value of $\Delta G^{\neq}$. This procedure was used to analyze different reaction paths for the hydrolysis of phenyl esters by β-CD.

One of the most "popular" reactions in organic chemistry is the Diels–Alder (DA) reaction between a diene and a "dienophile" (scheme 1).

Virtually thousands of works have been devoted to the study of DA reactions. From a theoretical point of view, DA reactions are also a sort of paradigm used as a test of the reliability of any new methodology. It is not surprising then that the first quantum-chemistry study of reactivity inside CD [50] was devoted to the DA reaction of cyclopentadiene with several dienophiles.

The TSs for the DA reactions were located with the AM1 method. The interaction between the TSs and the β-CD was represented by a Lennard-Jones potential keeping the internal coordinates of both host and guest fixed. For each position of the center of mass of the host, the Euler angles were minimized and the lowest energy geometry picked as the TS for the encapsulated DA reaction. It was assumed that one of the two reagents, the diene (cyclopentadiene), is preferentially included into the cavity and the dienophile approach this complex to form the TS–CD activated complex. Within this methodology notable differences in the stabilization energies induced by the presence of CD were seen depending on the substituents of the diene.

Other groups have studied the effect of CD-encapsulation on DA reactions. Houk and colleagues [51] used a force field (MM2) to analyze the catalytic role of β-CD on the DA reaction between cyclopentadiene and diethyl fumarate. MD simulations constraining all bond lengths were carried out to approximately localize the mimimum energy structures of reactants, TSs and products. The relative energy of reactants, TSs and products was obtained through B3LYP quantum-electronic calculations. A modest reduction of the $\Delta G^{\neq}$ upon encapsulation was seen. This fact is mainly due to the "entropy trap" effect that favors the reaction inside CD because the usual entropy barrier associated to a bimolecular reaction (the two reagents have to collapse into only one transition state) is not present here as an unimolecular complex is previously formed inside the cavity.

Li, Chung and Chao [52] carried out a similar study on the DA reaction where different regioproducts are possible. The gas-phase TSs were located within the HF (6-31G*) *ab initio* procedure and the different stabilization of these frozen structures when encapsulated inside β-CD were studied by means of a force field. The best host–guest conformation was sought through restricted MD runs. In this manner the binding behavior of the guest molecule in both the secondary hydroxyl and primary hydroxyl rims of the CD was examined. Simulations showed that the best binding site was at the bottom rim with the primary hydroxyl groups. This implies that when a guest molecule contains polar functional groups, hydrogen-bonding interactions with the primary hydroxyl groups may be an important factor in determining the stability of the overall complex. MD simulations of the different structures also showed that CD-mediated reactions can occur with just shallow binding at the periphery of the CD in the boundary between the highly polar aqueous medium and the relatively hydrophobic cavity.

In all the works cited up to now in this subsection, the reactivity of the entire macromolecular system was not considered at a full MO level but the effect of the guest was always modelled using some sort of force field. The use of hybrid methods to such a problem has been recently undertaken in a series of works by Casadesús *et al.* [37,53,54]. The analyzed system is the intramolecular H-atom transfer in 2-(2′-hydroxyphenyl)-4-methyloxazole (HPMO) in gas phase and embedded in β-CD (Scheme 2).

This reaction is known to be photoactivated, that is, it does not occur in the ground electronic state but in an excited electronic state. A previous theoretical

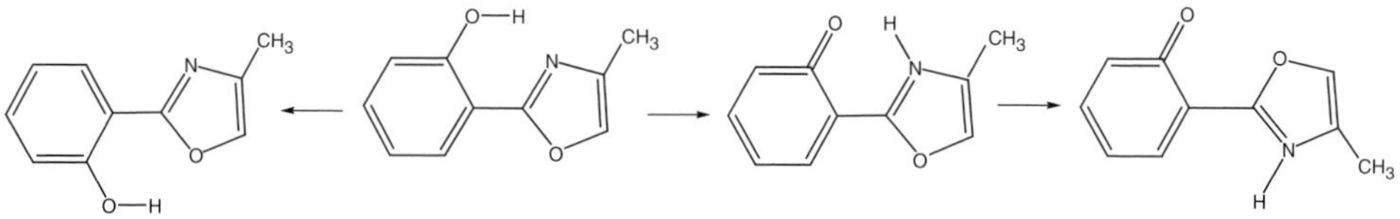

| Enol rotamer (ER) | Enol (E) | Keto (K) | Keto rotamer (KR) |

Scheme 2.

treatment [55] showed that a high energy barrier and a large endoergicity impeded the reaction to occur in the ground state (S_0) but in the first singlet excited electronic state (S_1) the excited state intramolecular proton transfer (ESIPT) was an exoergic process with just a moderate energy barrier.

The first work [54] considered the reaction in Scheme 2 with and without β-CD so that the possible catalytic effect of β-CD could be analyzed. Both the ground and first singlet excited electronic states were considered so that this work is not only pioneering the study of reactivity inside CD using pure quantum chemistry methods but it is also the first one to consider the photoreactivity inside CD. The ONIOM hybrid method was used to locate the geometries and energies of the reactant (enol form), product (keto form) and the TS. The high level includes just the HPMO guest (HF/6-31G* for S_0 and CIS/6-31G* for S_1) whereas the β-CD is included in the lower level and dealt with the semiempirical PM3 method. A previous search revealed two energetically accessible conformations of the initial guest (enol) inside the cavity, which are labeled **E1** and **E2**. Both conformations differ in the relative orientation of the host and, as their energies are quite similar, an equilibrium between both is predicted to exist at regular temperatures. The two initial conformations energy profiles are depicted in Fig. 4 and the geometries of the located stationary points (**E**, **TS** and **K** structures) are given in Fig. 5.

The dashed vertical arrows in Fig. 4 indicate the vertical transition (i.e., the point initially accessed upon electronic excitation assuming the Franck–Condon principle). In a photoinduced reaction, it is more significant to measure the energy barrier from this point than from the absolute enol minima in S_1. Comparing these energy barriers in S_1 with the ones obtained in gas phase at the same theoretical level, it is seen that for **E₁** the energy barrier is lowered whereas **E₂** increases it a little bit so that if thermal equilibrium is present a very minor effect is expected for the ESIPT reaction encapsulated in β-CD.

From the experimental side this photoreaction has been recently studied using ultrafast (femtochemistry) techniques. The first results apparently pointed to a slowing down of the ESIPT reaction when HPMO was encapsulated inside β-CD [56]. However, another work by the same authors pointed to the direct ESIPT reaction being almost unaffected when HPMO is placed inside a CD nanocavity [57]. In this work Douhal *et al.* observed two groups of time-resolved

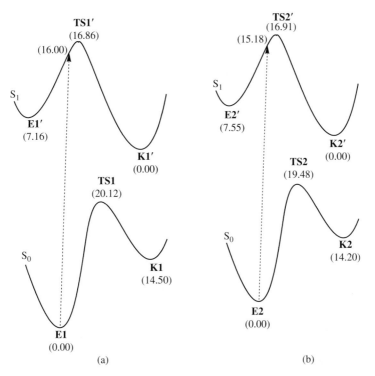

Fig. 4. Schematic energy profiles for the intramolecular proton transfer of HPMO inside β-CD showing the relative energy of each stationary point in kcal/mol as obtained within the ONIOM method (see text). The dashed line indicates the vertical transition. (a) and (b) correspond, respectively, to the process taking place from **E1** and **E2** structures (see Fig. 5). A prime is used to denote the first electronically excited singlet state S_1.

fluorescence transitions overlapping at 430 nm. At short wavelengths all the transients showed fast decays after the initial raise whereas at longer wavelengths the transient showed competition between rise and decay. This was interpreted as the presence of two mechanisms for the ESIPT reaction. A direct (very fast) one in which the proton transfer takes place directly and a slower one where the system evolves along two different coordinates: the proton-transfer motion at earlier times and the twisting motion of the heterocyclic moieties later on. This last motion would account for the slower rise component of fluorescence transitions. These interesting results prompted Casadesús *et al.* to enlarge the theoretical study by taking into account the internal rotational processes of HPMO and the effect of encapsulation on them [54]. The ONIOM hybrid method was used as before at the same level of calculation for S_0. For S_1 the CIS/6-31G* method was still used to find the stationary points but the energies were recalculated at the more reliable TDDFT level. Calculations indicated that the enol rotamer ER is out of reach as a large energy barrier is to be overcome. On the

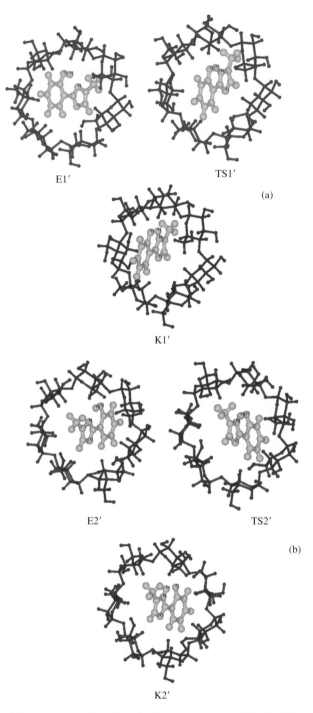

Fig. 5. Structures of the enol, transition state and keto structures of the HPMO molecule inside β-CD as obtained within the ONIOM method (see text). A prime is used to denote the excited state.

contrary, the keto rotamer KR can be easily accessed once the ESIPT process has taken place. In Fig. 6 the potential energy profile for the E → K → KR process in S_1 is depicted both for the isolated HPMO (gas phase) and encapsulated in β-CD. To be noted again a very minor effect of encapsulation on the energy barrier but a more clear difference in the final stability of the KR rotamer that is clearly more stable inside β-CD (this can be easily explained in terms of better inter-molecular host–guest interactions for the rotamer).

In order to evaluate the differences in the dynamics of the whole process in S_1 when encapsulation takes place, statistical RRKM calculations (i.e. TS theory) were carried out. Rotational degrees of freedom were not considered and the number of vibrational states was obtained through a direct count procedure. Two elementary reactions were considered: the E → K intramolecular proton-transfer reaction (IPT) and the subsequent inter-ring rotation (IRR) K → KR. For the proton-transfer process the tunneling was accounted for by using a simple one-dimensional Eckart potential. The rate constants at different energies were obtained this way for both processes in gas phase and inside β-CD. Figure 7 compares the results. Given the energy schemes of Fig. 6, it is not surprising to see that IPT is clearly faster than IRR and almost unaffected by the presence of β-CD. Conversely, the internal rotation IRR is dramatically slowed down upon encapsulation (compare Fig. 7a and b), the rate constant for the IRR process

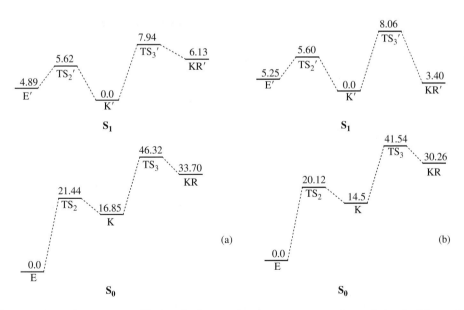

Fig. 6. Schematic energy profile for the intramolecular proton transfer and the C–C inter-ring rotation of HPMO in the ground state S_0 and the first electronically excited singlet state S_1. Energies are given in kcal/mol. A prime is used to denote the excited state. (a) Isolated (gas phase) results; (b) encapsulated inside a β-CD molecule.

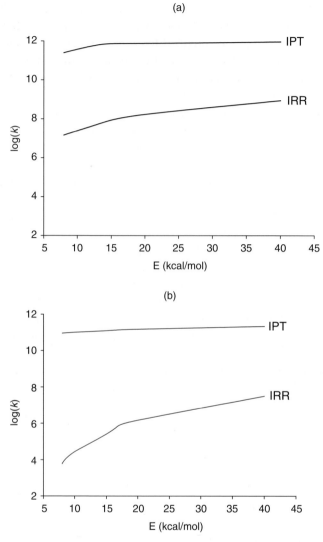

Fig. 7. RRKM rate constants (in s⁻¹) for the intramolecular proton transfer (IPT) and internal ring rotation (IRR) of the keto rotamer in the first electronically excited singlet state of HPMO. Energies are relative to the keto structure. (a) Isolated (gas phase) results; (b) encapsulated inside a β-CD molecule.

falling by more than two orders of magnitude in the encapsulated system. Such a big difference may come as a surprise giving the similar energy profiles in gas phase and CD shown in Fig. 6. The differences in rate constants come from the differences in the frequencies of the TSs for the internal rotation that are globally larger in the host–guest complex, therefore accounting for a higher Gibbs free-energy barrier for the process inside β-CD. In a more colloquial language, it can be said that rigidity increases upon encapsulation. These higher frequencies account for a higher energy barrier in terms of the zero point energy (not considered in the potential energies of Fig. 6) but also leads to a higher entropic term. Both factors contribute almost equally to the slowing down of the IRR process when the substrate is placed inside β-CD.

A more modest attempt to analyze photoreactivity inside CD can be found in the work of S. Monti *et al* [58]. Here the effect of β-CD on the excited-state decarboxylation reaction of two benzolthiophene derivatives is analyzed using the MM3 force field and a MC statistical search carried out to find the best conformations. For the lowest-energy structures so localized, semiempirical CNDO/S single-point calculations are carried out to obtain the energies of several excited states and the corresponding dipolar transition moments.

In concluding this subsection a quite recent work by E.A. Castro and colleagues [59] has to be cited as this is, as far as I know, the only pure MO treatment of reactivity inside CD apart from the HPMO work explained above. In this work the different rates of hydrolysis measured for *N*-phenylphthalamide (Ph-phta) and *N*-adamantylphthalamide (Ad-phta) in presence of β-CD is analyzed by means of the semiempirical PM3 method. In particular, the theoretical work just considers the equilibrium of the reactive amide with a zwitterionic structure and the eventual formation of a tetrahedral intermediate that is known to lead to the hydrolysis through a considerably lower-energy barrier. Scheme 3 depicts these structures.

PM3 calculations show that the most stable 1:1 host–guest complex has the phenyl or adamantyl substituent inside the β-CD whereas the phthalamide group is well outside the cyclodextrin cavity as seen in Fig. 8. The calculated PM3 energies indicate that, in agreement with experiment, there is a stronger driving force toward complexation for the Ad-phta system. Once the reactive host–guest complexes are formed, theoretical data also indicate that the way toward the zwitterion is made easier for Ad-phta and more difficult for Ph-phta. After the zwitterion structures have been attained, Ph-phta would be kept more strongly anchored (via two host–guest hydrogen bonds) than Ad-phta (only one hydrogen bond) so that it would be easier for the latter to attain the proper

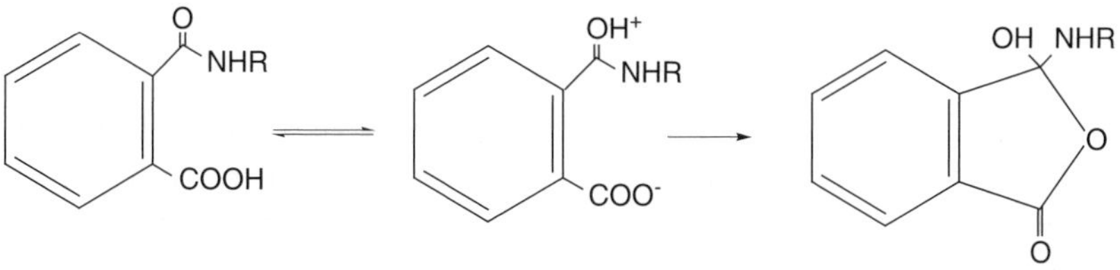

Scheme 3.

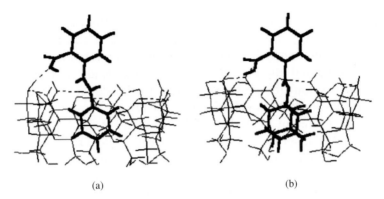

(a)   (b)

Fig. 8. (a) Phenyl- and (b) adamantyl-phtalamide/β-CD complexes at the optimized (PM3) geometry. From Ref. [59].

geometry for generating the intermediate. These results are in qualitative agreement with experimental evidence indicating that the hydrolysis of ph-phta in presence of β-CD is strongly inhibited whereas ad-phta hydrolysis is just slightly constrained when β-CD is present.

4. Future trends and perspectives

To end this overview, let me say a few words about the likely evolution of the calculations devoted to theoretically analyze the stability and reactivity of cyclodextrin complexes. Hybrid methods have recently opened the door for the most sophisticated *ab initio* and DFT methods to deal with molecular systems with hundreds or even thousands of atoms if the reactive center is well localized within a particular zone of the whole macromolecular system. As computers keep on growing in capacity and theoretical methods are progressively more well established, theoretical work on host–guest complexes will be ready to make a jump from the qualitative predictions, already available now, to the quantitative results. This will provide a powerful tool for the experiments as theory will be helpful not just to explain reactions already known but to predict new reactions and effects of the supramolecular environment that have been until now out of reach of precise measurements (as intermolecular forces are quite tenuous, their theoretical evaluation is subject to noticeable relative errors). On the other hand, it is to be noted that electronic calculations alone are not able to disclose the rich molecular dynamics of such a large molecular systems. Up to now MC and MD procedures have been restricted to find the more stable conformations but they are able to tell us more details about the mechanism of the inclusion process and reactivity inside CD and other large guests. The applicability of these methods to such problems has also been prevented because only MM methods can be used to perform the huge number of potential energy calculations needed. As QM and QM/MM methods become cheaper, the real dynamics of such systems shall be analyzed from first-principle methods. From this moment, reliable theoretical predictions will be routinely done in the more complex fields of molecular sciences such as nanotechnology, molecular machines and the large-scale biologically relevant molecules.

Acknowledgments

Financial support from the Spanish "Ministerio de Educación y Ciencia" and the "Fondo Europeo de Desarrollo Regional" through project CTQ2005–2007115/BQU and from the "DURSI de la Generalitat de Catalunya (2005SGR00400)" is acknowledged.

References

[1] W. Saenger, Angew. Chem. Int. Ed. Eng. 27 (1988) 89.
[2] W. Saenger, J. Jacob, K. Gessler, T. Steiner, S. Daniel, H. Sanbe, K. Koizumi, S.M. Smith, T. Takaha, Chem. Rev. 98 (1998) 1787.
[3] L. Barr, P.G. Dumanski, C.J. Easton, J.B. Harper, K. Lee, S.F. Lincoln, A.G. Meyer, J.S. Simpson, J. Incl. Phenom. Macro. Chem. 50 (2004) 19.
[4] J.J.P. Stewart, MOPAC 2002, Fujitsu Limited, Tokyo, 1999.
[5] C. Lee, W. Yang, R.G. Parr, Phys. Rev. B 37 (1988) 785.
[6] A.D. Becke, J. Chem. Phys. 98 (1993) 5648.
[7] J.B. Foresman, M. Head-Gordon, J.A. Pople, M.J. Frisch, J. Phys. Chem. 96 (1992) 135.
[8] K. Andersson, P.-Å. Malmqvist, B.O. Roos, J. Chem. Phys. 96 (1992) 1218.
[9] E. Runge, E.K.U. Gross, Phys. Rev. Lett. 52 (1984) 997.
[10] R.E. Stratmann, G.E. Scuseria, M.J. Frisch, J. Chem. Phys. 109 (1998) 8218.
[11] S. Dapprich, I. Komáromi, K.S. Byun, K. Morokuma, M.J. Frisch, J. Mol. Struct. (THEOCHEM) 461–462 (1999) 1.
[12] C. Betzel, W. Saenger, B.E. Hingerty, G.M. Brown, J. Am. Chem. Soc. 106 (1984) 7545.
[13] K.A. Connors, Chem. Rev. 97 (1997) 1325.
[14] N.S. Bodor, M.J. Huang, J.D. Watts, J. Pharm. Sci. 84 (1995) 330.
[15] V.G. Avakyan, V.B. Nazarov, M.V. Alfimov, A.A. Bagatur'yants, N.I. Voronezheva, Russ. Chem. Bull. Int. Ed. 50 (2001) 206.
[16] C.S. Nascimento Jr., H.F. Dos Santos, W.B. De Almeida, Chem. Phys. Lett. 397 (2004) 422.
[17] M.A.F.O. Britto, C.S. Nascimento Jr., H.F. Dos Santos, Quim. Nova. 27 (2004) 882.
[18] C.S. Nascimento Jr., C.P.A. Anconi, H.F. Dos Santos, W.B. De Almeida, J. Phys. Chem. A 109 (2005) 3209.
[19] P. Bonnet, I. Bea, C. Jaime, L. Morin-Allory, Supramol. Chem. 15 (2003) 251.
[20] J. Pozuelo, F. Mendicuti, E. Saiz, Polymer 43 (2002) 523.
[21] H. Dodziuk, O. Lukin, Polish J. Chem. 74 (2000) 997.
[22] L.M.A. Pinto, M.B. De Jesus, E. De Paula, A.C.S. Lino, J.B. Alderete, H.A. Duarte, Y. Takahata, J. Mol. Struct. (THEOCHEM) 678 (2004) 63.
[23] D.J. Barbiric, E.A. Castro, R.H. De Rossi, J. Mol. Struct. (THEOCHEM) 532 (2000) 171.

[24] D.J. Barbiric, R.H. de Rossi, E.A. Castro, J. Mol. Struct. (THEOCHEM) 537 (2001) 235.

[25] S. Monti, G. Marconi, F. Manoli, P. Bortolus, B. Mayer, G. Grabner, G. Köhler, W. Boszczyk, K. Rotkiewicz, Phys. Chem. Chem. Phys. 5 (2003) 1019.

[26] N. Balabai, B. Linton, A. Napper, S. Priyadarshy, A.P. Sukharevsky, D.H. Waldeck, J. Phys. Chem. B. 102 (1998) 9617.

[27] A.M. Da Silva, J. Empis, J.J.C. Teixeira-Dias, J. Incl. Phenom. Macrocycl. Chem. 33 (1999) 81.

[28] V.G. Avakyan, V.B. Nazarov, M.V. Alfimov, A.A. Bagatur'yants, Russ. Chem. Bull. 48 (1999) 1833.

[29] V.B. Nazarov, V.G. Avakyan, T.G. Vershinnikova, M.V. Alfimov, Russ. Chem. Bull. 49 (2000) 1699.

[30] H.F. Dos Santos, H.A. Duarte, R.D. Sinisterra, S.V. De Melo Matos, L.F.C. De Oliveira, W.B. De Almeida, Chem. Phys. Lett. 319 (2000) 569.

[31] M. Oana, A. Tintaru, D. Gavriliu, O. Maior, M. Hillebrand, J. Phys. Chem. B 106 (2002) 257.

[32] W.A. Adeagbo, V. Buss, P. Entel, J. Incl. Phenom. Macro. Chem. 44 (2002) 203.

[33] V.B. Nazarov, V.G. Avakyan, M.V. Alfimov, T.G. Vershinnikova, Russ. Chem. Bull. Int. Ed. 52 (2003) 916.

[34] R.V. Belosludov, H. Sato, A.A. Farajian, H. Mizuseki, Y. Kawazoe, Mol. Cryst. Liquid Cryst. 406 (2003) 195.

[35] R.V. Belosludov, H. Sato, A.A. Farajian, H. Mizuseki, Y. Kawazoe, Thin Solid Films 438–439 (2003) 80.

[36] R.V. Belosludov, H. Sato, A.A. Farajian, H. Mizuseki, K. Ichinoseki, Y. Kawazoe, Jpn. J. Appl. Phys. 42 (2003) 2492.

[37] R. Casadesús, M. Moreno, A. González-Lafont, J.M. Lluch, M.P. Repasky, J. Comput. Chem. 25 (2004) 99.

[38] V.B. Nazarov, V.G. Avakyan, S.P. Gromov, M.V. Fomina, T.G. Vershinnikova, M.V. Alfimov, Russ. Chem. Bull. Int. Ed. 53 (2004) 2525.

[39] R.V. Belosludov, A.A. Farajian, H. Mizuseki, K. Ichinoseki, Y. Kawazoe, Jpn. J. Appl. Phys. 43 (2004) 2061.

[40] W.A. Adeagbo, V. Buss, P. Entel, Phase Trans. 77 (2004) 53.

[41] K.B. Lipkowitz, Chem. Rev. 98 (1998) 1829.

[42] M. Zubiaur, C. Jaime, J. Org. Chem. 65 (2000) 8139.

[43] I. Pastor, A. Di Marino, F. Mendicuti, J. Photochem. Photobiol. A 173 (2005) 238.

[44] X. Li, L. Liu, Q. Guo, S. Chu, Y. Liu, Chem. Phys. Lett. 307 (1999) 117.

[45] Y. Cao, X. Xiao, R. Lu, Q. Guo, J. Mol. Struct. 660 (2003) 73.

[46] Y. Cao, X. Xiao, S. Ji, R. Lu, Q. Guo, Spectrochim. Acta A 60 (2004) 815.

[47] N. Blidi Boukamel, A. Krallafa, D. Bormann, L. Caron, M. Canipelle, S. Tilloy, E. Monflier, J. Incl. Phenom. Macrocycl. Chem. 42 (2002) 269.

[48] V. Luzhkov, J. Åqvist, J. Am. Chem. Soc. 120 (1998) 6131.

[49] V. Luzhkov, J. Åqvist, Chem. Phys. Lett. 302 (1999) 267.

[50] E. Alvira, C. Cativela, J.I. García, J.A. Mayoral, Tetrahedron Lett. 36 (1995) 2129.

[51] S.P. Kim, A.G. Leach, K.N. Houk, J. Org. Chem. 67 (2002) 4250.

[52] W.-S. Li, W.-S. Chung, I. Chao, Chem. Eur. J. 9 (2003) 951.

[53] R. Casadesús, M. Moreno, J.M. Lluch, Chem. Phys. Lett. 356 (2002) 423.

[54] R. Casadesús, M. Moreno, J.M. Lluch, J. Photochem. Photobiol. A 173 (2005) 365.

[55] V. Guallar, J.M. Lluch, M. Moreno, F. Amat-Guerri, A. Douhal, J. Phys. Chem. 100 (1996) 19789.

[56] A. Douhal, T. Fiebig, M. Chachisvilis, A.H. Zewail, J. Phys. Chem. A 102 (1998) 1657.

[57] D. Zhong, A. Douhal, A.H. Zewail, Proc. Natl. Acad. Sci. USA 97 (2000) 14056.
[58] S. Monti, S. Encinas, A. Lahoz, G. Marconi, S. Sortino, J. Pérez-Prieto, M.A. Miranda, Helv. Chim. Acta 84 (2001) 2452.
[59] A.M. Granados, R.H. De Rossi, D.A. Barbiric, E.A. Castro, J. Mol. Struct. (THEOCHEM) 619 (2002) 91.

Cyclodextrin Materials Photochemistry, Photophysics and Photobiology
Abderrazzak Douhal (Editor)
DOI 10.1016/S1872-1400(06)01008-9

Chapter 8

Fast and Ultrafast Dynamics in Cyclodextrin Nanostructures

Abderrazzak Douhal

Departamento de Química Física, Sección de Químicas, Facultad de Ciencias del Medio Ambiente, Universidad de Castilla-La Mancha, S.N. 45071 Toledo, Spain

1. Introduction

Cyclodextrins (CDs) are effective nanocavities to encapsulate different molecules, and this book contains ample information on the behaviour of several inclusion systems. They have the ability to modify the reaction or relaxation coordinates of the guest depending on the nature and size of the cavity. Several contributions have shown the effect of CD on the spectroscopy and dynamics of the guest and they have been reviewed [1–3]. Therefore, in this chapter we will focus on those published during the 2004–2005 period, and show the advances in the field related to fast and ultrafast (pico- and femtosecond (fs) timescales) dynamics of aromatic guest within CD nanocavities. The molecular systems, which we will consider here, show elementary key events in solution and upon caging: proton (or H-atom) transfer, charge transfer or twisting motion. Detailed information is in the original contributions, and therefore to save space I will mainly refer to these works.

2. Concept of femtochemistry in nanocavities

Femtosecond studies of caged molecules in nanocavities provide direct information on the relationship between time and space domains of molecular relaxation [3]. Therefore, simple and complex molecular systems have been studied using CD's proteins, micelles, pores and zeolites as nanohosts, demonstrating the confinement effect of the hydrophobic nanocavities on the spectroscopy and dynamics of the guests [2,3]. Relevant information on the induced ultrafast dynamics of caged wavepackets involving breaking/making chemical bonds and solvation has been acquired.

The fs experiments studying CD complexes measure the ultrafast time of a chemical reaction or solvation and subsequent events of trapped guest, which may be converted to a trapped photoproduct. Upon caging a molecule at the ground state, a confined nanostructure is formed and may have interesting physical and chemical properties. The nanocaging changes the degrees of freedom available for the embedded molecule to move along the reaction coordinates, and confines the wavepacket in a small area of propagation [4]. Therefore, reducing the space for molecular relaxation makes the system robust and immune to transferring heat over long distances to the surrounding solvent. Within CD, the dielectric reaction field around the guest changes.

The caged system has a limited freedom to move along the reduced potential-energy surface (PES). The neighbouring water molecules located at the gates of CD may affect its evolution. We note that in semiconductors a similar situation occurs where the conduction electrons are not only particles, but waves. Trapped electrons can only have energies dictated by the present wave patterns that will fit in this small space. Interestingly, cooling (or vibrational relaxation) process due to a fast (picosecond regime) exchange of heat with the environment (caging medium) might be also controlled by changing the nature or the size of the cage. It is well known that the nature of solvent plays a key role in the issue of a chemical reaction, in bulk solvent and in a cavity. Furthermore, these phenomena are reminiscent to those involving enzymes and zeolites in their efficient catalysis. In addition to that the solvation of water confined in a nanospace, such as those offered by CDs, is slower than that found in bulk medium and the results are important for a better understanding of protein biochemistry and biophysics [2,3,5,6].

3. Few examples

3.1. Methyl 2-amino-4,5-dimethoxy benzoate (ADMB)

ADMB is an aromatic molecule with two groups able to form an intramolecular H-bond (Scheme 1), and might be considered as analogous to the anaesthetic procaine and tetracaine. The intramolecular bond is comparable to that found in methyl salicylate, where the hydroxyl group has been substituted by the amino one [7]. The position of the maxima in absorption spectra depends on the polarity and H-bonding properties of the solvent [8]. The binding and related processes of ADMB within CD might be considered as a chemical model to mimic the interactions of drugs with hydrophobic pockets of CDs and

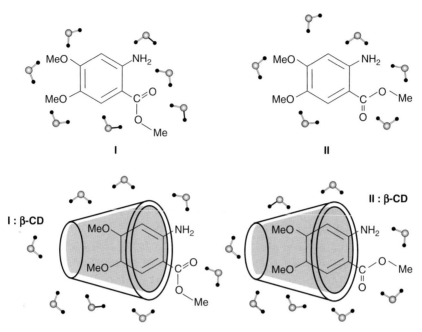

Scheme 1. Molecular structures of ADMB under **I** and **II** conformations, and the suggested inclusion complexes with CD.

biological substrates. In our group, we have studied this molecule in several solvents and in CD cavities [9]. We have shown that the addition of CD to a neutral aqueous solution of ADMB does not lead to any change in the UV-visible absorption spectrum in which the first transition is located at 340 nm [7]. The results indicate that the absorption coefficient of ADMB does not change when going from water into CD cavity. However, the emission spectra show a large increase in intensity. From the emission intensity change with the concentration of CD, we obtained an equilibrium constant of $12\,\mathrm{M}^{-1}$ and a complex stoichiometry of 1:1 at 293 K. The emission results, supported by theoretical calculations, indicate that upon electronic excitation, the $NH_2 \cdots O = C-O$ intramolecular H-bond becomes stronger at S_1 leading to an emission with a large Stokes shift ($\sim 7400\,\mathrm{cm}^{-1}$). The shift is explained in terms of an intramolecular charge-transfer (ICT) reaction from the amino group to the phenyl moiety, and not due to an excited-state intramolecular proton-transfer (ESIPT) reaction, as it occurs in the analogous molecule anthranilic acid. Thus, the increase in the emission intensity upon addition of β-CD is not due to the formed phototautomers, but of an enhanced emission of the caged intramolecular H-bonded ADMB molecule within β-CD. The calculated emission quantum yield of the complex, $\phi = 0.29$, is much larger than in pure water, $\phi = 5 \times 10^{-3}$. The presence of efficient non-radiative processes in water due to intermolecular H-bonding interactions with water molecules or due to twisting motion of the amino groups of ADMB is the main reason for the low-emission quantum yield. This explanation is supported by the strong dependence of the fluorescence lifetimes on viscosity, polarity and H-bonding character of the solvent [8]. The polarity effect of the alcohols is a consequence of the involvement of conformational changes due to twisted configurations of ADMB. The efficiency of these non-radiative channels is reduced due to the confinement effects on the ADMB into β-CD.

The ps emission dynamics of ADMB in neutral water and in presence of β-CD was examined (Fig. 1). In water solution, the emission decay at 450 nm is fit by a bi-exponential function with time constants of 47 ps (93%) and 110 ps (7%). These times were assigned to conformers II (with a larger dipole moment) and I, respectively. The calculations also suggest that the non-radiative processes due to twisting motion are expected to be faster in II than in I, leading to a shorter emission lifetime for II as compared to I. In presence of β-CD, the decay was best fitted to a four-exponential function. For example at [β-CD] = 15 mM, and gating the emission at 450 nm, the lifetimes are: 40 ps (67%), 110 ps (22%), 0.8 ns (3%), and 3.7 ns (8%). The ps components are assigned to free I and II conformers, respectively. The two other ns components, whose pre-exponential factors increase with the [β-CD] are due to ADMB:CD complexes. The

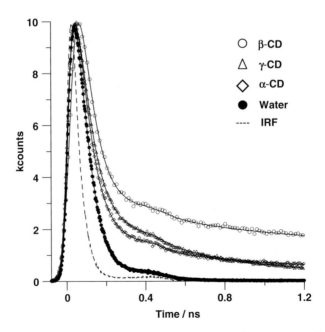

Fig. 1. Magic-angle fluorescence decays of ADMB in water and in presence of 15 mM of α-, β- and γ-CD. The instrumental response function (IRF) is ∼65 ps.

amplitude change upon addition of CD indicates that the complexation process involves conformer II rather than I. It should be mentioned that the confinement effects on the photodynamics of the molecule lead to longer lifetimes of the trapped ADMB. It is worthy, however, to notice that depending on the gated emission wavelength, the ns component varies between 0.6–1.1 and 3.4–3.9 ns, respectively. The heterogeneous environment of the complexes having different conformations of the guest and different interactions with water molecules is the source of the rich dynamics. Thus, the longest ns component corresponds to more protected complex than the shortest one. The latter is more exposed to water interactions that favour the twisting of the ester group, like in the II conformer. The observed caging effects in a smaller nanocavity (α-CD) and in a larger one (γ-CD) are not so significant. In the case of the complex with α-CD, the ADMB molecule does not penetrate enough into the cavity. The amino and ester groups are still exposed to water molecules. For the case of the complex with γ-CD, the twisting motion is highly favoured due to its larger cavity giving shorter decays.

Using time-resolved anisotropy measurements of ADMB in β-CD solutions, the decays give two rotational times: $\phi_1 = 53$ ps and $\phi_2 = 510$ ps (Fig. 2). ϕ_1 is assigned to the rotational relaxation time of the caged guest. This component also contains a contribution of the rotational time of free ADMB, as measured in pure water ($\sim$97 ps). ϕ_2 is assigned to the overall rotational time of the

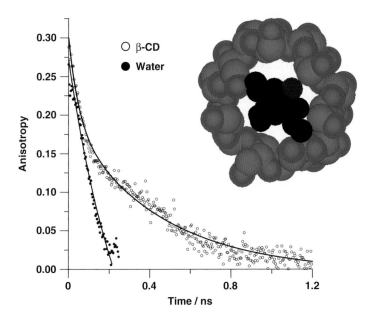

Fig. 2. Time-resolved anisotropy decay observed at 460 nm of ADMB emission in water and in presence of 15 mM of β-CD.

confined complex of ADMB:β-CD, which is close to 520 ps obtained using the Debye–Stokes–Einstein hydrodynamic theory under stick conditions.

3.2. Nile Red

Recently, ICT and solvation dynamics of Nile Red (NR) in the nanocavity of CD have been studied using ps emission spectroscopy [9]. The addition of CD enhances the week fluorescence of NR in water, and the observed change in steady-state emission spectra depends on the nature of cavity. While for β-CD, the increase of the emission band intensity was observed without a significant change in the wavelength of the emission band maximum. For the largest cavity, a large shift to lower wavelength (from 600 to 540 nm) of emission upon addition of 15.4 mM of γ-CD clearly indicates the effect of CD [9]. To better understand the nature of the complexes, the effect of urea on the emission of NR:CD solutions was studied. The experiments show that for β-CD complexes, the entity is rather robust and it is attributed to an inclusion complex involving 1:1 stoichiometry. However, for γ-CD complexes, the authors suggest 1:1 and 2:1 (NR:CD) capped complexes but involving hydrogen bonding between the NR and the OH groups of CD (Scheme 2). Time-resolved anisotropy experiments giving rotational relaxation times in water (50 ps), for β-CD (92 ps and 0.89 ns) and γ-CD (90 ps) solutions support these explanations. From the point of view of emission dynamics, the results show a Stokes shift in β-CD water solution, while it is absent for the γ-CD one. In addition to that, when compared to water solution, the non-radiative rate for twisted ICT process is almost 2.5 times slower than that observed in water. The γ-CD solution shows a weaker retardation. These differences were explained in terms of different types of complexes (inclusion entity for β-CD and capped for γ-CD), as it is suggested by the anisotropy experiment and urea effect on the emission behaviour.

3.3. Coumarin 153

Recently, the UV-visible absorption and emission spectra of Coumarin 153 (C-153) were studied in water and in presence of dimethyl- and trimethyl-β-CD (DMβ-CD and TMβ-CD) [10,11]. At room temperature, 1:1 and 1:2 (guest:-host) complexes were observed. The apparent equilibrium constant for 1:2 complex formation is 3350 and 38,500 M^{-2} for DMβ-CD and TMβ-CD, respectively. Those involving 1:1 complexes are 220 and 1220 M^{-1}, respectively. The fluorescence anisotropy decay was fitted using a bi-exponential model to give rotational times of 1 and 2.5 ns. These times were assigned to 1:1 and 1:2 complexes, respectively. From these times, the estimated hydrodynamic radius of the complexes are $\sim$9 and $\sim$11.5 Å for 1:1 and 1:2 entities, respectively. These

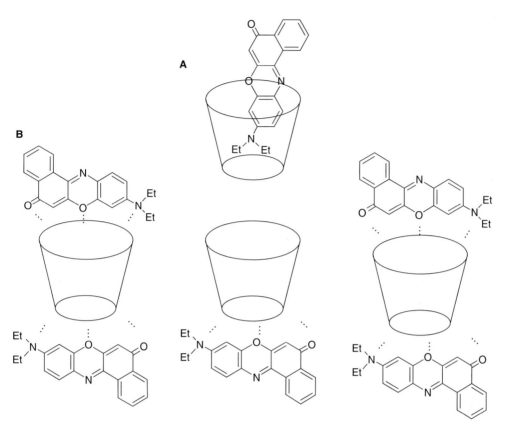

Scheme 2. Nanostructure of the inclusion and H-bonded complexes between Nile Red and CD. A for β-CD and B for γ-CD complexes.

values suggest that for the former, part of the guest is located outside the cavity, while for the latter the guest is almost completely protected by the macrocapsule formed by the two CD cavities (Scheme 3). In another study, the same authors report on the formation of aggregates of C-135:γ-CD complex in water solution [11]. A large number (more than 53) of CD entities was proposed to form the CD nanotube containing C-153. The average solvation time of these aggregates decreases from 680 to 160 ps when the temperature changes from 278 to 318 K. In a parallel way, and in the same temperature range, the Stokes shift was found to decrease from 800 to 250 cm^{-1} [11]. The observed temperature effect on the solvation time and its ultraslow component were attributed to a dynamical exchange between free and bound water molecules to the CD. This exchange is reminiscent to the behaviour of biological water in which free and bounded molecules have different response to a change in their rotational and translational component when adsorbed on or very near to the biological protein

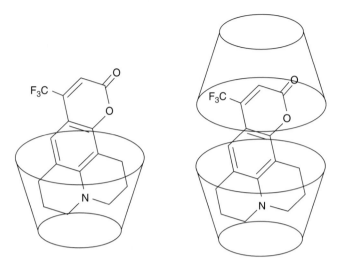

Scheme 3. Nanostructure of the inclusion complexes between Coumarin 153 and DMβ-CD or TMβ-CD. Adapted from Ref. [10].

molecule. The authors also reported a fs emission study using DMβ-CD cavity [11]. The results show that the solvation process involves a fast component having time of 2.4 ps and a slower component with times of 50 and 1450 ps. For TMβ-CD solution, the solvation dynamics is longer and the times are ~10, 240 and 2450 ps. The non-availability of a large number of water molecules around the methoxy groups of TMβ-CD might be the reason of the slowing of the longer solvation time [11].

3.4. Pyranine

Pyranine (PY) is a large aromatic molecule (Scheme 4) which shows excited-state intermolecular proton-transfer reaction with water molecules [12]. Several studies have been reported on the intermolecular proton transfer from PY to water. The interaction at the excited-state results in an excited-state proton-transfer (ESPT) reaction producing an anionic form, which emits a greenish-yellow fluorescence band. Here, we will consider only those of CD complexes using ultrafast spectroscopy [13]. In presence of CD, a 1:1 complex is formed and the normal emission of PY in water due to the enol form (E, 440 nm) increases, while that due to the anionic (A, 550 nm) structure decreases (Scheme 4). The change clearly shows the effect of CD on the emission behaviour. Gating the emission of E (440 nm) and of A (550 nm), the fs study reported a 0.8 ps component assigned to solvation of locally excited (LE) enol prior to proton transfer, and a 2–3 ps component attributed to solvent-assisted interconversion of LE to

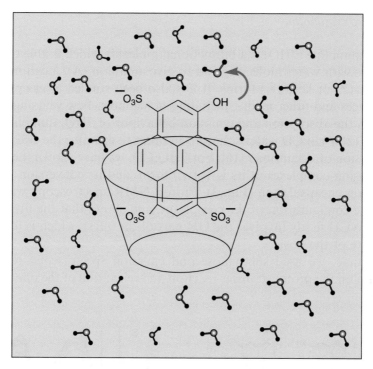

Scheme 4. Nanostructure of inclusion entity between pyranine (PY) and γ-CD surrounded with water molecules. The arrow between the OH group of caged PY and water indicates the observed intermolecular proton transfer reaction.

charge-transfer state before the occurrence of proton-transfer reaction to give A. These assignments are based on the results of previous experimental and theoretical works [14,15]. The proton-transfer reaction within the hydrogen-bonded complex occurs in a longer time, ~90 ps, and is followed by ion pair (IP) recombination to give back E or by dissociation to produce A. In presence of γ-CD, the ESPT becomes slower, and this was observed in the decay of E and rise of A emissions. While the first component (0.8 ps) assigned to solvation of LE state does not change upon encapsulation, that of solvent-assisted conversion of LE to charge transfer changes from 2–3 to 8 ps. Furthermore, the rise of caged A emission shows two longer components having times of 160 ps and 1.4 ns. Following the analysis of the data supposing equilibria at the electronically excited state, the results suggest that the most affected rate constant by encapsulation is that corresponding to recombination of the geminate IP for which the time increases by a factor of 3.5 [13]. In a parallel way, the rate of dissociation to give caged A is slowed down 1.5 times. These results clearly show the confinement effect on intermolecular proton transfer of a guest, and reflect the dynamics of special water molecules found at the gates of CD.

3.5. 7-Hydroxyquinoline

7-Hydroxyquinoline (7HQ) is a bifunctional molecule which is able to exchange two protons with water molecules and to give an anion (A), a cation (C) and a zwitterionic (Z) or keto (K) forms. It has also been studied in gas phase, polymeric matrices and other media, including bi-alcohols. Few years ago, we have reported on the absorption and emission behaviour of 7HQ, quinoline (Q) and 2-naphthol (2NP) in CD water solutions in order to examine the effect of caging on the emission of Z tautomer [16]. For 7HQ:CD, we have shown the formation of 1:1 inclusion complexes of its K (or Z) form, and its conversion to the enol (E) one upon encapsulation (Fig. 3). Proton NMR spectroscopy was used for studying the configuration of the complexes, and shows that the fitting process of 7HQ into CD firstly involves the OH part of the guest, which is found at the smallest gate of the host.

At that time, we used sub-nanosecond emission spectroscopy to get some information on the proton-transfer reaction and relaxation of the phototautomer

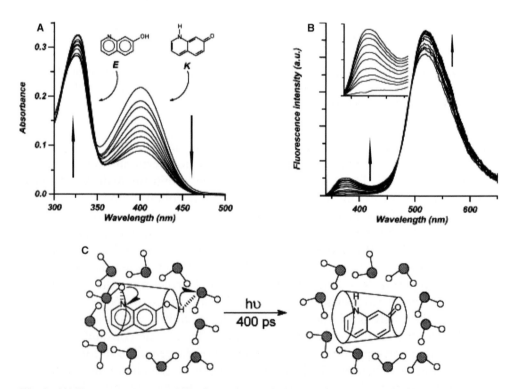

Fig. 3. (A) Room-temperature UV-absorption and (B) emission spectra of 7HQ in presence of DMβ-CD in neutral water solutions. The wavelength of excitation is at 330 nm. (C) Photoinduced double proton-transfer reaction of caged E to give caged K. Adapted from Ref. [16].

caged K form. We were the first to observe a retardation of these processes when the bifunctional molecule is trapped within CD nanocavities. The double proton-transfer reaction in caged E takes place in almost 400 ps, while in pure water solution the reaction occurs in few tens of ps (Fig. 3C). We explained the results in terms of a slowering of the proton-transfer dynamics of water molecules found at both gates of the nanocavity. For comparison, we have studied CD-caged 2NP (and of Q) and found that the caged proton-transfer reaction for 7HQ is almost 14 times faster than that of caged naphthol. The rate constant of naphtholate structure formation is reduced to $2 \times 10^8 \, s^{-1}$ upon CD encapsulation [16]. In 2001, we have reported on ps and fs dynamics of 7HQ in water and in presence of various CDs [17]. We showed how both proton-transfer reactions of caged E to give caged A and K forms are slower than those occurring in the bulk solution. We observed times changing from few hundreds of fs to few hundreds of ps. Recently [18], a study using ps emission spectroscopy was reported and it agrees with our previous published results (Scheme 5).

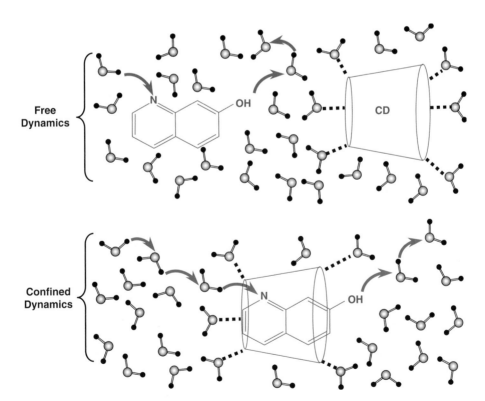

Scheme 5. Confined nanostructure of the complex between the E form of 7HQ and DMβ-CD, and conversion of K in water to caged E. Adapted from Ref. [16].

3.6. 3-Hydroxyflavone

3-Hydroxyflavone (3HF) is a widely studied molecule that undergoes an ESIPT reaction from its first excited electronic state to give a Z tautomer [19–22]. In hydrogen-bonding media, H-bonded structures to the solvent molecules give normal emission at blue side of the emission spectrum. Detection of anionic forms at the S_0 and S_1 states has been reported [23–26]. Due to its extreme sensitivity to water, the effect of CD on UV-visible absorption and emission characteristics was studied in *N,N*-dimethyl formamide (DMF) [27].

Upon addition of γ-CD to a DMF solution containing 3HF, the UV-visible absorption spectrum shows a new band at ~430 nm, indicating the formation of the anion (A) of the dye and a decrease of its E absorption band (~340 nm) (Fig. 4). The presence of 1.5 M of water does not significantly change the spectrum. Therefore, we explained the change at 430 nm region in terms of formation of A inclusion complex of 3HF into the γ-CD nanocage (Scheme 6). The OH group of the guest has been deprotonated due to an increase in its acidity helped by the H-bonding network of the OH groups of γ-CD. Using β-CD, the increase in the absorption intensity of the A band is very weak (5–10%). The changes in absorption spectra suggest the formation of a 1:1 inclusion complex between 3HF and CD, and the obtained equilibrium constants at 293 K are 48 M^{-1} (γ-CD) and 0.2 M^{-1} (β-CD). These values indicate a larger stabilization into a larger (γ-CD) nanocavity.

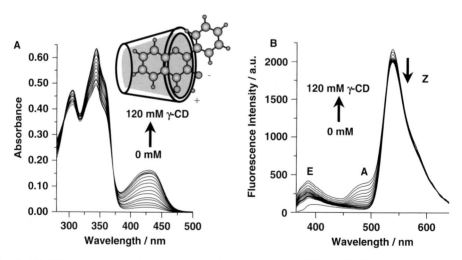

Fig. 4. (A) UV-visible absorption and (B) emission spectra of 3HF in *N,N*-dimethyl formamide upon addition of different concentrations of γ-CD. The excitation wavelengths was 340 nm. E, A and Z within the figure indicate the emission bands corresponding to enol, anion and zwitterionic structures, respectively.

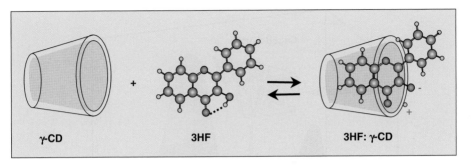

Scheme 6. Conversion of enol form of 3HF to its anion upon encapsulation by γ-CD.

When the solution is excited at 320 nm, we observed three emission bands at 380, 480 and 520 nm. We assigned these bands to 3HF bonded to γ-CD (more fluorescence than those bonded to DMF), A and Z structures, respectively. This one is due to a fast intramolecular proton-transfer reaction within the dye [28]. However, upon excitation at 420 nm the emission of A shows an increase in intensity upon addition of CD due to the formation of a more fluorescent inclusion molecule as a consequence of confinement effects on the caged anion. The excitation spectra suggest a larger relaxation of the guest inside CD, allowing a larger electronic conjugation between the aromatic moieties of anionic 3HF.

The emission decay of 3HF in pure DMF at 480 nm gives time constants of $\sim$10 ps (99%), $\sim$273 ps (0.7%) and $\sim$1.9 ns (0.3%) assigned to E, Z and A, respectively. Upon addition of 120 mM of γ-CD, these values change to $\sim$12 ps (68%), $\sim$270 ps (7%) and $\sim$2.14 ns (25%). The increase in the contribution of the ns component is because the population of caged A is largely excited at this wavelength. Gating the decay at 560 nm, a 12-ps rise component is recorded and it is close to the one assigned to the proton-transfer motion to produce Z, according to the value obtained in [28]. Figure 5 shows ps time-resolved emission spectra of 3HF water solution in presence of γ-CD, and the result indicates the absence of ESIPT reaction within the cavity and a longer lifetime of the confined anionic structure.

The observed rotational times obtained from the time-resolved anisotropy experiments reveal structural information on the formed complexes. While in pure DMF, the anisotropy decay gives a rotational time $\phi = \sim$85 ps. In presence of CD, we obtained rotational times $\phi_1 = \sim$170 ps, $\phi_2 = \sim$475 ps for 3HF:γ-CD complexes, and $\phi_1 = \sim$175 ps, $\phi_2 = \sim$440 ps for 3HF:β-CD ones. In both cases, ϕ_1 is two times longer than that of free 3HF in DMF indicating a docking of A into the cavity, ϕ_2 corresponds to the overall rotational time of the complexes, and accordingly it is larger for the larger (in size) complex.

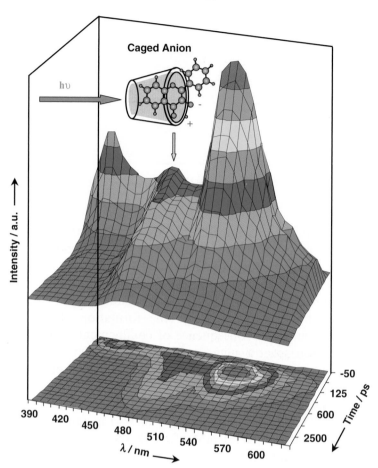

Fig. 5. Picosecond time-resolved emission spectra of 3HF in DMF in presence of γ-CD.

3.7. Orange II

Orange II (OII, Fig. 1) is an aromatic molecule which has O–H $\cdots$ N and N = N bonds. It shows an ESIPT and *trans–cis* isomerization reactions. It is widely used in the dyeing of textiles, food and cosmetics, making it present in the wastewaters of the related industries. OII exists under azoenol (AZO) and ketohydrazone (HYZ) forms (Scheme 7). Their relative populations depend on the medium. In a water solution for example, the shift of the H-atom within the O–H $\cdots$ N intramolecular H-bond to the nitrogen site makes HYZ structure the most stable one (about 95%). While in a more polar solvent like dimethyl sulfoxide (DMSO), the HYZ population decreases to 70%, and in solid state it becomes the only populated structure.

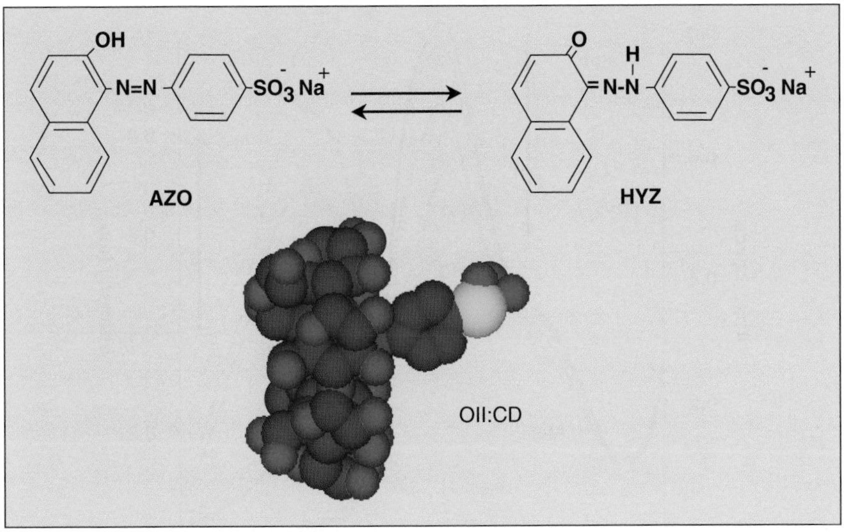

Scheme 7. Orange II (OII) in its AZO and HYZ structures, and its confined nanostructures with β-CD.

Recently, we have shown the occurrence of photoinduced proton-transfer reaction in AZO to give HYZ structure with a very low-emission quantum yield [29]. The results of studying the photodynamics of OII might be used for a better understanding of other chemical and biological systems undergoing isomerization reactions like retinal rhodopsin, for example. The inclusion complexes of OII and CDs have been studied by several groups showing the formation of a 1:1 entity. Ultrafast transient lens (UTL) technique with a response function of about 300 fs has been used to study the dynamics of OII in CD [30]. The slowing of the longest component in CD (13 ps in water, 28 ps in β-CD and 140 ps in γ-CD solutions) observed in the UTL signal was explained in terms of restriction to *trans–cis* isomerization of AZO tautomer due to H-bond formation between CD and the guest [30]. In 2005, we have reported on ps and fs emission studies of *trans* OII in water, few organic solvents, CD and human serum albumin (HSA) protein [29]. The emission of the photoformed *trans* OII was gated at several wavelengths upon excitation of the AZO structure.

Upon encapsulation by CDs, the hydrophobic nature of the interior of the cage converts *trans* HYZ into *trans* AZO-caged tautomer. The complex formation is reflected by a change in the relative absorption intensities of *trans* AZO (400 nm) and *trans* HYZ (485 nm) forms (Fig. 6). Previous ^{1}H and ^{13}C NMR experiments have suggested the inclusion of OII under its *trans* AZO form into CD through the naphthol unit [31,32]. The degree of penetration of naphthol moiety depends on the size of the cavity, being more important in a larger one.

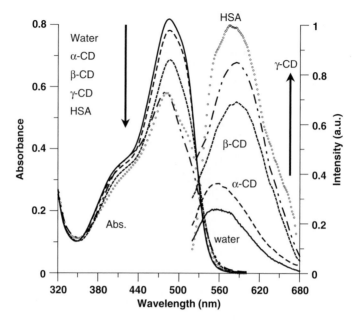

Fig. 6. UV-visible absorption and emission spectra of OII in water and in presence of CD and HSA protein. From Ref. [29].

In absence of CD, the emission intensity is very weak, peaking at 550 nm. The addition of β- and γ-CD induces an increase in emission intensity and a red shift of the maximum (from 560 to 590 nm). The emitting structure (HYZ form) originates from the encapsulated *trans* AZO one. Therefore, in presence of CD the ground-state AZO–HYZ equilibrium shifts to the *trans* AZO side due to the encapsulation of this form. The large Stokes shift ($\sim$8700 cm^{-1}) of the emission (590 nm) relative to the absorption of the encapsulated *trans* AZO form (390 nm) is explained in terms of production of an ESIPT reaction in the caged *trans* AZO structure to produce an encapsulated *trans* HYZ form. The increase in emission intensity in presence of CD is due to a restriction in motion of caged and photoproduced HYZ structure. For OII/CD complexes, the phenyl part is located outside the molecular pocket, and therefore torsion/twisting or rotation around N–N or C–N bond is not completely restricted. These motions enhance the rate constant of non-radiative processes leading to a low-fluorescence quantum yield. Furthermore, ^{13}C–T$_1$ measurements have shown that for some populations of OII:β-CD complexes, the guest molecule still can move within the cavity. For α-CD solution, the change is weaker due to a lower equilibrium inclusion constant mainly dictated by the small size of the host cavity.

Picosecond emission decay of OII in neutral water gives a time of $\sim$12 ps, and adding β-CD, the decay becomes longer (Fig. 7). The multi-exponential functions fit gives times of 25–45 ps (non-relaxed encapsulated *trans* HYZ), 150–175 ps

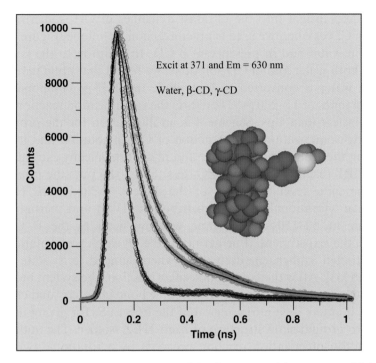

Fig. 7. Picosecond emission decays of OII in water and in presence of CD.

(encapsulated *cis* HYZ) and ~0.8 ns. At the red side of emission spectrum, the very small contribution of the 0.8 ns component does not change, while those of 25–45 ps and 150–175 ps show clear decrease and increase, respectively. The reason for this behaviour is the presence of different nanostructures having different degrees of penetration/confinement of the naphthol part, and thus different location of the N = N bond of AZO respect to the cavity. The experiment does not show any ps-rise component at longer wavelengths of observation. This indicates the absence of a common ps channel connecting these structures. The same trend was observed using γ-CD. The relatively small shortening in the ps components at red region, due to a faster relaxation in γ-CD, reflects the involvement of torsion/twisting motions in both cavities. The results show also that photoinduced intramolecular proton-transfer reaction in confined *trans* AZO occurs leading to caged *trans* HYZ. In these nanostructures, probably with different angles of inclusion of OII into CD, the restriction to twisting should be larger, and the emitting structures at the green region have different relaxation times. It is worthy to note that the fast *trans–cis* isomerization governs the short emission lifetime of OII. The heterogeneity in the complexed populations leads to a large distribution of the confined geometries at S_0 with different routes for ultrafast relaxation from S_1, and produces different twisted and encapsulated *cis* HYZ rotamers.

To get a deep insight into the ultrafast dynamics of OII isomerization in solution and in CD cavities, we used fs up-conversion technique. Figure 8 shows the fs transient in water and in presence of β-CD. In addition to the fs component observed in both water and in CD solutions, we found that upon nanocaging the 1 ps time in water is converted to 4 ps, while the 2–15 ps component becomes longer giving times of 15–200 ps. Therefore, the isomerization reaction within CD takes place in less than 1 ps. Scheme 8 is an illustration for the proton-transfer and *trans–cis* isomerization photoreaction of OII in solution and in CD nanocavities along the PES of both ground and first electronically excited states. The ultrafast ESIPT reaction in *trans* AZO (less than 30 fs) puts the *trans* phototautomer at the same level of energy of the directly excited *trans* HYZ forms. Intramolecular vibrational-energy redistribution (IVR) and inertial solvent response occur in 50–150 fs. This time window includes the fs contribution observed in the experiment. Furthermore, the dynamics of rotation-free and rotation-restricted azobenzene derivatives when pumped to S_2 state show comparable times [33]. After the fs-intramolecular reaction, the system becomes more flexible due to a loosening of the N = N double bond character and change in the bond order. It enters then into a region of the PES at S_1 for giving the *cis* HYZ or relaxing to ground-state structure of *trans* HYZ isomer. The authors suggest that this photo-isomerization reaction proceeds by a rotational/twisting mechanism in solution and through an inversion mechanism in CDs and in a protein (human serum albumin, HSA) nanocavity, as it was suggested for flexible

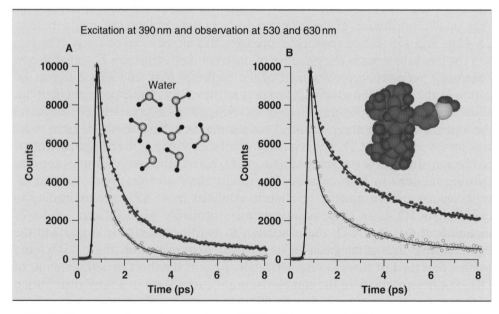

Fig. 8. Femtosecond emission transient of OII in (A) water and (B) in presence of β-CD.

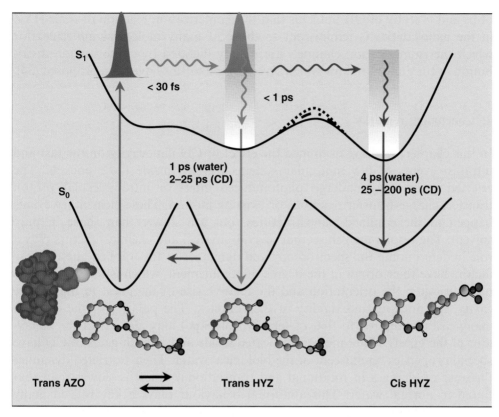

Scheme 8. Illustration of the potential-energy curve at S_0 and S_1 states of OII *trans* and *cis* isomers and tautomers. The inset shows the recorded times for the fast and ultrafast experiments in water and in CD nanocavities. Adapted from Ref. [29].

and rigid azobenzene derivatives [33]. The barrier to cross the *cis*-isomer region should be very small. Confined nanostructures with a stronger docking have lower probability to rotation (around N–N and C–N bonds) inside the cavity, and most probably, the inversion mechanism should be the operative one.

Because of CD confinement, the time scale for the C–N and C–C coupling, key event for the isomerization reaction, may change when compared to that of OII in water solution [3,4]. As we previously pointed out, the space restriction makes the twisting/rotation channel less favourable. Therefore, the *trans*-isomer wavepacket may take longer time to enter the isomerization region, and the phenyl inversion occurs. We note that the switching between rotation/twisting and inversion mechanisms for OII photodynamics in solution and in the nano-cavities involves the co-existence of three structures: two tautomers and one rotamer. The confined structures in the cavities have different topologies of the PES and therefore exhibit rich dynamics. One of the main interests in the results of ps and fs study of OII in CD is that the isomerization reaction of *trans* HYZ in the nanocavities is reminiscent to those of many biological molecules for which the isomerization channels are mainly dictated by conformation distribution of the guests, and for which several mechanisms have been proposed [34].

4. Concluding remarks

In this chapter, we have examined the effect of CD nanocavity on the fast and ultrafast events of few molecular systems trapped inside these pockets. The selected guests may undergo photoinduced inter- or intramolecular proton transfer, charge-transfer reactions or twisting motion. These elementary events happen in the confined nanostructures, but are slower than those in pure solvent. The results then show that several physical and chemical factors play a role in determining the spectroscopy and dynamics of the trapped guest. These factors have their origin in the degree of confinement, which is reflected by the nanostructure, the orientation and the relative size of the guest to that of the cavity, the docking and rigidity of the complex. The polarity of the cage (in many cases compared to that of tetrahydrofurane) may also affect the behaviour of the guest. Water molecules located inside and at both gates of CD have special properties reminiscent of the biological water. Their restricted dynamics (because of changes in rotational and translational times) is slower than that found in normal water. This abnormal behaviour plays a key role in many chemical and biological phenomena, and one can think about taking advantage of this relatively slow response of caged water to explore new directions for research and potential applications in nano- and biotechnology. As a final remark, and already opening the window to other directions of research in nanocaging, besides the few CD reports examined here, several studies have been carried out using other nanospaces provided by membranes, normal or reverse micelles, polymers, lipid vesicles, liquid crystals, sol–gels, dendrimers, proteins, DNA, zeolites and nanotubes. It is clear that understanding the fast and ultrafast events (solvation and chemical reactions) in these chemical and biological nanospaces will give a wealth of information for developing important fields of research like those of nano- and biotechnology. The use of time and space domain techniques, based on ultrafast laser technology and high-resolution microscopy, will play a key role in the conception, development and shaping these fields of research.

Acknowledgements

This work was supported by the MEC and the JCCM through the projects: MAT2002-01829, CTQ2005-00114 and SAN-04-000-00.

References

[1] For a Recent Literature on Photochemistry and Photophysics of Cyclodextrin Inclusion Complexes; see for example Special Issue of J. Photochem. Photobiol. A: Chem. 173 (2005) 229.

[2] N. Nandi, K. Bhattacharyya, B. Bagchi, Chem. Rev. 100 (2000) 2013.

[3] A. Douhal, Chem. Rev. 104 (2004) 1955.

[4] A. Douhal, Science 276 (1997) 221.

[5] S.K. Pal, A.H. Zewail, Chem. Rev. 104 (2004) 2099.

[6] B. Bagchi, Chem. Rev. 105 (2005) 3197.

[7] L. Tormo, J.A. Organero, A. Douhal, J. Phys. Chem. B 109 (2005) 17848.

[8] L. Tormo, Ph.D. Dissertation, University of Castilla-La Mancha, September 2004.

[9] P. Hazra, D. Chakrabarty, A. Charkraborty, N. Sarkar, Chem. Phys. Lett. 388 (2004) 150.

[10] D. Roy, S.K. Monda, K. Sahu, S. Ghosh, P. Sen, K. Bhattacharyya, J. Phys. Chem. A 109 (2005) 7359.

[11] P. Sen, D. Roy, S.K. Mondal, K. Sahu, S. Ghosh, K. Bhattacharyya, J. Phys. Chem. A 109 (2005) 9716.

[12] K.K. Smith, K.J. Kaufmann, D. Huppert, M. Gutman, Chem. Phys. Lett. 64 (1979) 522.

[13] S.K. Mondal, K. Sahu, P. Sen, D. Roy, S. Ghosh, K. Bahattacharyya, Chem. Phys. Lett. 412 (2005) 228.

[14] L.T. Genosar, B. Cohen, D. Huppert, J. Phys. Chem. A 104 (2000) 6689.

[15] T.-H. Tran-Thi, C. Prayer, P. Millie, P. Uznanski, J.T. Hynes, J. Phys. Chem. A 106 (2002) 2244.

[16] I. Garcia-Ochoa, M.A. Diez Lopez, M.H. Viñas, L. Santos, E. Martinez Ataz, F. Sanchez, A. Douhal, Chem. Phys. Lett. 296 (1998) 335.

[17] A. Douhal, Ultrafast Single and Double Proton Transfer Reactions in Solution and in Caging Media, VI Femtochemistry Conference, M. Martin and J.T. Hynes (Organizers) July 2003, Paris.

[18] H.J. Park, O.-H. Kwon, C.S. Ah, D.-J. Jang, J. Phys. Chem. B 109 (2005) 3938.

[19] A.J.G. Strandjord, P.F. Barbara, J. Phys. Chem. 89 (1985) 2355.

[20] M. Itoh, H. Kurokawa, Chem. Phys. Lett. 91 (1982) 487.

[21] S.J. Formosinho, L.G. Arnaut, J. Photochem. Photobiol. A: Chem. 75 (1993) 21.

[22] S.M. Dennison, J. Guharay, P.K. Sengupta, Spectrochim. Acta A 55 (1999) 903.

[23] M. Itoh, K. Tokumura, Y. Tanimoto, Y. Okada, H. Takeuchi, K. Obi, I. Tanaka, J. Am. Chem. Soc. 104 (1982) 4146.

[24] A.J.G. Strandjord, S.H. Courtney, D.M. Friedrich, P.F. Barbara, J. Phys. Chem. 87 (1983) 1125.

[25] P.T. Dzygan, J. Schmidt, T.J. Artsma, Chem. Phys. Lett. 127 (1986) 336.

[26] A. Douhal, M. Sanz, L. Tormo, J.A. Organero, ChemPhysChem 6 (2005) 419.

[27] L. Tormo, A. Douhal, J. Photochem. Photobiol. A: Chem. 173 (2005) 358.

[28] S. Ameer-Beg, S.M. Ormson, R.G. Brown, P. Matousek, M. Towrie, E.T.J. NIbbering, P. Foggi, V.R. Neuwahl, J. Phys. Chem. A 105 (2001) 3718.

[29] A. Douhal, M. Sanz, L. Tormo, Proc. Natl. Acad. Sci. USA 102 (2005) 18807.

[30] H. Yui, M. Takei, Y. Hirose, T. Sawada, Rev. Sci. Instrum. 74 (2003) 907.

[31] M. Suzuki, Y. Sasaki, Chem. Pharm. Bull. 27 (1979) 1343.

[32] M. Suzuki, Y. Sasak, M. Sugiura, Chem. Pharm. Bull. 28 (1979) 1797.

[33] Y.-C. Lu, E.-W.-G. Diau, H. Rau, J. Phys. Chem. A 109 (2005) 2090.

[34] R.S.H. Liu, G.S. Hammond, T. Mirzadegan, Proc. Natl. Acad. Sci. USA 102 (2005) 10783 and references therein.

Cyclodextrin Materials Photochemistry, Photophysics and Photobiology
Abderrazzak Douhal (Editor)
DOI 10.1016/S1872-1400(06)01009-0

Chapter 9

Structural Characterization of Colloidal Cyclodextrins: Molecular Recognition by Means of Photophysical Investigation

Antonino Mazzaglia[a,b], Norberto Micali[c], Luigi Monsù Scolaro[b]

[a]*Istituto per lo Studio dei Materiali Nanostrutturati, ISMN-CNR, Salita Sperone 31, 98166, Messina, Italy*
[b]*Dipartimento di Chimica Inorganica, Chimica Analitica e Chimica Fisica, Salita Sperone 31, 98166, Messina, Italy*
[c]*Istituto per il Processi Chimico Fisici, IPCF-CNR, Via la Farina 237, 98123 Messina, Italy*

1. Cyclodextrin colloids

Over the past decade the aggregation properties of cyclodextrins (CD) have been widely investigated. Non-covalent interactions such as hydrogen and ionic bonding, hydrophobic and van der Waals interactions are supposed to be the driving forces in the formation of CD supramolecular aggregates. Furthermore, molecular shape of natural or modified CD influences the recognition of identical or different CD structures which can interact "head to head" or "tail to tail" in short or long aggregates [1,2]. The spontaneous self-assembly finds potential applications in the search for "smart" nanomaterials and, when therapeutics are encapsulated, as drug delivery systems. Self-aggregation of natural CD in polydispersed stable nanoaggregates (with size of 200–300 nm for β-CD) has been recently studied by Bonini *et al.* [3] by means of scattering techniques, the diameters being in agreement with those previously reported by Coleman *et al.* [4]. On the other hand, modified CD, such as random methylated-β-CD (RMβ-CD) and hydroxypropyl-β-CD (HP-β-CD), display a better water solubility than natural CD. The aggregation process in these systems is prevented by disruption of hydrogen bonding network on both CD faces following substitution. In this respect it was evident that the aggregation behaviour involves the hydrophilic rims of CD. Moreover, Polarz *et al.* [5] employed CD-based silica materials to investigate the driving force for the self-assembly of the CD molecules to ordered aggregates.

The aggregation of CD molecules in water along single aliphatic polymeric chains has been first reported by Harada and Kamachi [6]. The kinetic aspects of these phenomena have been investigated by Becheri *et al.* [1], who reported that threading of CD molecules is cooperative and driven by enthalpic interactions between participating species. Moreover, addition of neutral or ionic species affects the process by modifying the hydrogen-bonding network of CD and by the adsorption of these solutes at the polymer and/or CD surface. In analogy, a large number of rotaxanes and polypseudorotaxanes possessing CD as ring components has been reported and the kinetic control of threading has been investigated [7].

The presence of the primary and secondary hydrophilic face surrounding the hydrophobic cavity bestows properties of amphiphile molecules to the natural CD. By modifying the upper and lower rims of CD with hydrophilic or hydrophobic groups self-assembly processes can be promoted either in solution or on surfaces.

In this chapter, we will focus on the self-organization of amphiphilic CD which form self-assembled colloidal systems with specific functionalities. In particular, the potential application of CD nanoaggregates and their nanostructured complexes with different guests in drug delivery will be mentioned. Light scattering and fluorescence techniques will be presented in order to characterize in detail their structure in aqueous solution. In this respect heterotopic colloids of amphiphilic cyclodextrins and photosensitizers will be shown as potential novel systems for targeted photodynamic therapy of tumours (PDT).

1.1. Amphiphilic CDs as building blocks for self-assembled carrier systems

Complete or selective modification at one or both faces of CD can lead to a change in physico-chemical properties of these molecules. The introduction of hydrophobic groups (such as long-chain ethers, esters, amines, thiols or amides) at only one face offers a range of amphiphilic molecules widely studied as supramolecular assemblies [8]. In this respect, it has been shown that amphiphiles of hydrophobically modified CD can form a variety of monolayers, multilayers, and Langmuir–Blodgett films at the air–water interface [9–11]. Amphiphilic CD possessing lyotropic and thermotropic properties are particularly interesting. As an example, CD modified with hydrophobic and hydrophilic faces (the so-called Janus CD) were studied under different medium conditions and have been used as calibration standards for the determination of molecular weights [12].

Amphiphilic CD can be admixed to phospholipid monolayers [13] as well as lipid vesicles [14,15], and they can be dispersed into nanospheres, showing promising properties for drug encapsulation [16]. In the early 1990s Lehn's group synthesized modified CD named "bouquets", bearing multiple poly(oxyethylene) or polymethyene chains on both the primary and secondary rim. These systems have been shown to be incorporated into lipid vesicles where they act as trasmembrane ion channels [14]. In the same period, different French groups demonstrated promising properties of amphiphilic CD organized materials for drug encapsulation and delivery. In Saclay, R. Auzely Velty, F. Dyedaini-Pilard, B. Perly, and T. Zemb observed that amphiphilic monosubstituted CD derivatives (i.e. cholesteryl-CD) can be efficiently incorporated in membranes retaining their inclusion properties towards external guests [15,17]. Recently the insertion and behaviour of modified amphiphilic CD in suspended bilayer lipid membrane have been investigated by electrophysiological methods, observing that CD molecules build single well-defined ionic channels [18].

Moreover, it has been shown by Wouessidjewe, Duchene, Skiba, and Lesieur groups that "skirt-shaped CD" bearing fatty acyl chains on their secondary hydroxyl groups can form nanospheres. This colloidal carrier system has been studied in the area of pharmacotechnical aspects of drug association [16]. However, with few exceptions, these material have poor water solubility and the (mixed) aggregates in water are systems out of thermodynamic equilibrium [16]. On the other hand, highly soluble amphiphilic CD forming spherical micelles in water have been produced [17]. The properties of these compounds have been studied in detail in the absence and in the presence of a synthetic phospholipidic matrix by NMR and scattering techniques (light, X-rays, neutrons) [19]. Recently, Darcy group found that counterbalancing hydrophobic substitution of the primary side of β-CD by the introduction of oligo (ethylene glycol) on the secondary side in a graft synthesis leads to amphiphilic molecules with high solubility in water [20,21].

1.2. Improving the water solubility of carrier systems based on supramolecular CD aggregates

In these last years, the design of novel carrier systems with increased solubility in water has still been a significant challenge. In Dublin, Ravoo and Darcy [20] synthesized the first example of vesicles composed entirely of amphiphilic CD.

These particles, by combining the properties of liposomes and macrocyclic host molecules, created new possibilities for the development of host–guest carriers and drug delivery systems. Heptakis (6-alkylthio-6-deoxy)-β-CD are less water-soluble precursors of CD molecules which, upon further substitution, are able to form vesicles. This family of compounds displays thermotropic mesophases and forms monolayers at the air–water interface [22]. The poor water solubility of these CD is probably due to the strong intramolecular hydrogen bonding of the secondary hydroxyl groups as in native β-CD. In the case of native β-CD, even a random and low degree of substitution with hydrophilic hydroxyethyl or oligo (ethylene glycol) groups disrupts the hydrogen bond network and dramatically improves water solubility. This solubilizing effect had already been observed on the reaction of β-CD with ethylene carbonate [23] or with poly(oxyethylene) [24].

The introduction of hydrophilic oligo(ethylene glycol) units on the lower rim of alkylthio-β-CD significantly increases the water solubility of these molecules. Differently from polymeric micelles and dendrimers, these amphiphilic CD display a relatively definite and low number of hydroxyl groups (seven terminal OH groups on the oligo-ethylene glycol chains), which are chemically versatile for an easy design of new systems [21,25,26]. Depending on the balance between their hydrophobic and hydrophilic portions, the new CD form different lyotropic phases in water. In this scenario, small micelles and micellar aggregates [27] or bilayer vesicles [20,25,28] have been reported as potentially less immunogenic (due to their oligo-ethylene oxide exterior [29]) and more versatile guest encapsulators with respect to a single CD molecule [28,30]. Such vesicles retain guest molecules even on dilution [20,28,30] and can be targeted by using receptor-specific groups such as galactosyl moieties [26]. The aggregates of these molecules display multiple complexation in binding polymer guest molecules [30] and lectins [26].

In another investigation, cationic vesicles of amphiphilic CDs were assessed for transfection [31,32]. In particular, a thiohexadecyl substituted CD bearing primary amine groups at the oligo(ethylene glycol) headgroups [25] increased transfection over unvectored DNA by a factor of 20,000, five times better that the simple polypyridylamino CD [32].

Chart 1 displays some of basic structures of amphiphilic CD studied in these years by Darcy, Ravoo, and Mazzaglia groups.

Another report shows that amphiphilic CD sulfates form anionic vesicles [33].

Nolan et al. [34] extended the use of vesicles and nanoaggregates of amphiphilic CD to other delivery systems by introducing a labile disulfide bond which links hydrophobic chains to the cavity. This labile bond could be readily cleaved upon endocytosis in acidic intracellular environment releasing the guest entangled in the hydrophobic moieties of the modified macrocycle.

Over the past 5 years the numerous investigations on the properties of amphiphilic CD depicted in Chart 1 have been carried out. These CD have many advantages as potential drug delivery systems and non-viral vectors. They

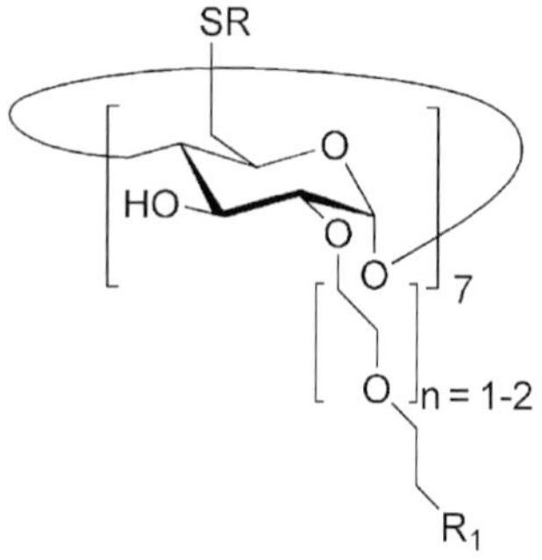

SR

$$1 \; R = C_2H_5 \quad R_1 = OH$$
$$2 \; R = C_6H_{13} \quad R_1 = OH$$
$$3 \; R = C_6H_{13} \quad R_1 = NH_3^+$$
$$4 \; R = C_{16}H_{33} \quad R_1 = OH$$
$$5 \; R = C_{16}H_{33} \quad R_1 = NH_3^+$$
$$6 \; R = C_6H_{13} \quad R_1 = SGal$$

Chart 1.

combine the properties of host macrocycle, their well-characterized pharmaceutical and toxicological profiles as drug delivery agents [35,36], including the ability to protect drugs from physical and chemical degradation, and the advantages of liposomes [37] and polymeric micellar containers [38], such as membrane permeability and low immunogenicity.

2. Molecular recognition exploiting optical spectroscopy (light scattering and time-resolved fluorescence anisotropy)

In order to fully understand the static interaction between molecules (and their aggregates) in solution, it is of basic importance to use spectroscopic techniques, which can characterize systems at thermodynamic equilibrium. On the other hand, techniques investigating interactions in a non-equilibrium regime, i.e. between molecules immobilized on surfaces and molecules in solution (e.g. surface plasmon resonance) could lead to scarcely comparable results in presence of aggregates in solution. Spectroscopic techniques do not require the probe molecules to be immobilized on solid surface and also eliminate the necessity of removing free molecules, usually required by other techniques for measuring interaction with biomolecules.

Several techniques are able to report the occurrence of a static interaction between molecules, including those based on detecting changes in the mass by measuring the translational (elastic light scattering, ELS, quasi-elastic light scattering, QELS, fluorescence correlation spectroscopy, FCS) and the rotational (time-resolved fluorescence anisotropy, TRFA) diffusion coefficient. Another useful spectroscopic technique to the same purpose is electrophoresis. In this latter case, it is possible to measure the mobility of the aggregates and the changes in the mass/charge ratio upon interaction.

Other spectroscopic techniques, such as circular dichroism, which is particularly devoted to chiral molecules, are not object of this report and will not be treated.

Quasi-elastic correlation spectroscopy and fluorescence correlation spectroscopy are suited to measure the diffusion coefficient of scatterers or fluorophores, reporting changes in size due to static interactions. Static light scattering gives analogous information through the measurement of the form factor or more simply of the gyration radii of the scattering particles.

TRFA gives insight on the rotational dynamics of fluorophores and it can detect a static interaction through changes of the inertia momentum.

Fluorescence techniques (FCS, TRFA) have the advantage of being selective (possibility of labelling with a specific fluorophore) and highly sensitive, together with the ability of detecting binding events between single molecules (they are incoherent techniques not limited by diffraction). Scattering techniques do not need a fluorescent tag, but they are not selective and much less sensitive (all the scatterers contribute to the observed signal).

In the following discussion, we will focus on some spectroscopic techniques able to give useful information of the static interaction between molecules and aggregated species in solution.

2.1. Elastic and quasi-elastic light scattering, and fluorescence correlation spectroscopy

In a QELS experiment, the measured intensity–intensity time correlation function $g_2(t)$ is related to the electric field correlation function, $g_1(t)$, by the Siegert relation [39]

$$g_2(t) = B(1 + f|g_1(t)|^2) \tag{1}$$

where B is the baseline and f is a spatial coherence factor. In the case of dilute solutions of monodispersed particles $g_1(t) = \exp(-\Gamma t)$, where the decay rate is $\Gamma = Dk^2$, D being the translational diffusion coefficient and k is the exchanged wavevector (being $k = [(4\pi n)/\lambda]\sin(\theta/2)$, where n is the refractive index of the sample and λ the wavelength of the incident light in vacuum). From the diffusion coefficient D, the mean hydrodynamic radius R_H of diffusing particles can be calculated using the Einstein–Stokes equation

$$R_H = \frac{k_B T}{6\pi\eta D} \tag{2}$$

where k_B is the Boltzmann constant, T the absolute temperature, and η the viscosity of the solvent.

For polydisperse scatterers, the field autocorrelation function may be expressed as the Laplace transform of a continuous distribution Γ of decay rates:

$$g_1(t) = \int G(\Gamma)\exp(-t\Gamma)\,d\Gamma \tag{3}$$

The effective diffusion coefficient, $D_{\text{eff}} = \bar{\Gamma}/k^2$ (where $\bar{\Gamma} = \int G(\Gamma)\,d\Gamma$ is the mean decay rate), can be obtained from the standard cumulant analysis [39]

$$g_1(t) = -\bar{\Gamma}t + \frac{1}{2!}\mu_2 t^2 - \frac{1}{3!}\mu_3 t^3 + \cdots \tag{4}$$

with μ_n the moments of the distribution $G(\Gamma)$ and $\sigma = \mu_2/\bar{\Gamma}^2$ the variance of the distribution. The polydispersity of the diffusion coefficient can be approximated to the variance for values of variance lower than 0.3.

For the ELS measurements the scattered intensity of the solvent is subtracted from that of the solutions; then the obtained angular profile is normalized using the scattered intensity of toluene a reference. The normalized scattered intensity can be written as [39]

$$I(k) = NKM_w cP(k)S(k) \tag{5}$$

where $P(k)$ and $S(k)$ are the normalized form and structure factor, respectively, N is the aggregation number, M_w the molecular weight of the molecule, c the mass concentration and K the optical constant. For the incident and scattered light orthogonally polarized with respect to the scattering plane it is $K = (4\pi^2 n^2)/(\lambda^4 N_A)(dn/dc)^2$ where n and dn/dc are the refractive index and the refractive index increment of the solution, respectively, and N_A the Avogadro number. For dilute samples, $S(k)$ becomes unity and the measured intensity $I(k)$ gives information on the static interaction through the apparent molecular weight NM_w and the form factor $P(k)$.

FCS monitors the random motion of molecules labelled with fluorescent tags inside a defined volume element irradiated by a focused laser beam. These fluctuations provide information on the rate of diffusion of a particle and this, in turn, is directly dependent on the particle's mass. As a consequence, any increase in the mass of a molecule, e.g. as a result of an interaction with a second molecule, can be readily detected as an increase in the particle's diffusion time. The intensity correlation function $g_2(t)$ in this case is related to the diffusion coefficient by means of the geometrical parameter dependent on the beam laser [39]:

$$g_2(t) \propto \left[N\left(1 + \frac{D_s}{\sigma_0^2}t \right) \right]^{-1} \tag{6}$$

where σ_o is the cross size of the focused light and D_s the self-diffusion coefficient that provides information on the static interaction.

For FCS, the laser wavelength must be able to excite the fluorescence of labelled molecules and the collecting optics must remove the scattered light and collects the fluorescence signal of the confocal image of the focused incident beam.

2.2. Time-resolved fluorescence anisotropy

Fluorescence anisotropy, $r(t)$, is defined using the following expression [40]:

$$r(t) = \frac{I_{VV}(t) - I_{VH}(t)}{I_{VV}(t) + 2I_{VH}(t)} = \frac{D(t)}{S(t)} \tag{7}$$

where the sum data $S(t)$ must be equal to the total intensity $I(t)$. In the first step $S(t)$ was analyzed using a reconvolution procedure based on the following multi-exponential model

$$S(t) = I(t) = I_0 \sum_i \alpha_i \exp(-t/\tau_i) \tag{8}$$

to obtain the parameters describing the intensity decay (α_i and τ_i). In the second step, holding constant the parameters obtained from the first step, the difference $D(t)$ was analyzed considering a multi-exponential decay of the anisotropy:

$$D(t) = S(t)r(t) = S(t)r_0 \sum_j g_j \exp(-t/\theta_j) = S(t) \sum_j r_{0\,j} \exp(-t/\theta_j) \tag{9}$$

where the parameters $r_0 = \Sigma_j r_{0\,j}$, θ_j and g_j represent, respectively, the limiting anisotropy in the absence of rotational diffusion, the individual rotational correlation times and the associated fractional amplitudes in the anisotropy decay ($\Sigma_j g_j = 1$). In the simple case of spherical molecules, each θ_j is related to the volume (V_j) of the rotating unit by the following equation [40]

$$\theta_j = \frac{\eta V_j}{k_B T} \tag{10}$$

where η is the micro-viscosity.

Finally, the decays of the parallel (I_{VV}) and perpendicular (I_{VH}) components of the emission were reconstructed using all the parameters obtained by the first two steps, on the basis of the following expressions:

$$I_{VV} = \frac{1}{3}I(t)[1 + 2r(t)] \tag{11}$$

$$I_{VH} = \frac{1}{3}I(t)[1 - r(t)] \tag{12}$$

When a direct determination of time-resolved fluorescence anisotropy is not possible, for example because of photo-bleaching, Perrin equation (in a system with a single lifetime and rotational correlation time)

$$r = \frac{\int_0^\infty I(t)r(t)\,\mathrm{d}t}{\int_0^\infty I(t)\,\mathrm{d}t} = \frac{\overline{I_{VV}} - \overline{I_{VH}}}{\overline{I_{VV}} + 2\overline{I_{VH}}} = \frac{r_0}{1 + \tau/\theta} \tag{13}$$

can allow for an estimation of the correlation rotational time θ from lifetime τ measuring the steady state fluorescence intensity $\overline{I_{VV}}$ and $\overline{I_{VH}}$.

A typical apparatus for time fluorescence measurements uses a time-correlated single-photon-counting (TCSPC) technique [41]. The fluorescence anisotropy data are then analyzed using the non-linear least-squares iterative reconvolution procedures. In this way, the instrumental resolution (corresponding to the minimum measurable time value) is about 50 ps. The sample compartment needs to be thermostated and eventually the sample is stirred (to minimize the effect of photo-bleaching).

3. Probing morphology and molecular recognition of photosensitizers in CD colloids

The interaction between drugs sensitive to the light and CD have been widely investigated in order to understand how the photostability and efficiency of drugs can be preserved. The structural properties of CD-drug materials can be studied by a combination of scattering, steady and time-resolved fluorescence techniques. In particular, scattering techniques can give information on the size, shape, charge, and interaction processes of CD aggregates. In this way, it is possible to modulate the recognition behaviour and the functionalities of these self-assembled systems. The entanglement of hydrophobic and hydrophilic photosensitizers into CD aggregates is strictly dependent on size and charge of the resulting hetero-colloids and is evident from changes in the stationary and time-resolved fluorescence properties.

3.1. Colloidal CD and photosensitizer drugs

Since long time CD as drug complexing agents have been the object of intense interest for both fundamental aspects and practical purposes [35,36]. Recently, this attention has turned to the problem of biological photosensitization by drugs [42]. Application of CD was suggested as a useful strategy to minimize the biological damage induced by drugs and increase drug photostability. Also the study of complexes between sunscreen agents and modified CDs has been described [43]. However, it is known that drug–CD complexes usually dissociate once injected into the body where there is also exposure to a wide range of endogenous species [44]. Polymeric or supramolecular aggregates of CDs have the ability to overcome these drawbacks. In this respect, Sortino *et al.* [45] provided the first example of a new cationic amphiphilic CD (**3** in Chart 1) able to influence significantly the photochemistry of a photosensitizing drug such as diflunisal, indicating oligo(ethylene glycol) amphiphilic CDs as an interesting system for increasing drug photostability and minimizing the biological damage.

3.2. Colloidal microenvironments and porphyrinoid sensitizers

The binding properties of complex systems formed by porphyrinoid sensitizers in colloidal microenvironments, such as liposomes, micelles, monolayers emulsions, are actively studied [46,47]. Over the last 20 years, liposomes interacting with hydrophobic porphyrins and metalloporphyrins have been employed to mimic membrane proteins [48] and to investigate distribution of sensitizers in cell compartments [46,49] in order to substantially improve the PDT (photodynamic therapy of tumours) efficacy and to preserve the photosensitizers [50]. Many investigations on porphyrins incorporated in neutral and charged micelles have been also reported [46,51], as far as their equilibrium or their kinetic behaviour are concerned [52]. The mechanism of interaction between photodynamic drugs, such as haematoporphyrins, protoporphyrins [53], glycoporphyrins [54], and cytotoxins [55] with liposomes has been widely investigated and it has been established that photophysical and photochemical parameters are quite sensitive to the microenvironment. Moreover, chiral recognition of micelles versus a chiral functionalized porphyrin was described [56]. Biopolymers [57] and calixarenes [58] constitute further micro-domains where chromophores can be localized.

A combination of non-covalent and electrostatic interactions influences the partition of dye in a lipid bilayer: hydrophobic guests are incorporated into the lipid region, while hydrophilic sensitizers interact mainly with the aqueous interface at the hydrated internal core of liposomes. As an example liposomal porphyrins could induce endocytoplasmatic damage, leading to change of mitochondria shape, while water-soluble haematoporphyrin mostly photosensitized the plasma membrane [55].

As a result of the localization of sensitizers inside the vesicles, monomers and self- aggregates of chromophores are formed. The presence of these supramolecular oligomers is due to high local concentration of sensitizer.

After being incorporated into the bilayer, the collision process of excited states and the rotational and diffusional freedom of the sensitizer molecule are consistently slowed down as a consequence of the increasing micro-viscosity [46].

For an efficient PDT, a greater amount of photosensitizer should be retained in tumour rather that in non-neoplastic tissues, upon systemic administration [59,60]. The behaviour into the cells of singlet oxygen, $O_2(a_1 \Delta_g)$, the actual photodynamically active species, has been recently investigated by time-resolved experiments with subcellular resolution [61]. These studies provide insight into the mechanisms of oxygen-dependent cell death by showing a longer lifetime of singlet oxygen (τ_Δ) in the nucleus than that suggested in other reports [62].

In this scenario, it is worth to noting as the different patterns of cell photodamage reflect the partition among specific sites and cell organelles, the distribution being mainly influenced by physico-chemical properties of the carrier/

drug complexes. Scattering techniques combined with steady-state and time-resolved fluorescence can be conveniently exploited to investigate both structural and dynamical features of these systems.

3.3. Heterotopic aggregates of CD and porphyrins

Cyclodextrins offer a hydrophobic microenvironment, which excludes water molecules and can include various kinds of photosensitizers. Systems of CD and porphyrins (Por) have been extensively studied as models for haemoproteins and some of these investigations have been focused on the hydrophobic environments around metalloporphyrins, similarly to the micro-domains of myoglobin and haemoglobin in which an iron-containing coordination site is involved. Hydrophilic porphyrin-β-CD conjugates and porphyrins bearing four covalently bound permethylated-β-CD were synthesized and characterized as soluble hosts having the advantage of multiple interactions with different substrates. The interaction between CD and green plant pigments and, fluorescent ratiometric methods, exploiting the coordinating properties of the metal in zinc(II)-porphyrin assembly in β-CD have been also described. Most of these topics concerning systems of CD/sensitizer and amphiphilic CD/sensitizer complexes are present in the literature cited in the Refs. [46,63,64]. Lang *et al.* [46] reported in detail some photophysical peculiarities of anionic and cationic porphyrins bound to CDs in aqueous solution as model sensitizers. It was evidenced that the binding constants (in the 10^3–$10^5 \, M^{-1}$ range) strongly depend on the cavity size and on the substituent groups grafted on the hydrophilic rims of CDs. Furthermore, the phototoxicity *in vitro* of these supramolecular sensitizers was described [65]. The non-covalent interaction between CD and anionic porphyrins affects the spectroscopic and photophysical properties of photosensitizers leading to red shift of Soret band and increase of triplet lifetime both in absence and in presence of oxygen. On the other hand, the quantum yield of singlet and triplet emission of the sensitizer and singlet oxygen formation does not undergo sensitive changes. These results suggested that stable complexes of CDs and porphyrins can prevent extensive aggregation (i.e. formation of J-aggregates) which would decrease the photodynamic efficacy. One of the main challenges is to achieve targeted PDT, that is to obtain the selective delivery of the photodynamic active drug directly to the action sites [66]. Our group is interested on the spectroscopic investigation of complex colloidal systems formed by porphyrin and amphiphilic CDs in order to characterize nanomaterials as "smart"carriers for specific recognition and cellular internalization [63].

Here we show our general approach to the study of such colloidal systems, reporting the case of an anionic water-soluble porphyrin (tetrakis-(4-sulfonatophenyl)porphyrin, TPPS) (Chart 2) and cationic amphiphilic CD **3**. The system

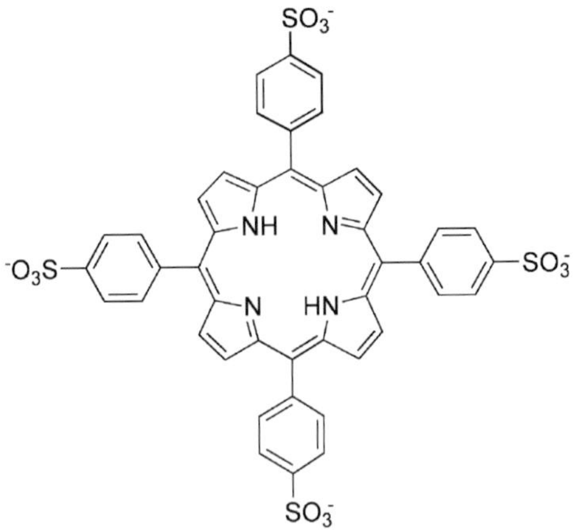

SO$_3^-$

$^-$O$_3$S

SO$_3^-$

SO$_3^-$

TPPS

Chart 2.

has been characterized using a combination of spectroscopic (UV/Vis absorption and fluorescence) and light scattering techniques (ELS, QELS, and electrophoretic mobility). Time-resolved fluorescence anisotropy experiments afford useful insights into the dynamics of porphyrins embedded within the nanocarriers.

Under neutral conditions this porphyrin is mainly present as a monomer, and dimerization has only been reported in few reports [63].

The ELS profiles of samples containing CD **3** aggregates and the system TPPS/CD **3** at 1:5 molar ratio are reported in Fig. 1.

The data were fitted using a single thin shell vesicles model with radius R [67] and a Gaussian size distribution $f(x)$ to take into account polydispersity, according to the relation

$$P(k) = \int \left[\frac{\sin[kR(x)]}{kR(x)} \right]^2 f(x)\, dx \qquad (14)$$

The slight discrepancy between fit and data can be attributed to the possible occurrence of multilamellar vesicles.

ELS analysis (Eq. 14) shows that CD **3** forms vesicles with $R = 290$ nm, and a polydispersivity of 30%, after short sonication and equilibration time (15–20 min). In a previous investigation, transmission electron microscopy (TEM) showed the occurrence of smaller CD **3** nanoparticles following longer sonication times [25]. On the other hand, interaction with TPPS (e.g. at porphyrin/CD 1:5 molar ratio) leads to a decrease of both R (140 nm) and polydispersivity (20%).

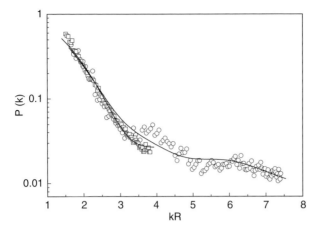

Fig. 1. Elastic light scattering profiles of CD **3** (○) and of TPPS/CD **3** (□) aggregates in phosphate buffer (10 mM, pH 7).

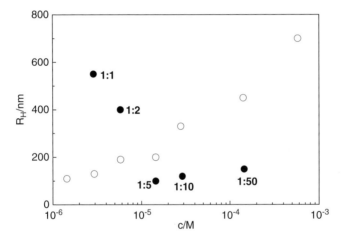

Fig. 2. R_H of vesicles versus CD **3** concentration in the absence (○) and in the presence of TPPS (3 μM, ●) at different porphyrin/CD molar ratios.

Figure 2 displays a comparison of the dependence of the hydrodynamic radii R_H as derived from QELS measurements (Eq. 2) on CD vesicles and porphyrin/CD systems as a function of CD concentration. In the absence of TPPS, mean R_H values for the vesicles increase from 120 to 1000 nm, as a function of the CD concentration (in the range 1×10^{-6}–6×10^{-4} M) thus confirming the lyotropic properties of CD. On adding TPPS (high values of the porphyrin/CD ratio), sensitive increase of the mean vesicle R_H occurs.

The observed increase of the aggregate size can be ascribed to the electrostatic nature of the interactions between positively charged CD and negatively

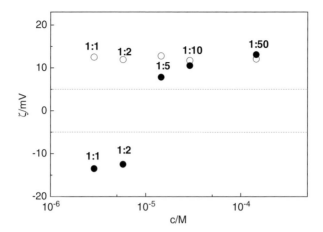

Fig. 3. Zeta potential (ζ) of the vesicles versus CD **3** concentration in the absence ($\bigcirc$) and in presence of TPPS ($\bullet$) at different porphyrin/CD molar ratios. The region marked between the dashed lines define the instability region for this system.

charged porphyrins. Consequently, the main features of the charge interaction process have been investigated by measuring the vesicles zeta potential (ζ) through laser Doppler electrophoresis experiments [64]. Figure 3 shows the zeta potential for the vesicles as a function of CD concentration in the absence and in the presence of TPPS.

In the absence of added porphyrin, at neutral pH the vesicles exhibit a zeta potential of $\zeta \approx +12.5\,\mathrm{mV}$, which holds nearly constant on changing concentration. A rather different behaviour is observed when the negatively charged TPPS is added to the CD system. At lower TPPS concentration (i.e. at 1:5, 1:10, and 1:50 molar ratios) most of the negative TPPS molecules are dispersed in the positive colloidal phase, due to their statistical interaction with more amphiphilic CD. On increasing TPPS concentration (i.e. at 1:1 and 1:2 molar ratio), the charges on the vesicle surface are balanced out via electrostatic interaction with the anionic guest. These results clearly suggest a deep structural reorganization as a consequence of charge and size modulation by the concentration of the components.

UV-Vis absorption and fluorescence emission features for this system under different experimental conditions are summarized in Table 1. As already reported in the literature [68], both a bathochromic shift and hypochromicity in the Soret band are indicative of porphyrins dispersed in colloidal phase.

The emission spectra of the unbound TPPS display a two-banded feature with a main higher energy peak at 642 nm. In the case of a 1:1 molar ratio, this component is located at about 642–643 nm and shows very low fluorescence anisotropy. At higher CD concentration, the maxima move to 650 nm, displaying a larger anisotropy.

Table 1
UV-Vis absorption and fluorescence emission maxima for TPPS porphyrin (alone) and in vesicles at different TPPS/CD **3** molar ratios[a]

	$\lambda_{max\ (abs)}$ (nm)	H (%)[b]	$\lambda_{max\ (em)}$ (nm)
TPPS	414	—	642
1:1	414	55	643
1:2	409, 420	72, 71	658
1:5	420	55	650
1:50	420	7	650

[a]UV-Vis absorption and fluorescence were measured in 10 mM phosphate buffer (pH 7, 298 K) and $\lambda_{ex} = 520$ nm.
[b]Hypochromicity percentage for TPP/CD**3** aggregates relative to the maximum intensity (Soret Band) for free TPPS.

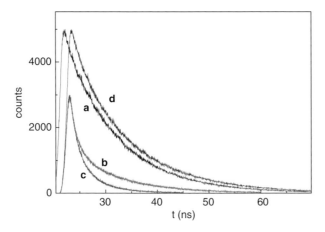

Fig. 4. Fluorescence decay traces ($\lambda_{ex} = 374$ nm) of TPPS (3 μM) in phosphate buffer (10 mM, pH 7, a) and in presence of CD **3** at 1:1 (b), 1:2 (c), and 1:50 (d) molar ratios.

The steady-state anisotropy can be related to the molecular motion by time-resolved fluorescence analysis, both reported in Fig. 4 and in Table 2.

As described by the Perrin's equation (13), the molecular rotational correlation time (θ) can be obtained by the values of steady-state anisotropy (r) and fluorescence lifetime (τ). An inspection of Fig. 4 and Table 2 shows that in buffered solution the steady state anisotropy and fluorescence lifetime of free TPPS are 0.01 and 9.7 ns, respectively. In the presence of high CD concentration, the increase of the steady-state anisotropy and of the fluorescence lifetime indicates that the rotational correlation time θ is increasing. These findings would suggest the entanglement of TPPS in the CD aggregates.

At low CD concentration, Eq. (13) does not allow certain conclusions about the rotational correlation time. However, the presence of short-living fluorescent

Table 2
Fluorescence lifetime parameters and steady-state anisotropy for TPPS porphyrin (alone) and in
vesicles at different TPPS/CD 3 molar ratios[a]

	τ_1 (ns)	τ_2 (ns)	τ_3 (ns)	$100\alpha_1$	$100\alpha_2$	$100\alpha_3$	r
TPPS[b]	9.7	—	—	100	—	—	0.01
1:1[c]	9.7	2.0	0.5	74	17	9	0.02
1:2[d]	8.8	3.0	0.8	20	60	20	—
1:10[e]	11.6	5.3	1.0	88	10	2	0.04
1:50[e]	11.6	5.0	0.8	89	9	2	0.07

[a]Fluorescence lifetimes were measured in 10 mM phosphate buffer (pH 7, 298 K) and $\lambda_{ex} = 374$ nm. α_i is the
amplitude of the intensity decay. [TPPS] = 3 μM.
[b]$\lambda_{em} = 642$ nm.
[c]$\lambda_{em} = 645$ nm.
[d]$\lambda_{em} = 654$ nm.
[e]$\lambda_{em} = 650$ nm.

species, probably due to porphyrins entangled in different hydrophobic compartments, or to small self-aggregated porphyrin oligomers were also detected in
these samples.

As already mentioned in Section 2.2, Perrin's equation allows for an indirect
measurement of the rotational correlation time in the case of systems having
both a single lifetime and rotational correlation time (assuming unaltered the
limiting anisotropy r_0).

In order to directly gain information on structural changes in the colloidal
microenvironment of TPPS, time-resolved fluorescence anisotropy experiments
at different CD/sensitizer have been carried out.

Unbound TPPS shows a completely depolarized fluorescence emission. The
size of this molecule is large enough (radius ≈ 10 Å) to exhibit a rotational
correlation time of about 0.9 ns in aqueous solutions (from Eq. 10). Different
from previous literature data obtained under diverse excitation conditions [68],
our experimental value for the initial anisotropy is zero, suggesting that the
relative angle between absorption and emission dipoles is close to the magic
angle (54.7°). In the presence of CD vesicles, TPPS fluorescence emission is not
completely depolarized, suggesting the static interaction between the fluorophore and the vesicles. Figure 5 reports a sequence of anisotropy decay curves for
TPPS/CD 3 system at different molar ratios. At low CD concentration (1:1 and
1:2 molar ratio), single-exponential curves describe the anisotropy decays with a
rotational correlation time $\theta \sim 1$–2 ns and $r_0 \sim 0.06$–0.08.

This experimental evidence is in full agreement with steady-state fluorescence
results and suggests that under these conditions most of the TPPS molecules are
unbounded. In this case, the measured anisotropy decay can be safely assigned
to the small percentage of TPPS molecules that are in different hydrophobic

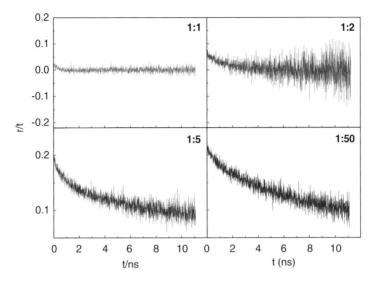

Fig. 5. Time-resolved fluorescence anisotropy [$r(t)$] of TPPS/CD **3** at different molar ratios.

compartments of the CD system and/or form supramolecular aggregates of chromophore.

On increasing CD concentrations, more than 90% of the total fluorescence emission can be accounted by TPPS molecules embedded in the colloidal phase. Consequently, the porphyrin rotation is slowed down and they are actually stabilized onto the vesicles surface. Indeed, the anisotropy decays evidence a double-exponential behaviour, in which the predominant "slow" components exhibit rotational correlation times in the range of 20–25 ns and r_0 values of *ca.* 0.15. Such values of rotational diffusion times (> 20 ns) are too long for the simple CD vesicles (corresponding to radii > 1000 Å). In a likely model TPPS molecules can be considered as embedded on the vesicles and the larger rotational correlation times correspond to the slow complete rotation of the CD/ porphyrin system and/or to local deformation of the colloidal wall [40].

The example discussed above demonstrates that molecular recognition can be driven by electrostatic interaction between photosensitizers and CD aggregates. Specific recognition of a protein with different non-ionic amphiphilic CDs can be also detected. Figure 6 displays fluorescence emission decays of tryptophan in lectin from *Pseudomonas Aeruginosa* (PA-I) in the absence (a) and in the presence (b) of galactosylated amphiphilic CD **6**, respectively.

As it is evident the interaction between lectin and CD **6** has sensitively affected fluorescence lifetime of fluorophores in the protein.

Photophysical investigations in combination with scattering techniques fully characterize the morphology in solution (size and shape) of a colloidal system interacting with cromophores species. These techniques are non-invasive and do

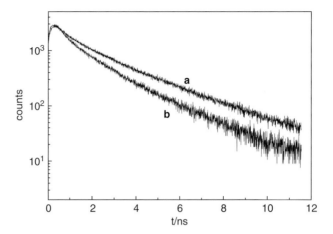

Fig. 6. Fluorescence emission decay of PA-I lectin ($10\,\mu M$) and PA-I/CD **6** aggregates at 1:16 molar ratio in aqueous solution (pH 7, $T = 298\,K$).

not affect the colloidal system (e.g. through immobilization or staining of the probe species) by assessing structural changes which accompany binding.

In particular, steady-state and time-resolved fluorescence as well anisotropy are powerful tools for studying function and conformation of sensitizer in vesicular membrane systems, in order to provide structural information on sensitizer–colloid binding or to evaluate protein–ligand interactions.

The application of these studies to the chemistry and pharmaceutic of cyclodextrins open the way to design of novel drug delivery sytems for specific cellular targeting (i.e. targeted PDT); [69].

References

[1] A. Becheri, P. Lo Nostro, B.W. Ninham, P. Baglioni, J. Phys. Chem. B. 107 (2003) 3979.

[2] M. Miyauchi, A. Harada, J. Am. Chem. Soc. 126 (2004) 11418.

[3] M. Bonini, S. Rossi, G. Karlsson, M. Almgren, P.Lo. Nostro, P. Baglioni, Langmuir 22 (2006) 1478.

[4] A.W. Coleman, I. Nicolis, N. Keller, J.P. Dalbiez, J. Incl. Phenom. 13 (1992) 139.

[5] S. Polarz, B. Smarsly, L. Bronstein, M. Antonietti, Angew. Chem. Int. Ed. 40 (2001) 4417.

[6] A. Harada, M. Kamachi, Macromolecules 148 (1990) 2821.

[7] T. Oshikiri, Y. Takashima, H. Yamaguchi, A. Harada, J. Am. Chem. Soc. 127 (2005) 12186.

[8] H. Parrot-Lopez, C.C. Ling, P. Zhang, A. Baszkin, G. Albrecht, C. de Rango, A.W. Coleman, J. Am. Chem. Soc. 114 (1992) 5479.

[9] M. Munoz, R. Deschenaux, A.W. Coleman, J. Phys. Org. Chem. 12 (1999) 364.

[10] M.H. Greenhall, P. Lukes, R. Kataky, N.E. Agbor, J.P.S. Badyal, J. Yarwood, D. Parker, M.C. Petty, Langmuir 11 (1995) 3997.

[11] H. Nakahara, H. Tanaka, K. Fukuda, M. Matsumoto, W. Tagaki, Thin Solid Films 284–285 (1996) 687.

[12] B. Hamelin, L. Jullien, A. Laschewsky, C. Hervé du Penhoat, Chem. Eur. J. 5 (1999) 546.
[13] A. Kasselouri, A.W. Coleman, A. Baszkin, J. Coll. Interf. Sci. 180 (1996) 384.
[14] L. Jullien, T. Lazrak, J. Canceill, L. Lacombe, J.M. Lehn, J. Chem. Soc. Perkin Trans. 2 (1993) 1011.
[15] J. Chem. Soc. Perkin Trans. 2 (1998) 2639.
[16] A. Geze, S. Aous, I. Baussanne, J. Putaux, J. Defaye, D. Woussidjewe, Int. J. Pharm. 242 (2002) 301.
[17] R. Auzély Velty, C. Péan, F. Djedaïni-Pilard, T. Zemb, B. Perly, Langmuir 17 (2001) 504.
[18] L. Bacri, A. Benkhaled, P. Guegan, L. Auvray, Langmuir 21 (2005) 5842.
[19] F. Djedaïni-Pilard, R. Auzély Velty, B. Perly, Rec. Res. Dev. Org. Biorg. Chem. 5 (2002) 41.
[20] B.J. Ravoo, R. Darcy, Angew. Chem. Int. Ed. 39 (2000) 4324.
[21] A. Mazzaglia, R. Donohue, B.J. Ravoo, R. Darcy, Eur. J. Org. Chem. (2001) 1715.
[22] C.-C. Ling, R. Darcy, W. Risse, J. Am. Chem. Soc. Chem. Commun. (1993) 438.
[23] R.B. Friedman, Proceedings of the 4th International Symposium on Cyclodextrins, in: O. Huber, J. Szejtli (Eds), A Chemically Modified Cyclodextrin, Kluwer, Dordrecht, 1988, p. 103.
[24] I.N. Topchieva, P. Mischnick, G. Kühn, V.A. Polyakov, S.V. Elezkaya, G.I. Bystryzky, K.I. Karezin, Bioconj Chem. 9 (1998) 676.
[25] R. Donohue, A. Mazzaglia, B. J. Ravoo, R. Darcy, Chem. Commun. (2002) 2864.
[26] A. Mazzaglia, D. Forde, D. Garozzo, P. Malvagna, B.J. Ravoo, R. Darcy, Org. Biom. Chem. 2 (2004) 957.
[27] D. Lombardo, A. Longo, R. Darcy, A. Mazzaglia, Langmuir 20 (2004) 1057.
[28] P. Falvey, C.W. Lim, R. Darcy, T. Revermann, U. Karst, M. Giesbers, A.T.M. Marcelis, A. Lazar, A.W. Coleman, D.N. Reinhoudt, B.J. Ravoo, Chem. Eur. J. 11 (2005) 1171.
[29] S. Salmaso, A. Semenzato, P. Caliceti, J. Hoebeke, F. Sonico, C. Dubernet, P. Couvreur, Bioconj Chem. 15 (2004) 997.
[30] B.J. Ravoo, J.C. Jacquier, G. Wenz, Angew. Chem. Int. Ed. 42 (2003) 2066.
[31] R. Darcy, B.J. Ravoo, A. Mazzaglia, R. Donohue, E. Kan, K. Gaffney, D. Forde, D. Nolan, C. M. O'Driscoll, S.-A. Cryan, A. McMahon, E. Gomez, Proceedings of the 12th International Symposium on Cyclodextrins, p. 665 Montpellier, 2004.
[32] S.-A. Cryan, A. Holohan, R. Donohue, R. Darcy, C.M. O'Driscoll, Eur. J. Pharm. Sci. 21 (2004) 57.
[33] T. Sukegawa, T. Furuike, K. Niikura, A. Yamagishi, K. Monde, S. Nishimura, Chem. Commun. (2002) 430.
[34] D. Nolan, R. Darcy, B.J. Ravoo, Langmuir 19 (2003) 4469.
[35] J. Szejtli, Chem. Rev. 98 (1998) 1743.
[36] K. Uekama, F. Hirayama, T. Irie, Chem. Rev. 98 (1998) 2045.
[37] D.D. Lasic, Angew. Chem. Int. Ed. Engl. 33 (1994) 1685.
[38] R. Haag, Angew. Chem. Int. Ed. 43 (2004) 278.
[39] B.J. Berne, R. Pecora, Dynamic Light Scattering, Wiley Interscience, New York, 1976.
[40] J.R. Lakowicz, Principles of Fluorescence Spectroscopy, Kluwer Academic, Plenum Publishers, New York, 1999.
[41] D.V. O'Connor, D. Phillips, Time-Correlated Single Photon Counting, Academic Press, New York, 1984.
[42] S. Monti, S. Sortino, Chem. Soc. Rev. 31 (2002) 287.
[43] S. Scalia, A. Molinari, A. Casolari, A. Maldotti, Eur. J. Pharm. Sci. 22 (2004) 241.
[44] V.J. Stella, V.M. Rao, E.A. Zabbou, V. Zia, Adv. Drug Del. Rev. 36 (1999) 3.
[45] S. Sortino, S. Petralia, R. Darcy, R. Donohue, A. Mazzaglia, New. J. Chem. 27 (2003) 602.
[46] K. Lang, J. Mosinger, D.M. Wagnerova, Coord. Chem. Rev. 248 (2004) 321 and references therein.

[47] G. Li, W. Fudickar, M. Skupin, A. Klyszcz, C. Draeger, M. Lauer, J.-H. Fuhrhop, Angew. Chem. Int. Ed. 41 (2002) 1828.

[48] M.C. Feiters, A.E. Rowan, R.J.M. Nolte, Chem. Soc. Rev. 29 (2000) 375 and references. therein.

[49] F. Ricchelli, G. Jori, S. Gobbo, M. Tronchin, Biochim. Biophys. Acta-Biomembr. 1065 (1991) 42.

[50] A.S.L. Derycke, P.A.M. de Witte, Adv. Drug Del. Rev 56 (2004) 17.

[51] C. Maiti, S. Mazumdar, N.J. Periasamy, Phys. Chem. B 102 (1998) 1528.

[52] J. Simplicio, K. Schwenzer, F. Maenpa, J. Am. Chem. Soc. 97 (1975) 7319.

[53] F. Ricchelli, S. Gobbo, G. Jori, C. Salet, G. Moreno, Eur. J. Biochem. 233 (1995) 165.

[54] I. Voszka, R. Galanti, P. Maillard, G. Csik, J. Photochem. Photobiol. B: Biol. 52 (1999) 92.

[55] C. Milanesi, F. Sorgato, G. Jori, Int. J. Radiat. Biol 55 (1989) 59.

[56] D. Monti, V. Cantonetti, M. Venanzi, F. Ceccacci, C. Bombelli, G. Mancini Chem. Commun. (2004) 972.

[57] R.F. Pasternack, C. Fleming, S. Herring, P.J. Collings, J. de Paula, G. De Castro, E.J. Gibbs, Biophys. J. 79 (2000) 550.

[58] L.D. Costanzo, S. Geremia, L. Randaccio, R. Purrello, R. Lauceri, D. Sciotto, F.G. Gulino, V. Pavone, Angew. Chem. Int. Ed. 40 (2001) 4245.

[59] Y.N. Konan, R. Gruny, E. Allemann, J. Photochem. Photobiol. B 66 (2002) 89.

[60] T. Hasan, A.C.E. Moor, B. Ortel, Cancer Med., 5th ed, B.C. Decker, Inc., Hamilton, ON, Canada, 2000.

[61] J.W. Snyder, E. Skovsen, J.D.C. Lambert, P.O. Ogilby, J. Am. Chem. Soc. 127 (2005) 14558.

[62] M. Niedre, M.S. Patterson, B.C. Wilson, Photochem. Photobiol. 75 (2002) 382.

[63] A. Mazzaglia, N. Angelini, R. Darcy, R. Donohue, D. Lombardo, N. Micali, V. Villari, M.T. Sciortino, L. Monsù Scolaro, Chem. Eur. J. 9 (2003) 5762 and references therein.

[64] A. Mazzaglia, N. Angelini, D. Lombardo, N. Micali, S. Patanè, V. Villari, L. Monsù Scolaro, J. Phys. Chem. B 109 (2005) 7258 and references therein.

[65] H. Kolárová, J. Mosinger, R. Lenobel, K. Kejlová, D. Jírová, M. Strnad, Toxicol. In Vitro 17 (2003) 775.

[66] A.C.S. Samia, X. Chen, C. Burda, J. Am. Chem. Soc. 125 (2003) 15736.

[67] J. Pencer, F.R. Hallett, Langmuir 19 (2003) 7488.

[68] C. Maiti, S. Mazumdar, N. Periasamy, J. Phys. Chem. B 102 (1998) 1528.

[69] S. Sortino, A. Mazzaglia, L. Monsù Scolaro, F. Marino Merlo, V. Valveri, M.T. Sciortino, Biomat. 27 (2006) 4256.

Cyclodextrin Materials Photochemistry, Photophysics and Photobiology
Abderrazzak Douhal (Editor)
© 2006 Elsevier B.V. All rights reserved
DOI 10.1016/S1872-1400(06)01010-7

Chapter 10

Functional Cyclodextrin Systems from Spectrophotometric Studies to Photophysical and Photochemical Behaviors

Yu Liu, Guo-Song Chen, Yong Chen

Department of Chemistry, State Key Laboratory of Elemento-Organic Chemistry, Nankai University, Tianjin 300071, P. R. China

1. Introduction

Cyclodextrins (CD) is a class of cyclic oligosaccharides mainly with six to eight D-glucose units linked by α-1,4-glucose bonds [1] (Fig. 1). Their most important advantage is to encapsulate various inorganic/organic guests within their hydrophobic cavities in both aqueous solution and the solid state. Therefore, they are extensively studied as not only excellent receptors for molecular recognition, but also functional building blocks to construct molecular devices through host–guest complexation or supramolecular assembly [2–5]. Generally, native CDs absorb below 300 nm. However, when large aromatic chromophores are attached to the CD rim or included in the CD cavity, the resulting CD-based systems always exhibit significant spectral behaviors. Therefore, the spectrophotometric measurement becomes a convenient and efficient method to investigate the inclusion complexation behaviors of CD systems. Moreover, benefiting from the good photophysical and photochemical properties of the attached/included chromophores, these CD systems can be used as not only spectral sensors for molecular recognition but also functional photochemical and photobiological materials.

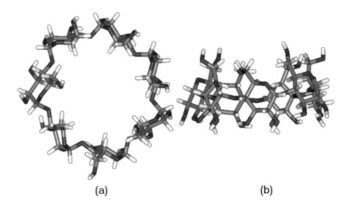

(a)   (b)

Fig. 1. Structure of β-CD from top (left) and side (right).

2. Spectrophotometric studies on the molecular recognition of CDs

CDs are well known to accommodate various guest molecules into their truncated cone-shaped hydrophobic cavity in aqueous media, which enables them to be successfully used in a wide variety of fields such as pharmaceuticals, artificial enzymes, and biomimetic materials. However, the molecular recognition abilities of native CDs are usually limited, which greatly unfavors their applications as molecular receptors and enzyme mimics. Therefore, a number of chemically modified CDs have been designed and synthesized to enhance the original binding abilities and the molecular selectivities of parent CDs. Recently, bridged CDs dimers, a new family of CD derivatives, are reported to exhibit the different inclusion complexation behaviors from those of mono-modified CDs, giving the significantly enhanced molecular recognition abilities through the cooperative binding of dual hydrophobic CDs cavities located in close vicinity. This section will concentrate on the recent developments of spectrophotometric studies on the molecular recognition of mono-modified CDs and bridged bis(CDs).

2.1. Spectrophotometric studies on the structural elucidation of CDs

2.1.1. Conformation analysis
The initial conformational analysis of CD derivatives is very important to understand their molecular recognition behaviors. It is well known that the elucidation of crystal structure is one of the most convincing methods of unequivocally illustrating the geometrical conformation of CD derivatives. However, the preparation of single crystals of CD derivatives that are suitable for X-ray crystallography is always difficult. Therefore, we try to elucidate the conformations of CD derivatives through a combinational analysis based on circular dichroism, 2D NMR and UV-Vis spectroscopy. Generally, if an achiral

guest/moiety is enclosed in or adjacent to a chiral environment, such as CD cavity, they will give either positive or negative induced circular dichroism (ICD) signals according to their location and orientation related to CD cavity [6–7]. If the guest molecule is located inside the CD cavity or perched on the edge of CD cavity, its electronic transition parallel to CD axis gives a positive ICD signal, whereas the perpendicular transition gives a negative signal, but this situation is reversed for a guest located outside the CD cavity. For example, the circular dichroism spectrum of mono {6-*O*-[4-((4-nitrophenyl)azo)phenyl]}-*β*-CD (**1**) shows a negative ICD signal at 381 nm and a positive ICD signal at 421 nm assigned to the π–π^* (parallel to the N = N bond) and n–π^* (perpendicular to the N = N bond) transition moments of azobenzene chromophore, respectively [8]. Therefore, we can deduce that the azobenzene moiety is deeply embedded in the *β*-CD cavity with an acclivitous orientation to form the self-included complexes (Fig. 2).

In addition, the circular dichrosim spectrum of diselenobis(benzoyl)-bridged bis(*β*-CD) **2** exhibits two positive Cotton effect peaks for the ^1L_b and ^1L_a transitions of phenyl group and a positive Cotton effect peak for the Se–Se transition. Therefore, we can deduce that the linker group in **2** is located at the exterior of CD cavity, where both the ^1L_a and ^1L_b transitions of phenyl chromophore are nearly perpendicular to the CD axis, resulting in the positive Cotton effect peaks. On the other hand, the Se–Se moiety is situated between the two narrow rims of dual CD cavities (Fig. 3). Hence, the transition moment of Se–Se bond is perpendicular to the CD axis, which consequently results in the positive Cotton effect [9].

UV-Vis spectroscopy is another useful method to investigate the conformation of CD derivatives. Recently, we compared the UV-Vis spectra of

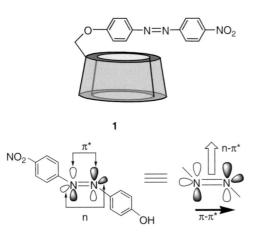

Fig. 2. Illustration of electric transition dipole moment for $\pi \rightarrow \pi^*$ (solid vector) or $n \rightarrow \pi^*$ (blank vector) transition of azobenzene moiety in **1**.

Y. Liu et al.

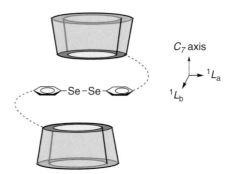

Fig. 3. Schematic structure of **2** deduced from circular dichroism spectrum.

2,2′-bipyridine-4,4′-dicarboxamide bridged bis(β-CD) **3** and its homologue 2,2′-bipyridine-3,3′-dicarboxamide bridged bis(β-CD) **4** to determine their conformations [10]. Generally, a coplanar bipyridine chromophore usually shows two absorption bands in the ultraviolet region. If two pyridine units twist along the central carbon–carbon bond, the absorption intensities of these two bands will quench, accompanied by the hypsochromic shifts of the absorption maximums, and sometimes only one absorption band can be observed. As we can see from Fig. 4, bis(β-CD) **3** shows two absorption bands at 241 nm and 294 nm, assigned to the absorptions of coplanar bipyridine chromophore. However, bis(β-CD) **4** only gives one absorption band at 270 nm. This phenomenon verifies the twist of bipyridine group in bis(β-CD) **4**. According to these spectral information, along with the molecular model studies, we deduce that the bipyridine group in bis(β-CD) **3** is coplanar and exists in a *cis* conformation. On the other hand, bis(β-CD) **4** adopt a *trans* conformation attributing to the steric hindrance between two carbonyl groups at 3,3′-positions of bipyridine. In this conformation, two pyridine rings in the linker group are not coplanar but twist with each other.

Besides circular dichroism and UV-Vis spectroscopy, 2D NMR spectroscopy is also an efficient method to study the conformation of CD derivatives since one can conclude that two protons are closely located in space if an NOE crosspeak is detected between the relevant protons in the NOESY or ROESY spectrum. Therefore, it is possible to estimate the orientation of the chromophore moiety in the CD cavity using the assigned NOE correlations, because if the chromophore moiety is self-included in the CD cavity, the NOE correlations between the protons of the chromophore moiety and the H3/H5 of the CD should be observed. For example, the 2D NOESY spectrum of 2,2′-biquinoline-4,4′-dicarboxamide-bridged bis(β-CD) **5** in D_2O (Fig. 5) shows six cross peaks between the linker's aromatic protons and the CD's protons. Among them, only cross-peak interactions with H3, H5, and H6 of CDs can be considered to analyze the results, because H2 and H4 are not facing the inner cavity and H1 is

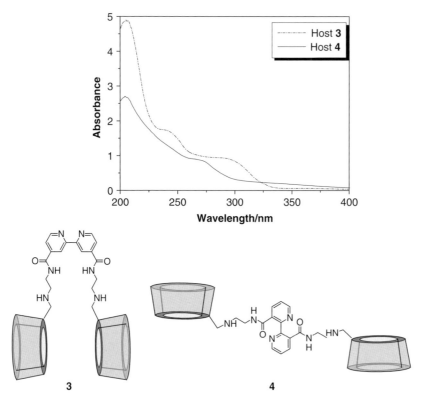

Fig. 4. UV-Vis spectra of bis(β-CD)s **3** and **4** in aqueous solution at 25°C ([**3**] = [**4**] = 0.9×10^{-4} mol dm^{-3}).

affected by D$_2$O. The cross peaks A, B, and C are respectively assigned to the NOE between 8,8′-, 7,7′-, and 6,6′-protons of biquinoline moiety and the CD's H5 protons. These results unambiguously indicate that the linker group is embedded in the CD cavity. Since the H5 protons are located at the primary side of the cavity, we may conclude that the biquinoline linker penetrates into the CD cavity from the narrower opening. However, no cross peaks are found between the biquinoline's 3,3′- or 5,5′-protons and the CD's interior protons. Hence, it is likely that the unsubstituted rings of biquinoline penetrate into the CD cavities shallowly from the primary side [11].

2.1.2. Conformation conversion

CD derivatives can change their conformations in different media, and these conformation conversions can greatly affect their molecular recognition abilities. For example, pyridine-2,6-dicarboxamide-bridged bis(β-CD) **6** displays quite different circular dichroism spectra at different pH values [12], which indicates

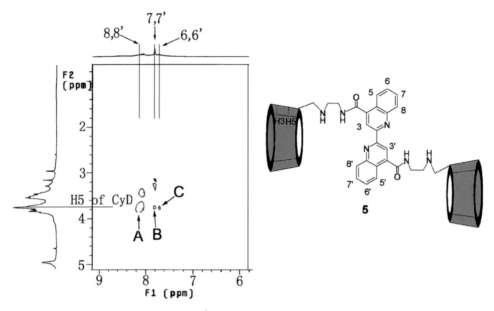

Fig. 5. Sectional ^{1}H NOESY spectra (300 MHz) of **5**.

that there should exist significant but different degrees of interactions between the linker group and the CD cavity. That is, the circular dichroism spectrum of **6** shows a strong positive Cotton effect peak around 213 nm in a pH 7.2 buffer, but displays a moderate negative Cotton effect peak around 220 nm and a weak negative Cotton effect peak around 270 nm in a pH 2.0 buffer. So we can deduce that the pyridine group in the linker of **6** is located outside the CD cavity in a pH 7.2 buffer, where the transition moment of ^1L_a band is nearly perpendicular to the CD axis and thus, induces the positive ICD signal. On the other hand, in a pH 2.0 buffer, the pyridine chromophore is shallowly included in the cyclodextrin cavity, where both the ^1L_a and ^1L_b transition moments are nearly perpendicular to the CD axis, resulting in the two negative Cotton effect peaks (Fig. 6). In addition, bis(β-CD) **6** emits stronger fluorescence in a pH 2.0 buffer than in a pH 7.2 buffer, which indicates that the pyridine chromophore in **6** is certainly located in a more hydrophobic environment in the pH 2.0 buffer. Therefore, the results of circular dichroism and fluorescence experiments jointly demonstrate a pH-switched conformation conversion of **6** as illustrated in Fig. 6.

When the organic solvent such as DMSO is added to the aqueous solution of azobenzene-modified β-CD with a self-inclusion conformation, the azobenzene substituent will gradually move out from the CD cavity. This process can be easily traced by UV-Vis spectrum. As can be seen in Fig. 7, the UV-Vis absorption maximum of **1** gradually decreases around 391 nm upon the addition of DMSO (lines a to e: DMSO ratio from 10% to 30%), and then increases

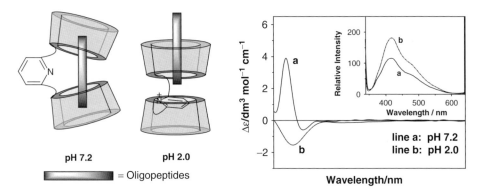

Fig. 6. (Left) pH-switched conformation conversion of bis(β-CD) **6**. (right) Circular dichroism spectrum of **6** (0.1 mM) in pH 7.2 and 2.0 buffers at 298 K. Inset: Fluorescence spectra of **6** (0.1 mM) in pH 7.2 and 2.0 buffers. The excitation wavelength is 330 nm.

around 374 nm by further increasing the DMSO ratio from 30% to 100% (e to i), accompanied by the obvious blue-shift (about 17 nm) of absorption peak [8]. These phenomena indicate that the CD chromophoric substituent enters the hydrophilic region from the hydrophobic cavity due to the binding of DMSO with CD cavity.

2.2. Spectrophotometric studies on the inclusion complexation of CDs

The host–guest inclusion complexations always lead to some accompanying spectral changes, which depend critically on the formation of new species, i.e. host–guest inclusion complex, showing the spectral enhancement or quenching. If the guest molecule is UV-visible responsive, when it includes into the CD cavity, its absorption generally decreases. If the CD chromophoric substitutent intramolecularly self-includes or intermolecularly includes into the hydrophobic cavity of CD, the absorption of the resulting CD system will gradually increase upon the addition of guest molecule due to the exclusion of the included chromophoric substitutent from CD cavity. When a fluorophoric guest molecule includes into the hydrophobic cavity of CD derivatives, its fluorescence intensity always shows remarkable enhancement. However, for the CD derivatives possessing a fluorophoric substitutent, which can self-include into its cavity, the fluorescence intensity will decrease upon the addition of a guest molecule. The complex stability constant (K_s) can be calculated by linear plot or nonlinear least squares curve-fitting method [13,14]. Ueno *et al.* [15,16] reported tens of chromophore-modified CDs that are sensitive to the changes in the binding state of CD cavity and consequently can be used as photosensors. That is, the excimer emission, absorption, and ICD intensities obviously change upon addition of a variety of organic guests. Similar results were

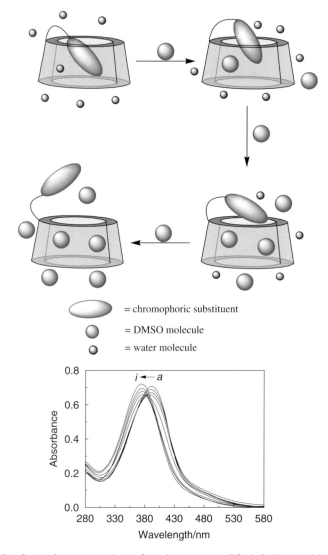

Fig. 7. (Top) Conformation conversion of azobenzene-modified β-CD **1** with the addition of DMSO. (bottom) UV-Vis spectra of **1** (2.6×10^{-5} mol dm^{-3}) in DMSO-H$_2$O solution at 298 K. DMSO ratio: a, 10%; b, 15%; c, 20%; d, 25%; e, 30%; f, 40%; g, 50%; h, 70%; i, 100%.

also found by Liu *et al.* from comprehensive studies on the binding behavior of L-tryptophan-modified and organoselenium-modified β-CDs [17]. These studies demonstrate that the spectral variations are due to the decomplexation of the chromophore moiety upon guest binding, and the induced-fit mechanism plays an important role in controlling the inclusion complexations between hosts and guests [18].

2.3. Spectrophotometric studies on the binding mode of CDs

Spectrophotometry is also an efficient method to investigate the binding mode between host and guest, because the different binding modes can result in the different spectral behaviors, such as the contrasting fluorescence. We have reported that bis(β-cyclodextrin-6-yl) 2,2'-bipyridine-4,4'-dicarboxylate is found to induce an unusual fluorescence enhancement of rhodamine B (RhB) upon complexation. This enhancement is attributed to the equilibium shift of RhB to the highly fluorescent carboxylate ion form induced by the cooperative binding of two appropriately preorganized CD cavities in the bis(β-CD) [19]. These studies demonstrate the formation of the sandwich-like complex between bis(β-CD)s and guest molecules. In addition, calix[4]arene-tethered mono- and bis(β-CD)s, which possess both CD and calixarene moieties in a single host molecule, are found to be able to induce the contrasting fluorescent behavior of acridine red (AR) upon inclusion complexation [20]. The fluorescence of AR is gradually quenched by the stepwise addition of calix[4]arene-tethered monoCD. In sharp contrast, the gradual addition of calix[4]arene-tethered bis(β-CD) to a dilute solution of AR significantly enhances the fluorescence intensity under the comparable conditions. Such contrasting fluorescence behaviors of AR indicate that the distinctly different binding modes are operative in the complexation of AR with calix[4]arene-tethered monoCD and bis(β-CD). For calix[4]arene-tethered monoCD, the calixarene cavity is actively incorporated in the inclusion complexation with AR, and the electrostatic interactions between the positively charged AR and the sulfonated calix[4]arene cavity result in the decreased fluorescence of AR. For calix[4]arene-tethered bis(β-CD), the calixarene moiety does not actively participate in the binding of AR but merely acts as a rigid tether. Therefore, the cooperative binding by two CD cavities increases the hydrophobicity around the AR fluorophore and thus, results in the enhanced fluorescence.

Many researches have demonstrated that the emission of fluorophore-appended CD is quenched upon inclusion complexation with a guest as a consequence of decomplexation of the initially self-included fluorophore moiety [21,22]. However, we find that the gradual addition of steroids can enhance the fluorescence of 2,2'-biquinoline-4,4'-dicarboxamide bridged bis(β-CD)s [9,23]. The enhancement of fluorescence intensity suggests that the biquinoline moieties in these bis(β-CD)s may undergo certain changes in location and orientation upon complexation, which consequently leads to the increased microenviromental hydrophobicity and/or steric shielding around the fluorophore. This unique fluorescence behavior and the 2D NMR results jointly indicate a cooperative "host-tether-guest" binding mode operative in the association of biquinoline-bridged bis(β-CD) with a guest molecule. According to this mode, the guest steroid is embedded into one hydrophobic CD cavity from the primary

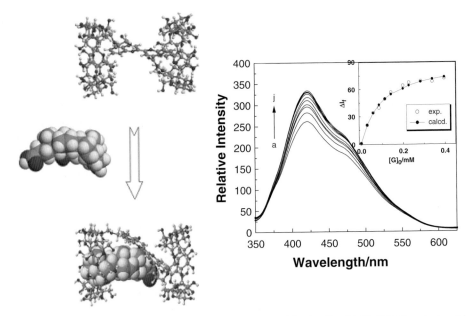

Fig. 8. The "host–tether–guest" binding mode between biquinoline bis(β-CD) **5** and steroids.

side upon complexation with bis(β-CD), while the tether group is partly self-included in the other cavity (Fig. 8).

3. Photophysical and photochemical behaviors of functional CD systems

3.1. Photophysical behaviors

3.1.1. Functional fluorophore/CD inclusion complex
Besides the preparation of CD derivatives containing chromophoric substituent, there is another strategy to obtain chromophoric CD hosts, i.e. to build chromophoric guest/CD inclusion complexes. These kinds of chromophoric CD systems can successfully respond to some nonchromophoric materials, like sugar [24] or ions [25]. The major role of CD in these systems is to create a hydrophobic and polar environment for the chromophoric moieties, making them much more sensible to the changes of their surroundings. Kano *et al.* [26] reported a very stable 2:1 inclusion complex of heptakis(2,3,6-tri-*O*-methyl)-β-CD (TMCD) with 5,10,15,20-tetrakis(4-sulfonatophenyl) porphinato iron(III). In this inclusion complex, the iron(III) center is located at a hydrophobic cleft formed by two face-to-face TMCD molecules. Various inorganic anions can coordinate to it to form five-coordinate high-spin conjugate, giving obvious Uv-vis absorption spectral changes.

   In addition to anions, this kind of CD-based inclusion complex can also bind cations [24]. Hayashita *et al.* reported that, when K^+ was added to a mixed solution of γ-CD and pyrene-modified benzo-15-crown-5 fluoroionophore, the significant fluorescence emission appeared at 370–410 nm. An equilibrium analysis of this system reveals that the major component responsible to the emission is a 2:1:1 complex composed of the crown ether fluoroionophore, K^+ and γ-CD. Moreover, they also reported that a pyrene-modified boronic acid fluorophore/β-CD complex could be used as a fluorescence sensor for sugar recognition due to a photo-induced electron transfer (PET) from the pyrene donor to boronic acid [25]. The fluorophore bearing pyrene moiety as fluorescent signal transducer exhibits no fluorescence emission initially according to its aggregation. After the addition of β-CD, its UV-Vis and fluorescence spectra greatly changed by forming the inclusion complex, and an efficient fluorescence emission response of the boronic acid fluorophore/β-CD inclusion complex upon sugar binding is found owing to the binding of boronic acid moiety with sugar.

   Recently, we investigated the pH-dependent fluorescence behaviors of pyronine Y/biquinolino-bridged bis(β-CD) **5** inclusion complex [27]. In an acidic or a neutral environment, pyronine Y is protonated. Because a positively charged group is difficult to be included in the CD cavity, the positively charged pyronine Y should be located outside the CD cavity and acts as an electron acceptor. Whereas the biquinolino group in bis(β-CD) **5** acts as an electron donor. Therefore, a PET process between the biquinolino group and the protonated pyronine Y results in the quenched fluorescence of pyronine Y. However, in a basic environment, the fluorescent group of pyronine Y can be efficiently shielded from the PET process by the inclusion complexation of the deprotonated pyronine Y with the CD cavity. Subsequently, the significantly enhanced microenvironmental hydrophobicity around the pyronine Y fluorophore, which comes from a cooperative contribution of two hydrophobic CD cavities, results in the fluorescence enhancement of pyronine Y. This property may enable the biquinolino-bridged bis(β-CD) **5** as an efficient chemical sensor for the protonation and deprotonation of xanthene dyes (Fig. 9).

3.1.2. CD-based assembly

If the fluorescent species are used as building blocks, the resulting assembly constructed from inclusion complexes of fluorescent species with CD will exhibit the fluorescence enhancement and peak shifts. For example, we have constructed a supramolecular polypseudorotaxane with conjugated polyazomethines from two inclusion complexes of β-CD with tolidine and phthaldehyde (Fig. 10) [28]. Different from that of azomethine (AZO) backbone **8** giving two maximums at 400.0 and 490.0 nm, the emission spectrum of azomethine/β-CD assembly **7** displays two maximums at 397.0 and 483.0 nm with an excitation wavelength of 290.0 nm, showing the significant hypsochromic shifts. On the

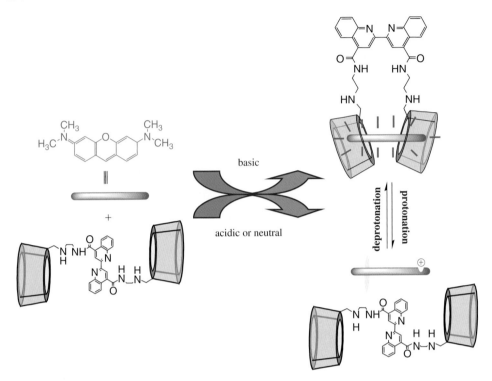

Fig. 9. The schematic fluorescence behaviors of pyronine Y/biquinolino-bridged bis(β-CD) **5** inclusion complex.

other hand, the photoluminescence spectrum (used a laser photosource) of azomethine/β-CD assembly **7** in DMF gives two maximum peaks at 438.9 and 713.6 nm with an excitation wavelength of 375.0 nm. These results indicate that β-CD can be used as the insulator for the azomethine backbone **8** and thus increases the stability of molecular electronic materials [29].

CD-based supramolecular assemblies always give the elongated fluorescence lifetimes as compared with their building blocks. Since the rates of complexation/decomplexation are much slower than that of the fluorescence decay, the decay profile of fluorescence intensity ($F(t)$) can be described as the sum of unimolecular decays for all fluorescing species present in the solution:

$$F(t) = \sum A_i \exp(-t/\tau_i) \qquad (i = 1, 2, \ldots) \tag{1}$$

where A_i and τ_i represent the initial abundance and lifetime of the ith species. As shown in Table 1, AZO-backbone **8** and AZO-assembly **7** give a short lifetime (τ_S) and a long lifetime (τ_L) respectively, and the fluorescence lifetimes of AZO-assembly **7** are longer than those of the backbone **8**, which indicates that the azomethine backbone **8** is located in the β-CD cavity to show a longer lifetime. Similar results are also found in the investigation of linear aggregation prepared

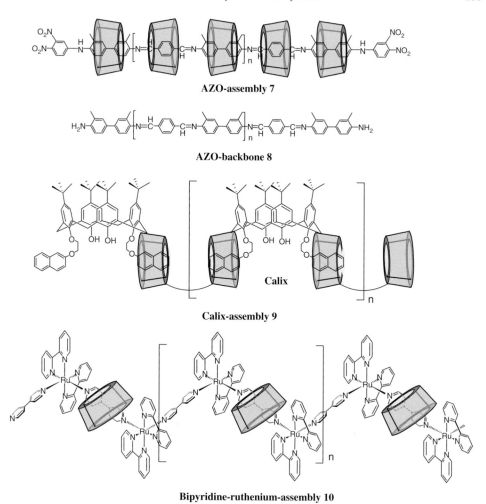

AZO-assembly 7

AZO-backbone 8

Calix

Calix-assembly 9

Bipyridine-ruthenium-assembly 10

Fig. 10. Structure of the polypseudorotaxanes **7**, **9**, **10** and backbone **8** listed in Table 1.

Table 1
Fluorescence lifetimes (τ) and relative quantum yields (Φ) in DMF at 298 K (S, short, L, long; λ_{ex}, excitation wavelength)

	8	7	Calix[4]arene	9
τ_S/(ns)	1.0	1.3	3.2	3.3
Φ_S/(%)	36.6	31.3	63.2	70.2
τ_L/(ns)	6.1	6.7	9.0	10.3
Φ_L/(%)	63.4	68.7	36.8	29.8
λ_{ex} (nm)	400	400	285	285
Reference	a	a	b	b

[a]Ref. [28].[b]Ref. [30].

by inclusion complexation of bridged bis(β-CD) with calix[4]arene derivative [30]. The naphthalene-modified calixarene in the mixed solution gives two different fluorescence lifetimes, indicating the presence of two independent fluorescing species in the solution. Judging from the two independent lifetimes (τ_S and τ_L), the short- and long-lived fluorescing species are most likely assigned to the calixarene framework and naphthalene groups. Because CD cavity can strongly bind naphthalene group, the calix-assembly **9** gives the prolonged long lifetime (from 9.0 to 10.3 ns), whereas the short lifetime do not show any significant change (from 3.2 to 3.3 ns), as compared with the parent calix[4]arene.

Bipyridine-ruthenium-CD assembly **10** can display a satisfactory luminescent behavior in both aqueous solution and the solid states benefiting from the fascinating photophysical property of bipyridine-ruthenium(II) units [31]. When excited at 456 nm, the obtained quantum yield of assembly **10** is 0.0921. Moreover, the assembly **10** can also emit strong fluorescence in the solid state. These luminescent properties will further its potential application as functional photophysical materials.

3.2. Photochemical behavior

3.2.1. Photoisomerism

The photo-induced *trans*/*cis*-isomerism of azobenzenes and/or stilbenes under the irradiation of UV/Vis lights can greatly affect their binding behaviors with CD. Huskens *et al.* [32] have prepared a CD dimer tethered by diarylethene 5,5′-dicarboxylic amide, which adopts the "open" or "close" conformation after irradiation with visible light ($\lambda > 460$ nm) or UV (313 nm) light, respectively. This switching process is completely reversible and can be easily detected through the appearance/disappearance of the absorption band at 528 nm in its UV-Vis spectrum. Because the binding ability of meso-tetrakis (4-sulfonatophenyl) porphyrin (TSPP) by the "open" dimer is a factor 35 higher than that by the "close" dimer, the photodirected release of the guest from the host is feasible and can be monitored by UV-Vis spectroscopy. This property enables this CD dimer to be an interesting candidate for the photocontrollable drug-delivery system and might find its application in photodynamic therapy (PDT).

4,4′-Azodibenzoic acid (ADA) can be isomerized from *trans* to *cis* conformer under irradiation with UV light (*ca.* 355 nm) and from *cis* to *trans* conformer under irradiation with visible light (*ca.* 430 nm). These photoisomerism behaviors can be used in the design of photoinduced molecular shuttle [33], because α-CD can only interact with *trans*-ADA but not interact with *cis*-ADA. Harada and his coworkers [34] have reported a kind of "sol-to-gel" and "gel-to-sol" transitions that can occur repeatedly by repetitive irradiations of UV and visible light. An aqueous solution of dodecyl (C_{12})-modified poly(acrylic acid) exhibited a gel-like behavior due to the polymer chains forming a network structure via hydrophobic associations of C_{12} side chains. When α-CD is added to the gel-like aqueous solution, the gel is converted to a sol mixture, because hydrophobic associations of C_{12} side chains are dissociated by the formation of inclusion complexes between α-CD and C_{12} side chains. Upon the addition of ADA just as received, where most of ADA molecules take the *trans*-form, to this binary sol mixture, α-CD dominantly bind with *trans*-ADA and then the hydrophobic associations of C_{12} side chains are restored, which consequently results in the sol-to-gel transition. When irradiating this ternary gel mixture with UV light, α-CD will dissociate with *cis*-ADA and bind with C_{12} side chains again. Therefore, the mixture undergoes a gel-to-sol transition. Further irradiating the ternary sol mixture with visible light, the mixture will undergo a sol-to-gel transition due to the photoisomerism of ADA from *cis* to *trans* conformer.

Tian *et al.* also reported a new [2] rotaxane as a light-driven molecular shuttle, where the α-CD ring can shuttle back and forth between the stilbene unit ("photo sensor") and biphenyl unit ("spacer") by alternating the irradiation frequency (see the contribution of Tian and Wang in this book), and this shuttling is accompanied by obvious changes in the intensity of fluorescence at 530 nm. In addition, this shuttle can be locked by simple acidification because of the hydrogen bonds between the primary hydroxyl groups of CD and the COO^- groups of the "lock".

3.2.2. Photo-induced DNA cleavage

As an important family of functional CD systems, CD-fullerene conjugate has an inherent advantage in the field of biochemistry. That is, CD cavity can selectively bind with the certain fragment of biological substrate and thus enable the site-specific interaction between fullerene and model substrate. For example, CD-C_{60} conjugate [35] or assembly [36,37] are reported to exhibit satisfactory DNA-cleavage ability under the visible-light irradiation at room temperature. One possible mechanism for this process may be that C_{60} can capture electron from oxygen under visible light, and make oxygen become the singlet oxygen ($^1O_2$). Then the sensitized $^1O_2$ reacts with the guanosines of DNA by either [4 + 2] or [2 + 2] cycloaddition to the five-membered imidazole ring of the purine base and thus, cleaves the DNA [38].

4. Conclusion and outlook

In recent decades, designing molecular devices containing host–guest recognition systems or supramolecular assembly to achieve a specific biological or physical function is one of the most important challenges for chemists. This chapter has covered some developments of the spectrophotometric studies on the molecular recognition of modified- or bridged-CDs bearing chromophoric substituents as well as the photophysical and photochemical properties of functional CD systems. We believe that, although the results and concepts described herein are drawn from a relatively limited systems based on CDs, they should be extended more generally to a wide variety of natural and synthetic supramolecular systems, and find their significance in material and life sciences.

Acknowledgment

This work was supported by NNSFC (No. 90306009, 20402008, 20421202, and 20572052).

References

[1] (a) J. Szejtli, Chem. Rev. 98 (1998) 1743; (b) W. Saenger, Angew. Chem. 92 (1980) 343; Angew. Chem. Int. Ed. Engl. 19 (1980) 344; (c) G. Wenz, Angew. Chem. 106 (1994) 851; Angew. Chem. Int. Ed. Engl. 33 (1994) 803.

[2] L. Szejtli (Ed.), Cyclodextrin Technology, Kluwer Academic, Dordrecht, The Netherlands, 1988.

[3] J.M. Lehn (Ed.), Supramolecular Chemistry, VCH, Weinheim, Germany, 1995.

[4] J. Szejtli, Chem. Rev. 98 (1998) 1743.

[5] J.L. Atwood, J.W. Steed (Eds), Encyclopedia of Supramolecular Chemistry, Marcel Dekker, New York, NY, 2004.

[6] D.A. Lightner, J.E. Gurst (Eds), Organic Conformational Analysis and Stereochemistry from Circular Dichroism Spectroscopy, Wiley-VCH, New York, 2000.

[7] M. Kajtár, C. Horvath-Toro, E. Kuthi, J. Szejtli, Acta Chim. Acad. Sci. Hung. 110 (1982) 327.

[8] Y. Liu, Y.-L. Zhao, H.-Y. Zhang, Z. Fan, G.-D. Wen, F. Ding, J. Phys. Chem. B 26 (2004) 8836.

[9] Y. Liu, Y. Song, Y. Chen, Z.-X. Yang, F. Ding, J. Phys. Chem. B 109 (2005) 10717.

[10] Y. Liu, X.-Q. Li, Y. Chen, X.-D. Guan, J. Phys. Chem. B 108 (2004) 19541.

[11] Y. Liu, Y. Song, Y. Chen, X.-Q. Li, F. Ding, R.-Q. Zhong, Chem. Eur. J. 10 (2004) 3685.

[12] Y. Liu, G.-S. Chen, Y. Chen, F. Ding, T. Liu, Y.-L. Zhao, Bioconjugate Chem. 15 (2004) 300.

[13] H.A. Benesi, J.H. Hildebrand, J. Am. Chem. Soc. 71 (1949) 218.

[14] Y. Wang, T. Ikeda, H. Ikeda, A. Ueno, F. Toda, B. Chem. Soc. Jpn. 67 (1994) 1598.

[15] T. Kuwabara, H. Nakajima, M. Nanasawa, A. Ueno, Anal. Chem. 71 (1999) 2844 and references therein.

[16] T. Aoyagi, H. Ikeda, A. Ueno, B. Chem. Soc. Jpn. 74 (2001) 157.

[17] (a) Y. Liu, B.-H. Han, S.-X. Sun, T. Wada, Y. Inoue, J. Org. Chem. 64 (1999) 1487; (b) Y. Liu, C.-C. You, T. Wada, Y. Inoue, J. Org. Chem. 64 (1999) 3630.

[18] Y. Liu, L. Li, H.-Y. Zhang, Induced Fit in Encyclopedia of Supramolecular Chemistry. Marcel Dekker, New York, NY, 2004, pp. 717–726.

[19] Y. Liu, Y. Chen, S.-X. Liu, X.-D. Guan, T. Wada, Y. Inoue, Org. Lett. 3 (2001) 1657.

[20] Y. Liu, Y. Chen, L. Li, G. Huang, C.-C. You, H.-Y. Zhang, T. Wada, Y. Inoue, J. Org. Chem. 66 (2001) 7209.

[21] Y. Liu, B.-H. Han, S.-X. Sun, T. Wada, Y. Inoue, J. Org. Chem. 64 (1999) 1487.

[22] H. Ikeda, M. Nakamura, N. Ise, N. Oguma, A. Nakamura, T. Ikeda, F. Toda, A. Ueno, J. Am. Chem. Soc. 118 (1996) 10980.

[23] Y. Liu, Y. Song, H. Wang, H.-Y. Zhang, T. Wada, Y. Inoue, J. Org. Chem. 68 (2003) 3687.

[24] A.-J. Tong, A. Yamauchi, T. Hayshita, Z.-Y. Zhang, B.D. Smith, N. Teramae, Anal. Chem. 73 (2001) 1530.

[25] A. Yamauchi, T. Hayashita, A. Kato, S. Nishizawa, M. Watanabe, N. Teramae, Anal. Chem. 72 (2000) 5841.

[26] K. Kano, H. Kitagishi, S. Tamura, A. Yamada, J. Am. Chem. Soc. 126 (2004) 15202.

[27] Y. Song, Y. Chen, Y. Liu, J. Photochem. Photobiol. A: Chem. 173 (2005) 328.

[28] Y. Liu, Y.-L. Zhao, H.-Y. Zhang, X.-Y. Li, P. Liang, X.-Z. Zhang, J.-J. Xu, Macromolecules 37 (2004) 6362.

[29] F. Cacialli, J.S. Wilson, J.J. Michels, C. Daniel, C. Silva, R.H. Friend, N. Severin, P. Samori, J.P. Rabe, M.J. O'Connell, P.N. Taylor, H.L. Anderson, Nature Mater. 1 (2002) 160.

[30] Y. Liu, H. Wang, H.-Y. Zhang, L.-H. Wang, Y. Song, Chem. Lett. 32 (2003) 884.

[31] Y. Liu, S.-H. Song, Y. Chen, Y.-L. Zhao, Y.-W. Yang, Chem. Comm. (2005) 1702.

[32] (a) A. Mulder, A. Juković, L.N. Lucas, J. van Esch, B.L. Feringa, J. Huskens, D.N. Reinhoudt, Chem. Comm. (2002) 2734;
     (b) A. Mulder, A. Juković, F.W.B. van Leeuwen, H. Kooijman, A.L. Spek, J. Huskens, D.N. Reinhoudt, Chem. Eur. J. 10 (2004) 1114.

[33] A. Stanier, S.J. Alderman, T.D.W. Claridge, H.L. Anderson, Angew. Chem. Int. Ed. 41 (2002) 1769.

[34] I. Tomatsu, A. Hashidzume, A. Harada, Macromolecules 38 (2005) 5223.

[35] Y. Liu, Y.-L. Zhao, Y. Chen, P. Liang, L. Li, Tetrahedron Lett. 46 (2005) 2507.

[36] Y. Liu, H. Wang, P. Liang, H.-Y. Zhang, Angew. Chem. Int. Ed. 43 (2004) 2690.

[37] Y. Liu, H. Wang, Y. Chen, C.-F. Ke, M. Liu, J. Am. Chem. Soc. 127 (2005) 657.

[38] (a) C. Sheu, C.S. Foote, J. Am. Chem. Soc. 115 (1993) 10446;
     (b) C. Sheu, C.S. Foote, J. Am. Chem. Soc. 117 (1995) 474.

Cyclodextrin Materials Photochemistry, Photophysics and Photobiology
Abderrazzak Douhal (Editor)
241

DOI 10.1016/S1872-1400(06)01011-9

Chapter 11

Supramolecular Photochirogenesis with Cyclodextrin

Cheng Yang[1], Yoshihisa Inoue[1,2]

[1]*ICORP Entropy Control Project, JST, 4-6-3 Kamishinden, Toyonaka 560-0085, Japan*
[2]*Department of Applied Chemistry, Osaka University, Suita 565-0871, Japan*

1. Introduction

Supramolecular photochirogenesis, an interdisciplinary field of science that integrates supramolecular chemistry, photochemistry and asymmetric synthesis, has in recent years rapidly developed and is a promising unique expansion of chiral photochemistry [1–3]. Controlled by non-covalent interactions, such as van der Waals, hydrogen bonding, π–π, dipole–dipole and hydrophobic interactions, inter- and intramolecular chirality transfer in the excited state proceeds with enhanced efficiency in chiral supramolecular environment. A fair number of supramolecular hosts or templates, including cyclodextrins (CDs) [4,5], chirally modified zeolites [6,7], Kemp's triacid derivatives [8–11], biomolecules [12–14] and chiral nanopores [15] have been explored as chiral environments or fields for asymmetric photochemical reactions. Among these chiral hosts or templates, native and modified CDs have hitherto been the most frequently employed by virtue of the fact that they are inherently chiral, readily available, UV-transparent, water soluble and, above all, capable of accommodating a wide range of organic guests in their different sized hydrophobic cavities.

CDs and organic guests are known to form inclusion complexes of varying stoichiometries both in solution and in the solid state, primarily through hydrophobic and van der Waals interactions [16]. It is generally believed that the inclusion complex of the ground-state substrate plays the major role in CD-mediated photoreactions, since the complexation/decomplexation processes are much slower than the decay kinetics of electronically excited substrate. Thus, supramolecular photoreactions usually proceed through two sequential steps, i.e. complexation followed by photoreaction. In the ground-state complex, the conformational freedoms of the included substrate are greatly reduced to give a limited number of conformers as a consequence of the steric confinement effect of chiral CD cavity. Irradiation of the complex leads to the formation of chiral photoproduct(s) with different quantum and chiral efficiencies hence completing the supramolecular photochirogenesis.

Compared to its thermal and non-chiral counterparts [4,5,17], chiral photochemistry with CD has only a short history and is still in its infancy, with CD-mediated enantio- and diastereodifferentiating photoreactions not reported until the end of the 1980s [18,19]. After a decade of slow growth, the study of CD-based supramolecular photochirogenesis started to develop more rapidly particularly since entering the 21st century. Recent studies have shown that photochirogenesis with CD not only provides us with novel, versatile and powerful methodologies for synthesizing chiral compounds and manipulating product chirality, but also functions as an indispensable tool for elucidating the nature of supramolecular interactions operating in the CD complex.

In this chapter, we will review the studies on CD-mediated photochirogenic reactions published by late 2005. We will also briefly refer to such works that employ CDs as chiral hosts but do not report the optical yields of photoproducts. However, we will not deal with the photochemical derivatization of CDs, although the products are certainly chiral [20–23].

2. Unimolecular asymmetric photoreactions

2.1. Geometrical photoisomerization

2.1.1. Photoisomerization of cycloalkenes

It is well documented that, upon direct or sensitized irradiation, (Z)-cyclooctene (**1Z**) undergoes geometrical isomerization to the (E)-isomer (**1E**), which is chiral and composed of a pair of planar chiral (R)- and (S)-**1E** enantiomers (Scheme 1) [24]. Nonetheless, the enantiodifferentiating photoisomerization of **1Z** sensitized by optically active benzenecarboxylates was only reported for the first time in 1978 [25], and has extensively been investigated to become a sort of "bench-mark" reaction for evaluating the ability of chiral photosensitizing systems [26,27].

The supramolecular enantiodifferentiating photoisomerization of **1Z** was first examined by direct photolysis at 185 nm of its β-CD complex in the solid state [28]. A stoichiometric 1:1 host–guest inclusion complex was prepared by precipitation by

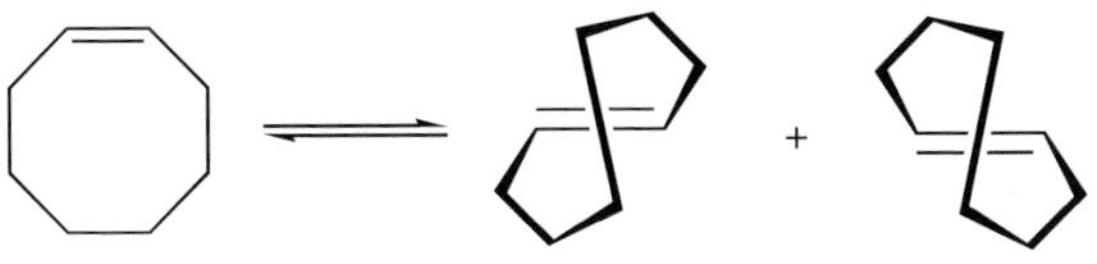

1Z   (R)-1E   (S)-1E

Scheme 1.

adding **1Z** to an aqueous solution of β-CD. The vacuum UV irradiation of the solid-state complex affords **1E** in an E/Z ratio of 0.47 at the apparent photostationary state (PSS), which is considerably smaller than that (0.96) obtained upon direct irradiation in pentane solution. The photoproduct obtained from the direct photolysis of **1Z**–β-CD complex was almost racemic (0.24% ee (enantiomeric excess)), demonstrating that the cavity interior of native β-CD is not sufficiently enantiodifferentiating for the successful photoisomerization of **1E**.

As an advanced strategy for achieving higher levels of photoenantiodifferentiation of **1E**, supramolecular enantiodifferentiating photosensitization was developed [29]. A series of CD derivatives with a chromophore appended to the primary rim were synthesized as chiral sensitizers (Scheme 2), and the photoisomerization of **1E** with these supramolecular sensitizers was studied over a wide range of solvents and temperatures [29–31]. The principle features of the supramolecular photosensitization system are outlined in Scheme 3. When the appended sensitizer moiety is included in the CD cavity (Scheme 3, left), the energy transfer to the substrate located in the bulk solution is interrupted by the CD wall. However, once the substrate is included within the chiral cavity, highly efficient energy transfer takes place to the substrate residing in the cavity immediately after the irradiation. Consequently, the undesirable, less-enantiodifferentiating sensitization outside the CD cavity can be prevented. Furthermore, the use of such a sensitizer-appended host allows the enantiodifferentiating photoisomerization to proceed in the presence of only a catalytic amount of the sensitizing host, so giving rise to photochemical chirality amplification.

High PSS E/Z ratios, $(E/Z)_{pss}$, of up to 0.8 are reported in the photosensitizations of **1Z** mediated by non-methylated CD derivatives **2a**, **3a–f** and **4a** in 1:1 methanol–water mixture. Since much lower $(E/Z)_{pss}$ values are obtained upon conventional photosensitization of **1Z** in organic solvents, these high ratios are characteristic of supramolecular photosensitization and hence may be attributed to the close contact, and efficient energy transfer to, of both isomers in the confined medium. The ee's of **1E** obtained upon photoisomerization of **1Z** in the presence of α and γ-CD benzoates, **2a** and **4a**, whose cavity sizes are not matched to that of **1Z**, are much lower than that obtained with β-CD analogue **3a**. The enantioselectivity of the reaction is also very sensitive to the structure of

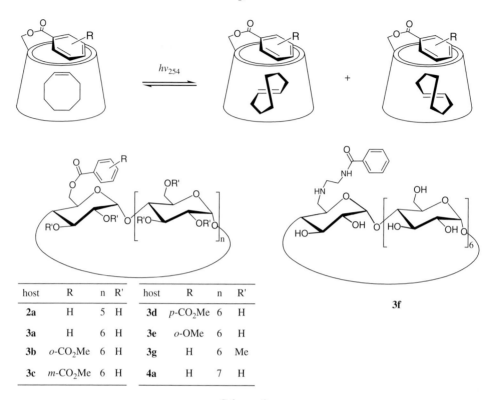

Scheme 2.

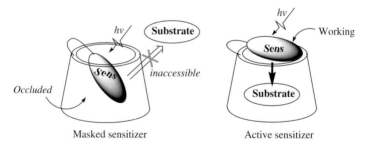

Scheme 3.

the appended sensitizer. The use of methyl phthalate derivative **3b** affords **1E** in good ee's of up to 24%. However, iso- and terephthalates, **3c** and **3d**, fail to give appreciable ee's, because the substituents at either *meta-* or *para*-positions disturb the inclusion of the aromatic moiety, and therefore the sensitization occurs outside the CD cavity [29,30]. The product ee obtained with **3e** depends critically on the solvent composition, as the major product is switched from (*R*)-(−)-**1E** (15% ee) in water to (*S*)-(-)-**1E** (5% ee) in pure methanol [31].

In marked contrast to the vital role of entropy observed in conventional asymmetric photosensitizations [26,27], the contribution of entropic factors to the photoisomerization mediated by non-methylated CD sensitizers seems insignificant. The ee values obtained with non-methylated CDs are not a simple function of reaction temperature and solvent, but are nicely correlated with host occupancy. Interestingly, the inherent low-entropy character of the CD cavity is dramatically modified, when the original hydrogen bond network connecting the adjacent 2- and 3-OHs are chemically modified by permethylation so as to make the CD skeleton flexible. As demonstrated in a recent study, the optical yield in the supramolecular photosensitization with **3g** exhibits a critical dependence on temperature [32]. The chiral sense of **1E** is reversed by performing the photoreaction above and below the equipodal temperature. The differential entropy change ($\Delta\Delta S^{\ddagger}_{S-R} = -11\,\mathrm{J\,mol^{-1}\,K^{-1}}$) obtained with **3g** is much larger than that obtained with non-methylated CDs, but quite similar to those reported for conventional sensitization, indicating a significant correlation between the degree of entropic contribution and the conformational freedom in the CD-based supramolecular system.

2.1.2. Photoisomerization of diphenylcyclopropane

Since the pioneering work of Hammond and Cole [33] on enantiodifferentiating photosensitized isomerization of 1,2-diphenylcyclopropane **5a** (Scheme 4), the chiral induction in this geometrical isomerization process has widely been studied by several groups [34–37]. Conventional chiral photosensitizers give only low ee's less than 10% for **6a** [37]. The enantiodifferentiating step is deduced to be the quenching of the chiral sensitizer rather than the decay from the intervening exciplex or radical–ion pair [38]. Recent studies on the asymmetric photo-isomerizations of diphenylcyclopropane derivatives within zeolite supercages and CD cavities have demonstrated that such immobilized systems are useful for improving the level of both enantio- and diastereoselectivity [39–44].

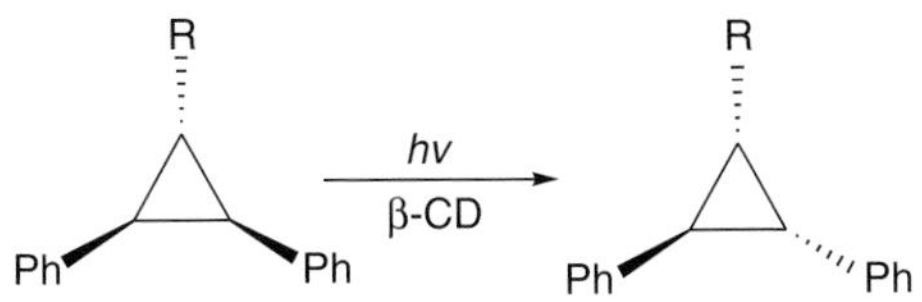

guests	— R	Sens*	% ee (de)
5a, 6a	—H	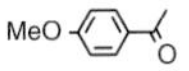 MeO-⟨⟩-⟨	13
5b, 6b	—C-OEt	no	13
5c, 6c	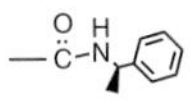 —C-N⟨⟩	no	28
5d, 6d	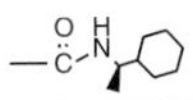 —C-N⟨⟩	no	30

Scheme 4.

Asymmetric photoisomerizations of β-CD-complexed diphenylcyclopropanes **5a–d** (Scheme 4) have been examined both in the solid state and aqueous solutions [45]. Solution-phase irradiation of the complexes does not show appreciable chirality transfer, typically giving the photoproduct in less than 2% ee or diastereomeric excess (de). β-CD inclusion complexes precipitated from aqueous solutions show a 1:1 host–guest stoichiometry for **5a** and **5b** but a 3:1 stoichiometry for **5c** and **5d**. The photoisomerizations of solid complexes of **5a–d** upon triplet sensitization or direct excitation affords the corresponding photoproducts with ee or de ranging from 13% to 30% (Scheme 4).

As shown in Scheme 5, the photoirradiation gives rise to a diradical intermediate via cleavage of the C_2–C_3 bond. Subsequent rotation around the C_1–C_2 or C_1–C_3 bond, followed by a ring closure, completes the isomerization between *cis* and *trans* isomers. Without a chiral inductor, the two modes of rotation proceed with equal probability to give racemic products. However, for the diphenylcyclopropane located in the chiral environment of the CD cavity, the efficiency of the two rotation modes should be different as a consequence of confinement in the chiral CD cavity and so give enantio- or diastereomerically enriched photoproducts.

β-CD exhibits a preference for accommodating one of the enantiomers upon complexation with **6a** or **6b**. Interestingly, the preferentially included enantiomer is exactly the same one that is produced by photoisomerization. Similar behavior has been reported for the enantiodifferentiating photoisomerization of (Z)-cyclooctene **1Z** included and sensitized by β-CD benzoate **3a** [29,30]. This phenomenon indicates a strong relationship between the relative affinity of the enantiomeric photoproducts upon complexation and the enantioselectivity upon photochirogenesis, and thus allows us to predict the enantiomer that will be produced in excess in the CD-mediated asymmetric photoreaction. If the general validity of such a relationship is further confirmed in future work, it will be a powerful tool for predicting the chiral outcome of supramolecular photochirogenic reactions.

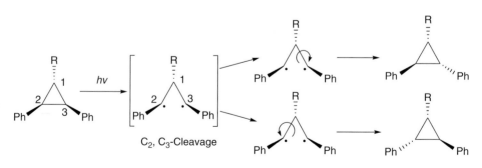

Scheme 5.

2.2. Photocyclization

2.2.1. Photocyclization of tropolone alkyl ethers

Upon photolysis, tropolone alkyl ether **7** undergoes intramolecular 4π electrocyclization to afford chiral bicyclo[3.2.0]heptadienone **8**, which rearranges to compound **9** upon prolonged irradiation (Scheme 6). The photocyclization of CD-complexed tropolone methyl ether **7a** was first studied by Takeshita and coworkers [46]. They found that the formation of CD complex improves the yield of the optically active product **8a**, but the ee value was not quantitatively determined.

Recently, Ramamurthy and coworkers systematically reinvestigated the stereo-differentiating photocyclizations of a series of tropolone derivatives **7a–g** (Scheme 6) within α-, β- and γ-CD cavities [47]. In line with the results reported in the photoisomerization of diphenylcyclopropanes [45], solution-phase photolyses in the presence of CDs produce **8a–g** with negligible ee's. CDs form inclusion complexes with tropolone alkyl ethers with 1:1 stoichiometries for **7a** and **7b** and with 2:1 host–guest stoichiometry for **7c**, **7d** and **7g** that possess much larger alkyl groups. Solid-state photoirradiation of the α-CD complex affords **8a** and **8b** in more than 20% ee, while the γ-CD complexes give only racemic cyclization products. Conversely, the optical yields obtained upon photocyclization of α-CD-complexed **7e** and **7f** are less than 5%, since the prochiral centers is located only on the rim of α-CD cavity due to the bulkiness of the alkyl groups in **7e** and **7f**. It was tentatively concluded that the size matching between the guest and the chiral cavity as well as confinement of the guest in the CD cavity are the most important factors controlling the stereoselectivity in this CD-mediated photocyclization.

Moderate ee's of 20–33% are obtained in the solid-state photolysis of the 2:1 β- or γ-CD complexes of **7c** and **7d**. As shown in Scheme 6, electronically excited tropolones may cyclize in two allowed disrotatory modes, 'in' or 'out' to give the

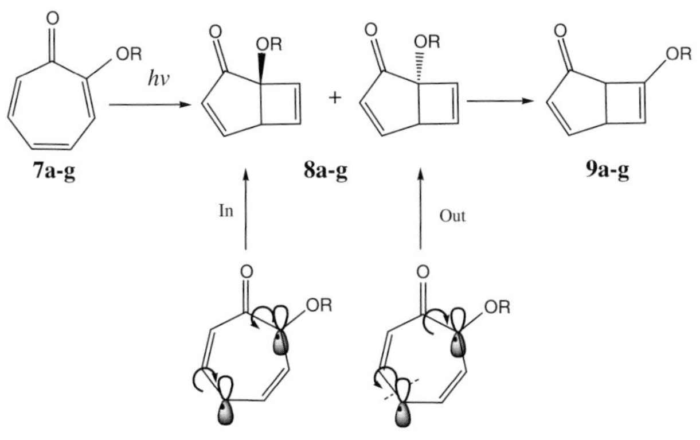

7a R = CH$_3$
7b R = CH$_2$CH$_3$
7c R = CH$_2$Ph
7d R = CH$_2$CH$_2$Ph
7e R = CH(CH$_3$)$_2$
7f R = (S)-CH$_2$ C*H(CH$_3$)CH$_2$CH$_3$
7g R = CH$_2$CH$_2$CH$_2$Ph

Scheme 6.

corresponding stereoisomers. When tropolone is included in a chiral CD cavity, it is likely that one of the two ring-closure modes is preferred to give the significant stereodifferentiation observed in this CD-complexed photocyclization.

2.2.2. Photocyclization of n-alkyl pyridones

A 4π cyclization can also be triggered upon excitation of N-alkylpyridones to yield chiral 2-azabicyclo[2.2.0]hex-5-en-3-ones [48]. Ramamurthy and coworkers investigated the effects of β-CD on the asymmetric photocyclization of N-methyl- and N-ethylpyridones **10a,b** (Scheme 7) [49]. It was found that even in the presence of a large excess amount of β-CD, photocyclization of **10a,b** in aqueous solution furnished the photoproduct in less than 5% ee. Since **10a,b** as well as their CD complexes are freely soluble in water, β-CD complexes of **10a,b** were prepared by mechanically grinding an equimolar mixture of the both components with pestle and mortar. Solid-state NMR and powder X-ray diffraction studies indicated that the resultant powder is not a physical mixture but an inclusion complex of CD and N-alkylpyridone.

Photocyclization of **10a,b** complexed with α- or γ-CD affords photoproducts **11a,b** as racemic mixtures. In contrast, solid-state photolysis of β-CD complexes yields **11a,b** with much enhanced enantioselectivities (of up to 59% ee). The optical yields of **11a,b** significantly depend on the content of hydrogen-bonding solvent in the host–guest complex. Respectable ee values ranging from 60% to 74% are obtained from the photolyses of the solid-state complex that is prepared in a methanol–hexane (5:95) mixture followed by air drying. This solid-state complex has similar inclusion structure to the sample obtained by mechanical grinding. Both the complexes contain 9% water, as revealed by NMR, X-ray and TGA analyses. Significantly, when the complexes are dried under vacuum and thus contain less water (2%), the product chirality is inverted and the antipodal product is given in 24% ee. Addition of a small amount of water or methanol to the vacuum-dried sample can recover the original enantioselectivity, but no effect is seen upon addition of non-hydrogen bonding solvent such as hexane.

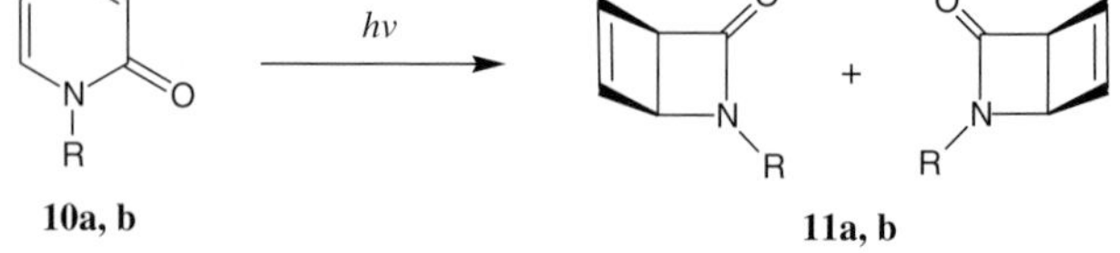

10a, b
10a: R = CH$_3$
10b: R = CH$_2$CH$_3$

11a, b

Scheme 7.

2.2.3. Photocyclization of phenoxyalkenes

The asymmetric intramolecular *meta* photocycloaddition of phenoxyalkenes **12a–g** (Scheme 8) included in β-CD have been examined by Eycken and coworkers [50]. The inclusion complexes of **12a–g** and β-CD are separated from the mixture of a warm aqueous solution of β-CD and substrates followed by centrifugation and decantation. The β-CD complexes thus obtained are of 1:2 host–guest stoichiometry with **12a** and **12d–f**, and 1:1 stoichiometry with the rest of phenoxyalkenes.

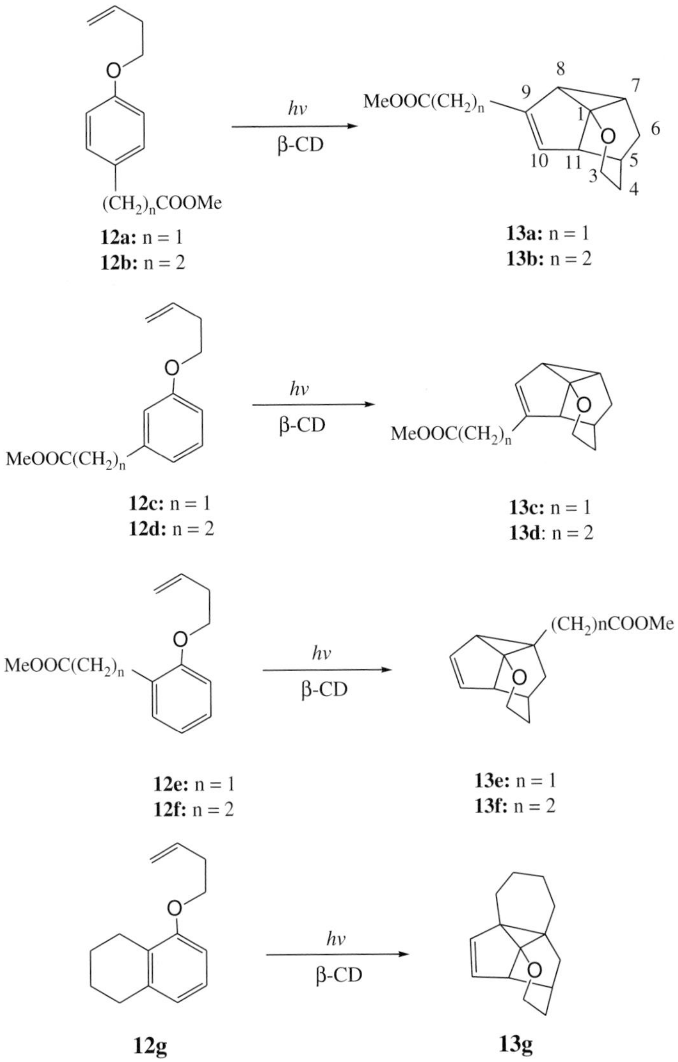

Scheme 8.

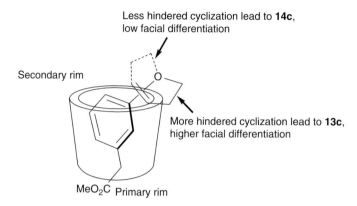

Fig. 1. Supramolecular photochirogenesis of phenoxyalkenes in CD cavity.

As shown in Scheme 8, intramolecular *meta* photocyclizations of phenoxyalkenes yield exclusively the 1,5-bridged dihydrosemibullvalenes. Photoirraditions of solid complexes furnish **12a–g** with 55–82% conversions, which are appreciably lower than that obtained in the solution-phase photoreaction. Photocyclization of phenoxyalkenes **12** gives, in general, only one regioisomer **13**, except for *meta*-substituted **12c,d**, which are converted to two regioisomers **13c,d** and **14c,d**. The **14c**:**13c** ratio increases from 1:1 for the free substrate to 3:1 for the β-CD complex, while **13d** and **14d** are produced in roughly equal amounts both in the presence and absence of β-CD. Photocyclizations of the 'linear' photosubstrate **12a,b** give insignificant ee's of less than 3%, for which the 'loose' complexation between **12a,b** and the CD, resulting from the parallel orientation of the guest with the CD axis, is thought to be responsible. In contrast, **13d**, **13f** and **13g** are obtained in better ee's of 9%, 10% and 13%, respectively.

NMR spectral studies suggest that the alkene tether of **12c** is positioned on the secondary rim of β-CD. The photocyclization of **12c** affords the minor product **13c** in the highest ee of 17%, but the major product **14c** in only 2% ee. As illustrated in Fig. 1, this phenomenon can be rationalized as a result of the trade-off between chemical and optical yield, as larger steric hindrance leads to low yield and high ee, while smaller steric hindrance results in high yield and low ee.

2.2.4. Photocyclization of dithienylethene

Diarylethenes are one of the most promising photochromic materials for such purposes as molecular optical memories and switches, because of their excellent thermal stability and fatigue-resistant nature [51]. A chiral dithienylethene has the potential for non-destructive readout by detecting optical rotation at wavelengths longer than its own absorption band. However, it is not a simple task to achieve appreciable stereoselectivity in solution-phase photocyclization of dithienylethenes, even for those with chiral substituents [52–54]. In this regard, a

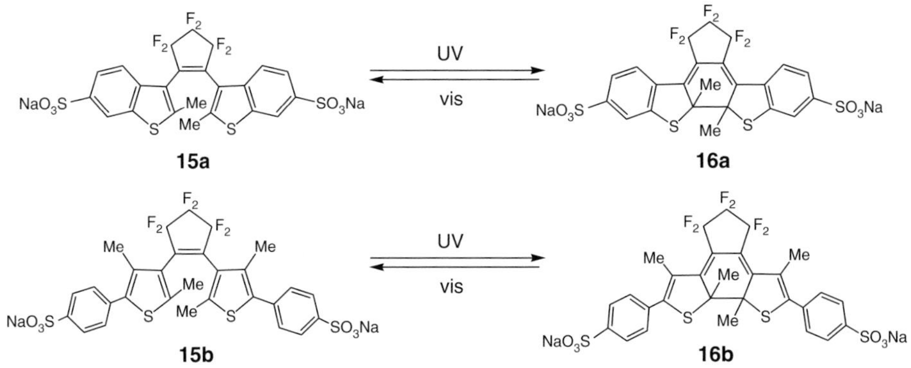

Scheme 9.

supramolecular photochirogenic approach was proven to be useful by recent work in which the closed form of dithienylethene is obtained in excellent diastereoselectivity (96% de) upon gel formation via supramolecular aggregation of the open form [52].

The effects of CDs on the photocyclization of **15a,b** (Scheme 9) have been studied by Irie and coworkers [55–57]. The open form of dithienylethene has two conformations, i.e. parallel and anti-parallel, of which only the rodlike anti-parallel conformer undergoes cyclization upon irradiation to give two closed-ring enantiomers (R,R and S,S). By the addition of a 10-fold excess of CD, the anti-parallel conformation of **15a** is significantly increased from 0.64 to 0.80 (γ-CD) and to 0.94 (β-CD). Accordingly, the cyclization quantum yield of **15a** is raised by a factor of 1.4 and 1.5 in the presence of γ- and β-CD, respectively. No effect is observed in the presence of α-CD, since its cavity is too small to include the guest. The complex of **15b** with β-CD exhibits a positive Cotton effect, which is opposite to that induced by γ-CD. Photochromic reaction of the CD complex of **15a,b** leads to remarkable changes in circular dichroism spectrum, which is ascribed to the change in direction of transition moments upon ring closure.

2.2.5. Photocyclization of N,N-dialkylpyruvamides

The Norrish Type II reaction of N,N-dialkylpyruvamides produces a mixture of β-lactams and oxazolidin-4-ones [58,59]. Solid-state photocyclization of N–benzyl-N-methylpyruvamide **17** (Scheme 10) was performed in the presence of β-CD or deoxycholic acid [18]. A 1:2 guest–host complex of **17** with β-CD was prepared by the precipitation method. The yield of product **18a** formed by methyl hydrogen abstraction is significantly improved from 3% in benzene to 17% in the β-CD complex, while the yields of oxazolidinone **18b** and the benzylic hydrogen

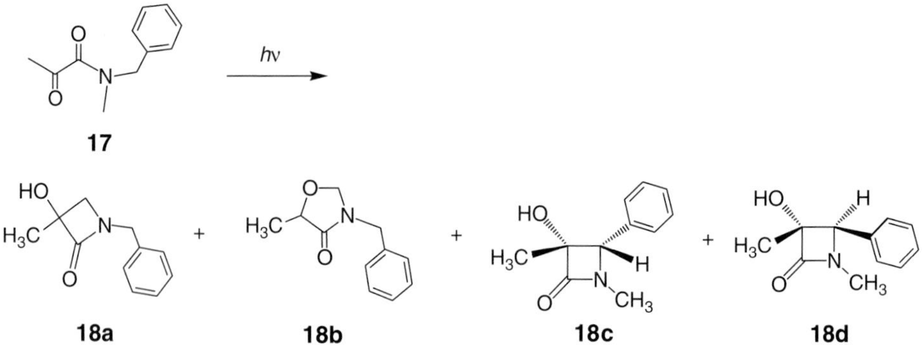

Scheme 10.

abstraction product **18c** are decreased from 15% and 33% to 6% and 12%, respectively. The minor product **18d** was given in 1% yield and 9% ee.

2.3. Photodecarboxylation of aryl ester

Photolysis of aryl esters yields a mixture of *ortho-* and *para*-phenolic ketones through the photo-Fries rearrangement. Complexation of aryl esters with CD can significantly modify the regioselectivity of the photo-Fries rearrangement. For example, the *para* product is greatly enhanced at the expense of the *ortho* product in the photo-Fries rearrangement of phenyl benzoate complexed with α- and β-CD [60]. Photoreactions of the γ-CD complexes of 1-naphthyl phenylacetates afford the corresponding 2-acylated 1-naphthols with very high selectivities of 94–98% among the possible eight rearrangement products arising from the photoreaction without CD [61].

However, the decarboxylation reaction predominates when the *ortho-* and *para*-positions of the aromatic ring are blocked by alkyl substituents [62]. Recently, Inoue and coworkers have studied the photodecarboxylation of chiral aryl esters accommodated in CD and polyethylene films [63–67]. In the photolysis of 2,4,6-trimethylphenyl-2-methylbutyrate **19**, two reaction pathways simultaneously open to afford alkylmesitylene **20** and 2,4,6-trimethylphenol **21**, respectively (Scheme 11). The yields of **20** and **21** obtained in aqueous β- and γ-CD solutions are much lower (0–11%) than those in organic solution, which is ascribed to the intermolecular reaction between CD and **19**. The β-CD-mediated photodecarboxylation of racemic **19** furnishes the (*R*)-enantiomer of **20** in up to 14% ee, suggesting that the reaction pathway differs in the presence and absence of β-CD and one of the enantiomers of **19** reacts faster than another in the β-CD cavity. When the photoreaction is carried out in the presence of γ-CD, only a very small amount of

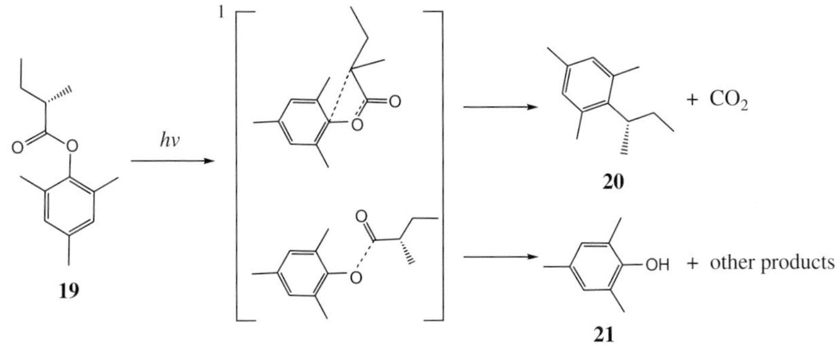

Scheme 11.

21 (<4%) is obtained and no **20** is detected, indicating that **19** reacts preferentially with the γ-CD to which it is bound.

2.4. Radical recombination of benzoin

Both thermal and photochemical formations of benzoin have been examined in the presence of CDs [19,68]. The aqueous cyanide-catalyzed benzoin condensation can be facilitated in the presence of γ-CD but inhibited by α-CD [68]. Rao and Turro studied the photochemical asymmetric induction in the synthesis of benzoin from β-CD complexed benzaldehyde [19]. Photoirradiation of α- and β-CD complexes of benzaldehyde in aqueous medium gives diphenylethane-1,2-diol as a major product with trace amounts of benzoin **22a** and 4-benzoylbenzaldehyde **22b** (Scheme 12). Benzaldehyde forms 1:1 complexes with α- and β-CD and a 1:2 complex with γ-CD. The α-CD complex is photochemically unreactive and yields no detectable product upon irradiation. Photoirradiation of the β-CD complex of benzaldehyde in the solid state affords **22b** in 24% yield and **22a** of 15% ee in 56% yield. Photolysis of γ-CD complex gives equal amounts of **22a** and **22b** in 78% combined yield. Upon photoirradiation, benzoin is formed first by the recombination of radical pairs arising from the hydrogen abstraction from a ground-state benzaldehyde by the excited triplet benzaldehyde [69,70]; then, the resultant benzoin undergoes the Norrish Type I cleavage to give a geminate radical pair, which recombines to regenerate benzoin [71,72]. Accordingly, the intracomplex rearrangement of the benzoin fragments should play a crucial role in the enantiodifferentiating process. As shown in Fig. 2, the complexation stoichiometries for α-, β- and γ-CDs are speculated to be 1:1, 2:2 and 1:2, respectively, on the basis of the chemo- and enantioselectivities observed with the relevant CDs. Thus, it was deduced that benzaldehyde is included, or individually 'insulated' in the cavity of α-CD and therefore lacks intermolecular reactivity. The

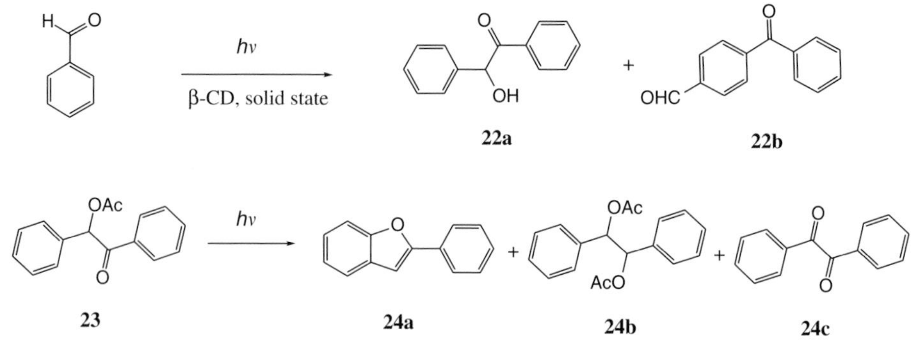

Scheme 12.

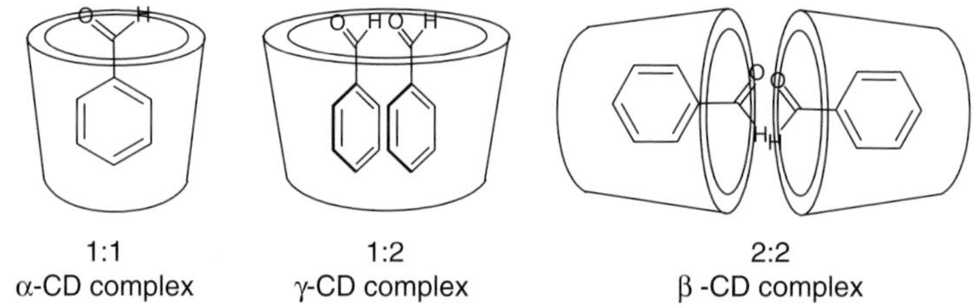

Fig. 2. Complexation modes of benzaldehyde with α-, β- and γ-CDs.

low enantioselectivity with the γ-CD complex was rationalized by the formation of a looser complex with γ-CD than with β-CD.

Photorearrangement of benzoin acetate **23** (Scheme 12) in the presence of β- or γ-CD highly favors the photocyclization product **24a** over the recombination products **24b** and **24c** originating from the Norrish Type I cleavage [73].

3. Bimolecular asymmetric photoreaction

A variety of bimolecular photoreactions that are suppressed in the solution phase proceed smoothly and selectively upon complexation with supramolecular hosts. Intermolecular photochemical reactions promoted by an achiral host or template, such as crown ether [74], cucurbituril [75,76] as well as self-assembled coordination cages [77], can occur with remarkable efficiency and chemoselectivity, for which the close proximity and the highly regulated orientation and special arrangement between substrates in the supramolecular aggregate are jointly responsible. The main advantage of CDs over the above-mentioned hosts is the inherent chirality, which makes them good cradles for asymmetric induction in bimolecular photoreactions.

3.1. [2+2] Photodimerization

3.1.1. Photodimerization of stilbene derivatives

Solution-phase photodimerizations of stilbene derivatives **25a–g** (Scheme 13) proceed in an inefficient and unselective fashion in the absence of CD, and the (Z)-isomers **26a–g** or the subsequent cyclized isomers **27a–g** are obtained as the major products. In contrast, the photodimerization of **25g** is completely inhibited in the presence of α- and β-CDs, but is dramatically promoted in the presence of γ-CD, selectively affording the *anti*-HH photodimer **30g** and the *syn*-HH photodimer **28g** in 79% and 19% yields, respectively [78]. Interestingly, upon photoirradiation of the solid-state complexes of γ-CD and **25a–f**, *syn*-HH and *syn*-HT dimers **28a–f** and **29a–f** are exclusively formed over the *anti* in combined yields ranging from 59% to 79% [79].

Complexation of *trans*-2-styrylpyridine with α or β-CD complex greatly suppresses the *trans–cis* photoisomerization in the solid state, and instead moderately facilitates the photodimerization to give the photodimers in low yields (<6%). However, complexation with γ-CD accelerates the photodimerization to give *syn*-HT dimer **29** in good yield [80].

3.1.2. Photodimerization of trans-cinnamic acids

The photochemical behavior of cinnamic acid derivatives **32a–g** (Scheme 14) has been investigated by employing γ-CD or CB [8] as photodimerization templates [76,81]. Solid-state complexes of **32a–g** with γ-CD cannot be prepared by the

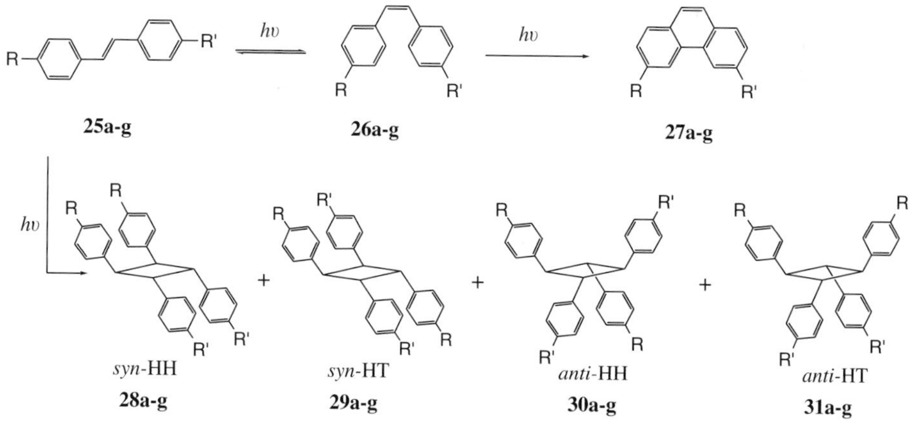

syn-HH	*syn*-HT
28a-g	**29a-g**

anti-HH	*anti*-HT
30a-g	**31a-g**

25a: R = R' = H   25e: R = OCH₃, R' = H
25b: R = R' = OCH₃   25f: R = CN, R' = H
25c: R = CH₃, R' = H   25g: R = R' = CH₂NMe₂
25d: R = Cl, R' = H

Scheme 13.

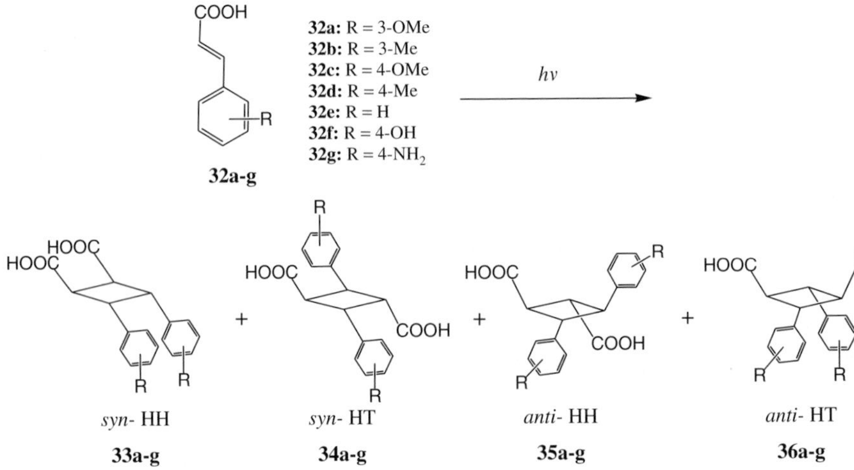

Scheme 14.

grinding technique but readily precipitate upon addition of these guests to an aqueous solution of γ-CD. In contrast to the photolysis of CB [8] complexes, where a mixture of *cis*-isomers, *syn*-HH dimers **33a–g** and *anti*-HT dimers **36a–g** are obtained, photoirradiations of solid-state γ-CD complexes afford exclusively *syn*-HH dimer **33a–g** in excellent yields (> 96%) [76].

3.1.3. Photodimerization of coumarins

The reactivity and selectivity of photodimerization of coumarins **37a–g** (Scheme 15) are significantly altered through complexation with β- and γ-CD. γ-CD and **37** form 1:2 inclusion complexes and yield, in general, *syn*-dimers **38** and **39** upon irradiation. However, no photodimer is obtained in the solid-state photolysis of β-CD complexes of **37b** and **37c**, suggesting that these substrates form 1:1 complexes with β-CD, in which each substrate molecule is isolated from its neighbor by the CD walls [82]. However, the 2:2 complexation with β-CD rather facilitates the photodimerization, and even switches the stereoselectivity from that observed for the photoreaction of γ-CD complexes. For example, photodimerization of the γ-CD–**37d** complex affords *syn*-dimers **38d** and **39d** in 57% and 28% yields, respectively, whereas *anti*-HT dimer **41d** is produced as the sole product upon irradiation of tetrad-type crystals of the 2:2 β-CD–**37d** complex [83,84].

3.1.4. Photodimerization of acenaphthylene

Photodimerization of acenaphthylene in aqueous solution is inhibited by β-CD, but leads to a quantitative conversion in the presence of γ-CD, giving **42a** and **42b** in a ratio of 99:1 (Scheme 16) [85]. The photocycloaddition of acrylonitrile to acenaphthylene included in β-CD affords only the *cis*-crossadduct **42c**.

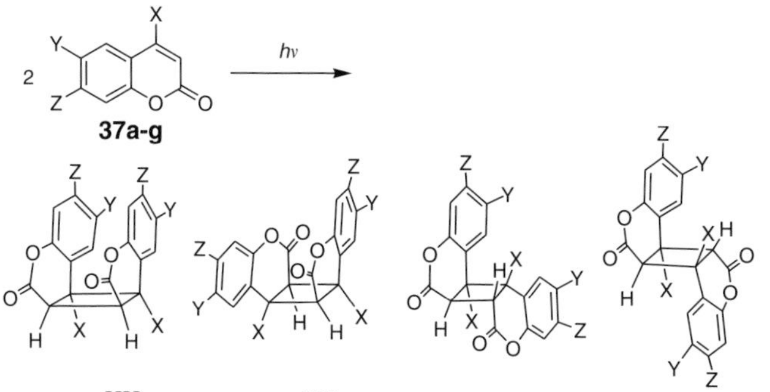

coumarins	X	Y	Z
37a	H	H	H
37b	H	H	CH$_3$
37c	H	H	OCH$_3$
37d	CH$_3$	H	CH$_3$
37e	CH$_3$	CH$_3$	H
37f	CH$_3$	H	OCH$_3$
37g	CH$_3$	H	OH

syn- HH	*syn-* HT	*anti-* HH	*anti-* HT
38a-g	**39a-g**	**40a-g**	**41a-g**

Scheme 15.

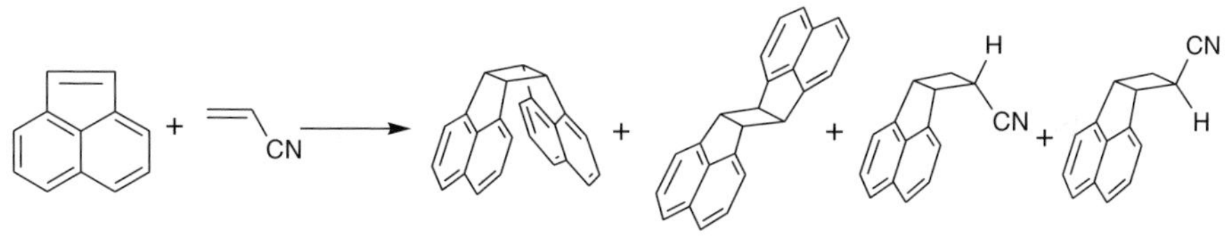

| **42a** | **42b** | **42c** | **42d** |

Scheme 16.

3.2. [4+4] Photocyclodimerization of anthracenecarboxylic acid

Tamaki *et al.* demonstrated that water-soluble anthracene derivatives, such as 2-anthracenesulfonate and 2-anthracenecarboxylate, form 2:2 complexes with β-CD and 2:1 complexes with γ-CD in aqueous solutions [86,87]. The [4+4] photocyclodimerizations of anthracene derivatives are greatly facilitated by more than eight times in the presence of β- or γ-CD.

Close examination of the photocyclodimerization of 2-anthracenecarboxylate **43** (Scheme 17) complexed by γ-CD derivatives has recently been done, as a consequence of the successful HPLC resolution of all of the photocyclodimers [88]. γ-CD forms a very stable 1:2 inclusion complex with **43** ($K_1 = 161 \, M^{-1}$ and $K_2 = 38,500 \, M^{-1}$ at 298 K). The chemical yield of *syn*-HT cyclodimer **44b** obtained upon photocyclodimerization is appreciably increased in the presence of γ-CD at the expense of HH cyclodimers **44c** and **44d**. The photocyclodimerization of **43** in the presence of γ-CD affords the major product **44b** in good ee's of up to 41%, while the ee of **44c** is consistently low (<5% ee).

Endeavors have been devoted to this photochirogenesis in order to improve the chemical and optical yields of the *anti*-HH cyclodimer **44c**. Similar approaches

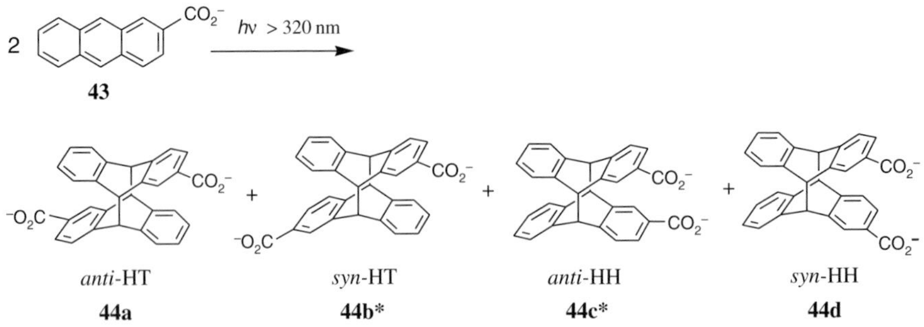

Scheme 17.

were employed independently by two groups to introduce two cationic groups on the primary face of γ-CD, so to control the orientation of the two guests accommodated in the CD cavity through electrostatic interactions [89–91]. Ikeda *et al.* reported that, by using dipyridinio-appended γ-CDs **45a–d** as chiral hosts (Scheme 18), the yield of **44c** is enhanced by a factor of 1.7 and the ee reached 13%, which is 10 times higher than that obtained with native γ-CD.

Cheng and Inoue examined the effect of solvent and temperature on the photocyclodimerization mediated by regioisomers of $6^A,6^X$-diamino-γ-CDs **46a–d** (Scheme 18). As shown in Table 1, the use of the diamino-γ-CDs greatly improves the yields of HH cyclodimers. By lowering the reaction temperature and solvent polarity, the yields of HH photocyclodimers are further increased as a result of the enhanced complexation and electrostatic interactions. By changing the host from **45a** to **45d**, the relative yield of **44c** is gradually increased with an accompanying reduction of the yield of **44d**, in good agreement with the increase in the inter-amino distance on the CD rim, thus exhibiting an excellent structure–function relationship (Fig. 3). Varying the temperature or solvent significantly affects the product's ee and even results in an inversion of the sense of product chirality. In an extreme case, i.e. the photocyclodimerization mediated by **46b** in 50% aqueous methanol, the ee of cyclodimer **44c** steadily decreases from 31% at 328 K to –4% at 263 K, and then to –27% at 228 K.

The temperature effect on the product's ee was analyzed for native and modified γ-CD-mediated photocyclodimerizations by using the differential Eyring equation, to give a good linear relationship in all cases as exemplified in Fig. 4. This result reveals that both the differential activation enthalpy and entropy are jointly responsible for the critical temperature dependence of the ee, and the dramatic temperature switching of product chirality. This is significantly different from the low entropy-dependence behavior observed for the enantiodifferentiating isomerization of cyclooctene included and sensitized by non-methylated β-CD derivatives [30,31], but can be reasonably interpreted in terms of the higher skeletal flexibility

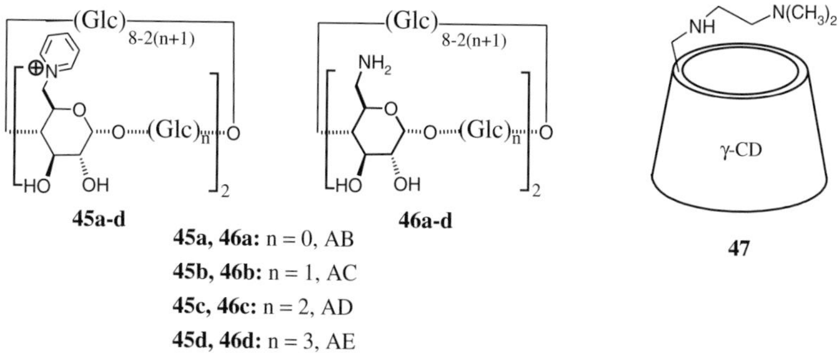

Scheme 18.

45a, 46a: n = 0, AB
45b, 46b: n = 1, AC
45c, 46c: n = 2, AD
45d, 46d: n = 3, AE

Table 1
Photocyclodimerization of **43** in the presence of γ-CD or **46a–d**

Host	Temp (K)	Solvent/vol(%)		Relative yield (%)				% ee	
		Water	MeOH	44a	44b	44c	44d	44b	44c
γ-CD	273	100	0	42.9	45.5	6.9	4.7	37.1	−0.8
46a	273	100	0	39.0	35.3	10.9	14.8	−4.0	4.2
46b	273	100	0	38.9	33.2	13.4	14.5	−1.5	−5.8
46c	273	100	0	37.4	36.2	14.9	11.5	15.4	−0.2
46d	273	100	0	37.3	37.1	16.5	9.1	20.3	−4.2
γ-CD	228	50	50	48.4	32.8	12.6	6.3	32.8	−11.9
46a	228	50	50	39.6	22.0	17.9	20.5	−9.9	−12.4
46b	228	50	50	38.7	19.3	21.6	20.4	−15.9	−26.9
46c	228	50	50	36.3	17.1	30.2	15.4	−0.7	−23.0
46d	228	50	50	35.2	21.8	31.9	11.0	9.2	−19.8

of γ-CD skeleton (relative to smaller α- and β-analogues), and the larger conformational freedoms in the termolecular complex with anthracenecarboxylates. It is noted that altering external entropy factors, such as temperature and solvent, does not influence the enantiodifferentiating mechanism, as demonstrated by the good compensatory relationship between the differential entropy ($\Delta\Delta S^{\ddagger}$) and enthalpy changes ($\Delta\Delta H^{\ddagger}$).

Intriguingly, when the photocyclodimerization was mediated by host **47** (Scheme 18), in which a flexible dicationic group are tethered to the primary face of γ-CD, the HH/HT selectivity is switched from 0.14 to 3.0, affording the HH dimers in 74.4% combined yield, and the ee of **44c** is also enhanced up to 41% [90].

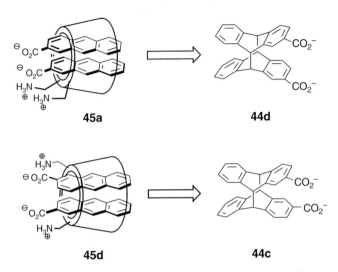

Fig. 3. Stereoselectivity controlled by inter-amino distances in 6^A, 6^X-diamino-γ-CDs.

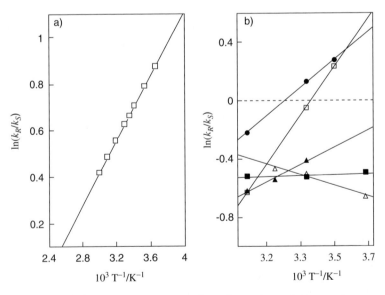

Fig. 4. Plots of the relative rate constants against the reciprocal temperature upon photo-cyclodimerization of **43** for the formation of (a) enantiomers of **44b** in the presence of native γ-CD in aqueous solution and (b) enantiomers of **44d** mediated by **46b** in aqueous solution containing 10% ($\bullet$), 30%($\square$), 50%($\blacktriangle$), 60%($\blacksquare$), and 80%($\triangle$) methanol.

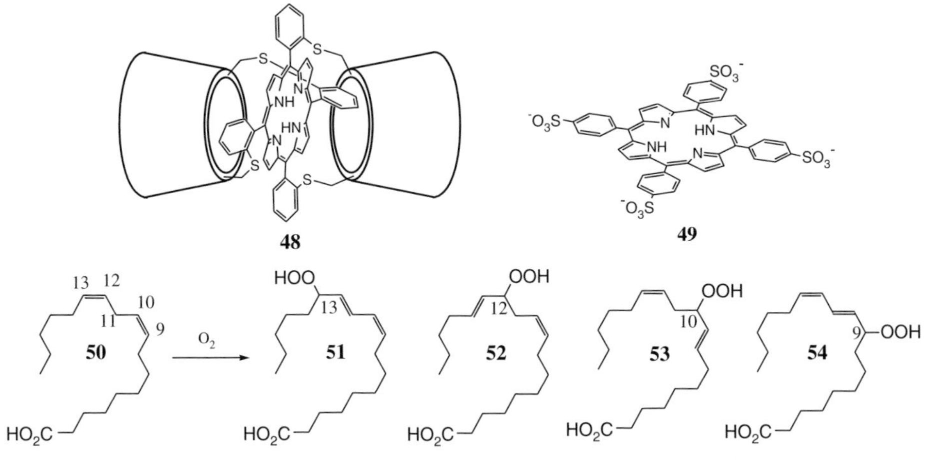

Scheme 19.

3.3. Singlet oxygenation of linoleic acid sensitized by CD-sandwiched porphyrin

Kuroda *et al.* studied the regio- and stereoselective oxygenation of linoleic acid by using the CD-sandwiched porphyrin **48** (Scheme 19) as a singlet oxygen generator [92]. Hydroperoxydienes **51–54** are produced as the major products upon the 'ene' reaction of singlet oxygen with linoleic acid. Photosensitized oxygenation of linoleic acid with tetrakis(*p*-sulfonatophenyl)porphyrin **49**, as a reference sensitizer, leads to a non-selective formation of regioisomeric **51/52** and **53/54**. Even in the presence of saturated *β*-CD, the oxygenation products obtained upon photosensitization with **49** are racemic. In contrast, the photo-oxygenation of **50** sensitized by an equimolar amount of **48** gives **51** in 20% ee and **52** in 12% ee with an 82% combined yield. The regio- and enantioselectivity in this supramolecular photochirogenic oxygenation was ascribed to the co-inclusion of singlet oxygen and **50** in the hydrophobic pocket of **48** during the ene reaction.

3.4. Paternó–Büchi reaction

The Paternó-Büchi reaction of 5-substituted adamantan-2-ones **55** with fumaro-nitrile to form oxetanes **56** and **57** (Scheme 20) has been studied by Turro and coworkers [93,94]. In the CD-free photocycloaddition performed in acetonitrile or water, the *anti*-oxetane **56** formed through the *syn*-face attack of fumaronitrile is slightly favored, with the **57/56** ratio varying from 47/53 to 36/64. When the photocycloaddition reaction is carried out in the presence of *β*-CD in aqueous solution, the stereochemical preference is switched to *syn*-oxetane **57** and the

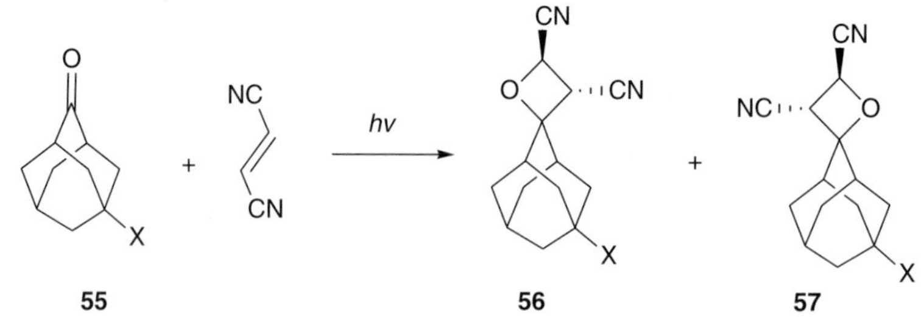

X= F, Cl, Br, OH etc.

Scheme 20.

57/56 ratio increases up to 86/14, as a result of the interference of the *syn*-face attack to **55** included in β-CD cavity [95]. In contrast, the use of α or γ-CD does not appreciably alter the product distribution.

4. Discussion and conclusions

Despite the relatively rapid growth particularly in recent years, supramolecular photochirogenesis with CDs has not yet been extensively explored in either depth or breadth. Readily available native CDs have been most frequently employed as chiral hosts, but the optical yields hitherto obtained with native CDs are only modest, in general. This implies that the chirality transfer controlled mainly by the non-directional hydrophobic interactions within the axially symmetrical CD cavity is not very efficient, considering the fact that there are more than 30 chiral centers symmetrically distributed around the CD's inner walls. The use of modified or functionalized CDs as chiral inductors has been demonstrated to be one of the most promising strategies for overcoming the limitations of native CDs. By adopting this approach, the size, shape, flexibility and chiral nature of the CD cavity are altered, distorted and made less symmetrical, as a result the chemo-, regio- and/or stereoselectivities are often improved indeed [29,30,89–92].

In most cases, the optical yield obtained in the solid-state photolysis is higher than that in solution. This can be accounted for in terms of two main reasons: first, in the solid-state complexes, particularly those prepared by the precipitating technique, almost all of the guest molecules are included in the CD cavities, and the unfavorable competitive reaction of free guests, leading to racemic products, can be entirely excluded. Second, in contrast to the larger positional and rotational freedoms allowed for the guest accommodated in a solution-phase CD complex, the more rigid alignment of guest in a solid-state CD complex leads to higher chemo/regio/ stereoselectivities. This is inline with the conventional concept that the more specific the recognition the higher the selectivity in subsequent reaction; and also with the wildly accepted concept in enzymatic reactions, in which the substrate is precisely positioned and oriented in a specifically designed chiral environment. However, our recent studies indicate that the *flexibility* of supramolecular photochirogenic systems offer a unique opportunity to control the product's ee and chiral sense [32,89]. In contrast to the rigid molecular recognition system, which is usually associated with a large enthalpy change, a flexible host accompanies a larger change in freedom upon complexation, and is therefore more entropy dependent. One can manipulate the entropy-related external factors, such as solvent, temperature and pressure, to modify or even invert the stereoselectivity of a reaction without using an antipodal chiral inductor or host, which would be beneficial when the antipodal host is not readily available, as is the case with CDs. In addition to well-documented entropy-controlled asymmetric photoreactions [2,3], CD-mediated supramolecular photochirogenic reactions are shown to exhibit vital entropic contributions in the ground- and excited-state interactions. Thus, as far as weak interactions are concerned, such phenomena may accelerate the entropy-oriented design of conformationally flexible supramolecular recognition and reaction systems.

References

[1] H. Rau, Chem. Rev. 83 (1983) 535.
[2] Y. Inoue, Chem. Rev. 92 (1992) 741.
[3] Y. Inoue, V. Ramamurthy, Chiral Photochemistry, Marcel Dekker, New York, 2004.
[4] P. Bortolus, G. Gràbner, G. Kohler, S. Monti, Coord. Chem. Rev. 125 (1993) 261.
[5] K. Takahashi, Chem. Rev. 98 (1998) 2013.
[6] A. Joy, V. Ramamurthy, Chemistry 6 (2000) 1287.
[7] J. Sivaguru, A. Natarajan, L.S. Kaanumalle, J. Shailaja, S. Uppili, A. Joy, V. Ramamurthy, Acc. Chem. Soc. 36 (2003) 509.
[8] T. Aechtner, M. Dressel, T. Bach, Angew. Chem. Int. Ed. 43 (2004) 5849.
[9] T. Bach, H. Bergmann, B. Grosch, K. Harms, J. Am. Chem. Soc. 124 (2002) 7982.
[10] A. Bauer, F. Weskamper, S. Grimme, T. Bach, Nature 436 (2005) 1139.
[11] Y. Inoue, Nature 436 (2005) 1099.
[12] T. Wada, M. Nishijima, T. Fujisawa, N. Sugahara, T. Mori, A. Nakamura, Y. Inoue, J. Am. Chem. Soc. 125 (2003) 7492.
[13] N. Sugahara, M. Kawano, T. Wada, Y. Inoue, Nucleic Acids Symp. Ser. 44 (2000) 115.
[14] V. Lhiaubet-Vallet, Z. Sarabia, F. Bosca, M.A. Miranda, J. Am. Chem. Soc. 126 (2004) 9538.
[15] Y. Gao, T. Wada, K. Yang, K. Kim, Y. Inoue, Chirality 17 (2005) 19.
[16] M.V. Rekharsky, Y. Inoue, Chem. Rev. 98 (1998) 1875.
[17] R. Breslow, S.D. Dong, Chem. Rev. 98 (1998) 1997.
[18] H. Aoyama, K. Miyazaki, M. Sakamoto, Y. Omote, Tetrahedron 43 (1987) 1513.
[19] V.P. Rao, N.J. Turro, Tetrahedron Lett. 30 (1989) 4641.
[20] C.A. Stanier, S.J. Alderman, T.D.W. Claridge, H.L. Anderson, Angew. Chem. Int. Ed. 41 (2002) 1769.
[21] S.S. Petrova, A.I. Kruppa, T.V. Leshina, Chem. Phys. Lett. 407 (2005) 260.
[22] V. Jullian, F. Courtois, G. Bolbach, G. Chassaing, Tetrahedron Lett. 44 (2003) 6437.

[23] M.G. Rosenberg, U.H. Brinker, J. Org. Chem. 68 (2003) 4819.
[24] A.C. Cope, C.R. Ganellin, H.W.J. Johnson, H.J.S. Winkler, J. Am. Chem. Soc. 85 (1963) 3276.
[25] Y. Inoue, Y. Kunitomi, S. Takamuku, H. Sakurai, Chem. Commun. (1978) 1024.
[26] Y. Inoue, T. Wada, S. Asaok, H. Sato, J.-P. Pete, Chem. Commun. (2000) 251.
[27] Y. Inoue, N. Sugahara, T. Wada, Pure Appl. Chem. 73 (2001) 475.
[28] Y. Inoue, S. Kosaka, K. Matsumoto, H. Tsuneishi, T. Hakushi, A. Tai, K. Nakagawa, L. Tong, J. Photochem. Photobiol. A: Chem. 71 (1993) 61.
[29] Y. Inoue, S.F. Dong, K. Yamamoto, L.-H. Tong, H. Tsuneishi, T. Hakushi, A. Tai, J. Am. Chem. Soc. 113 (1995) 2793.
[30] Y. Inoue, T. Wada, N. Sugahara, K. Yamamoto, K. Kimura, L.-H. Tong, X.-M. Gao, Z.-J. Hou, Y. Liu, J. Org. Chem. 65 (2000) 8041.
[31] Y. Gao, M. Inoue, T. Wada, Y. Inoue, J. Incl. Phenom. Macro. 50 (2004) 111.
[32] G. Fukuhara, T. Mori, T. Wada, Y. Inoue, Chem. Commun. (2005) 4199.
[33] G.S. Hammond, R.S. Cole, J. Am. Chem. Soc. 87 (1965) 3256.
[34] L. Horner, J. Klaus, Liebigs Ann. Chem. 1979 (1979) 1232.
[35] L. Horner, J. Klaus, Liebigs Ann. Chem. 1981 (1981) 792.
[36] M. Vondenhof, J. Mattay, Chem. Ber. 123 (1990) 2457.
[37] Y. Inoue, H. Shimoyama, N. Yamasaki, A. Tai, Chem. Lett. 20 (1991) 593.
[38] Y. Inoue, N. Yamasaki, T. Yokoyama, A. Tai, J. Org. Chem. 57 (1992) 1332.
[39] L.S. Kaanumalle, J. Sivaguru, R.B. Sunoj, P.H. Lakshminarasimhan, J. Chandrasekhar, V. Ramamurthy, J. Org. Chem. 67 (2002) 8711.
[40] J. Sivaguru, T. Shichi, V. Ramamurthy, Org. Lett. 4 (2002) 4221.
[41] J. Sivaguru, J.R. Scheffer, J. Chandarasekhar. V. Ramamurthy, Chem. Commun. (2002) 830.
[42] K.C. Chong, J. Sivaguru, T. Shichi, Y. Yoshimi, V. Ramamurthy, J.R. Scheffer, J. Am. Chem. Soc. 124 (2002) 2858.
[43] J. Sivaguru, R.B. Sunoj, T. Wada, Y. Origane, Y. Inoue, V. Ramamurthy, J. Org. Chem. 69 (2004) 5528.
[44] J. Sivaguru, T. Wada, Y. Origane, Y. Inoue, V. Ramamurthy, Photochem. Photobiol. Sci. 4 (2005) 119.
[45] S. Koodanjeri, V. Ramamurthy, Tetrahedron Lett. 43 (2002) 9229.
[46] H. Takeshita, M. Kumamoto, I. Kouno, Bull. Chem. Soc. Jpn. 53 (1980) 1006.
[47] S. Koodanjeri, A. Joy, V. Ramamurthy, Tetrahedron 56 (2000) 7003.
[48] E.J. Corey, J. Streith, J. Am. Chem. Soc. 86 (1964) 950.
[49] J. Shailaja, S. Karthikeyan, V. Ramamurthy, Tetrahedron Lett. 43 (2002) 9335.
[50] K. Vizvardi, K. Desmet, I. Luyten, P. Sandra, G. Hoornaert, E. Van der Eycken, Org. Lett. 3 (2001) 1173.
[51] M. Irie, Chem. Rev. 100 (2000) 1685.
[52] J.J. de Jong, L.N. Lucas, R.M. Kellogg, J.H. van Esch, B.L. Feringa, Science 304 (2004) 278.
[53] T. Kodani, K. Matsuda, T. Yamada, S. Kobatake, M. Irie, J. Am. Chem. Soc. 122 (2000) 9631.
[54] T. Yamaguchi, K. Uchida, M. Irie, J. Am. Chem. Soc. 119 (1997) 6066.
[55] M. Takeshita, M. Irie, Tetrahedron Lett. 40 (1999) 1345.
[56] M. Takeshita, N. Kato, S. Kawauchi, T. Imase, J. Watanabe, M. Irie, J. Org. Chem. 63 (1998) 9306.
[57] M. Takeshita, C. Choi, M. Irie, Chem. Commun. (1997) 2265.
[58] H. Aoyama, T. Hasegawa, M. Watabe, H. Shiraishi, Y. Omote, J. Org. Chem. 43 (1978) 419.
[59] H. Aoyama, T. Hasegawa, Y. Omote, J. Am. Chem. Soc. 101 (1979) 5343.
[60] R. Chenevert, N. Voyer, Tetrahedron Lett. 25 (1984) 5007.

[61] S. Koodanjeri, A.R. Pradhan, L.S. Kaanumalle, V. Ramamurthy, Tetrahedron Lett. 44 (2003) 3207.
[62] R.A. Finnegan, D. Knutson, J. Am. Chem. Soc. 89 (1967) 1970.
[63] T. Mori, M. Takamoto, T. Wada, Y. Inoue, Photochem. Photobiol. Sci. 2 (2003) 1187.
[64] T. Mori, T. Wada, Y. Inoue, Org. Lett. 2 (2000) 3401.
[65] T. Mori, Y. Inoue, R.G. Weiss, Org. Lett. 5 (2003) 4661.
[66] T. Mori, R.G. Weiss, Y. Inoue, J. Am. Chem. Soc. 126 (2004) 8961.
[67] T. Mori, H. Saito, Y. Inoue, Chem. Commun. (2003), 2302.
[68] E.T. Kool, R. Breslow, J. Am. Chem. Soc. 110 (1988) 1596.
[69] M. Cocivera, M.T. Anthony, J. Am. Chem. Soc. 92 (1970) 1772.
[70] G.L. Closs, D.R. Paulson, J. Am. Chem. Soc. 92 (1970) 7229.
[71] N.J. Turro, Pure Appl. Chem. 67 (1995) 199.
[72] N.J. Turro, M.H. Kleinman, E. Karatekin, Angew. Chem. Int. Ed. 39 (2000) 4436.
[73] N.R. Bantu, T.G. Kotch, A.J. Lees, Tetrahedron Lett. 34 (1993) 2039.
[74] D.G. Amirsakis, M.A. Garcia-Garibay, S.J. Rowan, J.F. Stoddart, A.J.P. White, D.J. Williams, Angew. Chem. Int. Ed. 40 (2001) 4256.
[75] S.Y. Jon, Y.H. Ko, S.H. Park, H-J. Kim, K. Kim, Chem. Commun. (2001) 1938.
[76] M. Pattabiraman, A. Natarajan, S.K. Lakshmi, V. Ramamurthy, Org. Lett. 7 (2005) 529.
[77] M. Yoshizawa, Y. Takeyama, T. Kusukawa, M. Fujita, Angew. Chem. Int. Ed. 41 (2002) 1347.
[78] W. Herrmann, S. Wehrle. G. Wenz, Chem. Commun. (1997) 1709.
[79] K.S. Rao, S.M. Hubig, J.N. Moorthy, J.K. Kochi, J. Org. Chem. 64 (1999) 8098.
[80] H.S. Banu, A. Lalitha, K. Pitchumani, C. Srinivasan, Chem. Commun. (1999) 607.
[81] F. Hirayama, T. Utsuki, K. Uekama, Chem. Commun. (1991) 887.
[82] J.N. Moorthy, K. Venkatesan, R.G. Weiss, J. Org. Chem. 57 (1992) 3292.
[83] T.J. Brett, J.M. Alexander, J.J. Stezowski, J. Chem. Soc. Perkin Trans. 2 (2000) 1095.
[84] T.J. Brett, J.M. Alexander, J.J. Stezowski, J. Chem. Soc. Perkin Trans. 2 (2000) 1105.
[85] Z. Tong, Z. Zhen, Youji Huaxue (1986) 44.
[86] T. Tamaki, T. Kokubu, K. Ichimura, Tetrahedron Lett. 43 (1987) 1485.
[87] T. Tamaki, T. Kokubu, J. Incl. Phenom. Macro. 2 (1984) 815.
[88] A. Nakamura, Y. Inoue, J. Am. Chem. Soc. 125 (2003) 966.
[89] C. Yang, G. Fukuhara, A. Nakamura, Y. Origane, K. Fujita, D.-Q. Yuan, T. Mori, T. Wada, Y. Inoue, J. Photochem. Photobiol. A: Chem. 173 (2005) 375.
[90] A. Nakamura, Y. Inoue, J. Am. Chem. Soc. 127 (2005) 5338.
[91] H. Ikeda, T. Nihei, A. Ueno, J. Org. Chem. 70 (2005) 1237.
[92] Y. Kuroda, T. Sera, H. Ogoshi, J. Am. Chem. Soc. 113 (1991) 2793.
[93] W-S. Chung, N.J. Turro, J. Silver, W.J.l. Noble, J. Am. Chem. Soc. 112 (1990) 1202.
[94] W.-S. Chung, N.J. Turro, S. Srivastava, H. Li, W.J. le Noble, J. Am. Chem. Soc. 110 (1988) 7882.
[95] A. Entrena, C. Jaime, J. Org. Chem. 62 (1997) 5923.

Cyclodextrin Materials Photochemistry, Photophysics and Photobiology
Abderrazzak Douhal (Editor)
DOI 10.1016/S1872-1400(06)01012-0

Chapter 12

Chemosensors for Detecting Molecules using Modified Cyclodextrins and Cyclodextrin–Peptide Conjugates

Hiroshi Ikeda, Akihiko Ueno

Department of Bioengineering, Graduate School of Bioscience and Biotechnology, Tokyo Institute of Technology, 4259-B-44 Nagatsuta-cho, Midori-ku, Yokohama 226-8501, Japan

1. Introduction

There has been considerable interest in the development of new methods of chemical sensing to recognize molecules or ions as important application of supramolecular chemistry [1–3]. Our group has prepared many kinds of chromophore-modified cyclodextrins (CD) as chemosensors for detecting molecules [4–8]. Colorless neutral molecules can be detected as changes in the intensities of the fluorescence, absorption, or circular dichroism using chemosensors based on chromophore-modified CD.

The discovery of the fluorescence variation of pyrene-modified γ-CD **1** upon addition of guests prompted the use of fluorophore-modified CD as chemosensors [9,10]. The pyrene-modified γ-CD forms an association dimer in aqueous solution and exhibits excimer emission around 470 nm. The addition of a guest into the solution of the pyrene-modified γ-CD causes a remarkable decrease in the excimer emission and an increase in the monomer emission around 378 and 397 nm. The guest-induced circular dichroism and absorption variations of **1** substantiated that this fluorescence variation occurs associated with the conversion from the association dimer to a 1:1 host–guest complex (Fig. 1). Since the

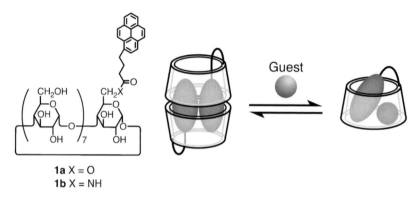

Fig. 1. Guest-induced conformational change of pyrene-modified γ-cyclodextrin.

guest-induced fluorescence variation is greatly influenced by the size and shape of the guest species, it is suggested that this system can be used as a molecular recognition-based sensory system for detecting various organic compounds. After this finding, our group has continued to develop improved chemosensors based on CD. This review presents the molecule sensing abilities of a number of modified CD, CD–peptide conjugates, and CD–protein conjugates.

2. Conformational change of dansyl-leucine-modified cyclodextrins

Ikeda *et al.* [11,12] prepared the fluorophore–amino acid–CD triad systems, *N*-dansyl-L-leucine-modified and *N*-dansyl-D-leucine-modified β-CD (**2**, **3**) and *N*-dansyl-L-leucine-modified and *N*-dansyl-D-leucine-modified γ-CD (**4**, **5**), and characterized their ability to act as fluorescent chemosensors for molecular detection (Fig. 2). Ikeda *et al.* examined the impact of the hydrophobic side chain of the amino acid and the leucine chirality on the guest binding. Conformational changes of dansyl-leucine-modified CD upon addition of guests were first investigated by fluorescence decay analysis and NMR techniques. A detailed summary of this investigation is given below, as it is a useful example to understand the sensing behavior of the fluorophore-modified CD.

Although the chromophore-modified CD adopts several conformations in aqueous solution, we can explain the conformational equilibrium by the simplified two-state model, consisting of a self-inclusion state and non-self-inclusion state (Fig. 3). Although the conformational interconversion occurs too rapidly to be followed by NMR spectroscopy, the analysis of the fluorescence decay of the fluorophore pendant can provide useful information with respect to the conformational features, as the fluorescence lifetimes of many fluorophores are in the range of a measurable timescale (nanoseconds). The dansyl (DNS) moiety is sensitive to the hydrophobicity around its local environment and has

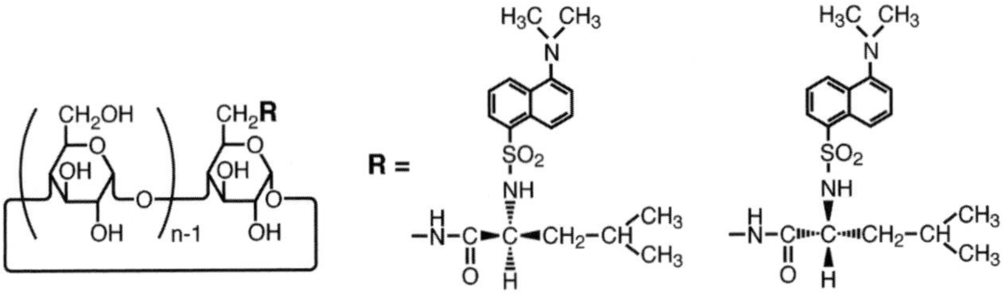

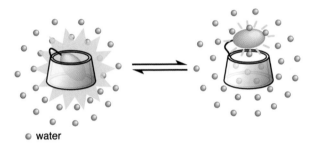

Fig. 2. Dansyl-leucine-modified cyclodextrins.

Fig. 3. Conformational equilibrium of modified cyclodextrin.

different lifetimes when located inside the cavity and in bulk water solution. All the fluorescence decays of the dansyl moiety could be expressed as a simple double-exponential function, and the results are summarized in Table 1. This means that there are two kinds of observable conformational isomers and these two species are in equilibrium as shown in Fig. 3. The longer and shorter lifetime species arise from the dansyl moiety being inside and outside the cavity, respectively.

The fluorescence decay ($D(t)$) is expressed by

$$D(t) = A_1\exp(-t/\tau_1) + A_2\exp(-t/\tau_2) \tag{1}$$

where A_i is a preexponential factor contributing to the signal at zero time and τ_i is the lifetime of the ith conformational state.

A_1/A_2 can be also written as

$$A_1/A_2 = (\varepsilon_1 C_1 \Phi_1/\tau_1)/(\varepsilon_2 C_2 \Phi_2/\tau_2) = (\varepsilon_1 C_1/\tau_{01})/(\varepsilon_2 C_2/\tau_{02}) \tag{2}$$

where the C_i is the concentration of the conformational state i, ε_i the molar extinction coefficient, Φ_i the quantum efficiency, and τ_{0i} the intrinsic fluorescence lifetime.

If $\varepsilon_1 = \varepsilon_2$ and $\tau_{01} = \tau_{02}$ \tag{3}

Table 1
Fluorescence lifetimes of **2–5** alone and in the presence of guest in aqueous solution[a]

Host	Guest	$\tau_1/(ns)$	A_1	$\tau_2/(ns)$	A_2	χ^2
2	—	5.7	0.33	17.7	0.67	1.32
3	—	6.9	0.23	17.8	0.77	1.38
4	—	7.3	0.46	13.1	0.54	1.13
5	—	7.9	0.78	13.5	0.22	1.06
2	1-AdOH	5.5	0.99	19.0	0.01	1.38
3	1-AdOH	5.9	0.97	17.9	0.03	1.07
4	(-)-Bor	5.5	0.57	13.7	0.43	1.23
5	(-)-Bor	5.7	0.86	13.2	0.14	1.03
4	c-HexOH	7.6	0.40	14.8	0.60	1.22
5	c-HexOH	7.3	0.45	14.1	0.55	0.95

[a]$[2-5] = 2 \times 10^{-5}$ M, $[1\text{-AdOH}] = 5 \times 10^{-3}$ M, $[(-)\text{-Bor}] = 3 \times 10^{-3}$ M, $[\text{c-HexOH}] = 5 \times 10^{-2}$ M. Decay curves were fitted to the equation: $D(t) = A_1 \exp(-t/\tau_1) + A_2 \exp(-t/\tau_2)$. χ^2 is the parameter for the goodness of the fit.

then

$$A_1/A_2 = C_1/C_2 \tag{4}$$

Since the fluorescence in the self-inclusion state and in the nonself-inclusion state arises from the same moiety, it is reasonable to consider the intrinsic fluorescence lifetimes as approximately equal [13–17]. Therefore, the equilibrium can be quantified from parameters obtained directly from the analysis of fluorescence decay curves. Although it is reported that τ_0 of the dansyl moiety is weakly dependent on solvent polarity and thus Eq. (4) is only an approximation, it is useful for discussion of the equilibrium semi-quantitatively [15].

The data in Table 1 reveal that the fractions of the longer lifetime forms of the β-CD derivatives are larger than those of the γ-CD derivatives. It suggests that the self-inclusion forms of the β-CD derivatives are more stable than those of the γ-CD derivatives and that the cavity of γ-CD is too large to form stable self-inclusion complexes of the DNS unit. The chirality of the leucine moiety also affects stability of the self-inclusion form. The self-inclusion form of **3** having the D-Leu unit is more stable than that of **2** having the L-Leu unit, while the self-inclusion form of **4** having the L-Leu unit is more stable than that of **5** having the D-Leu unit.

Upon addition of 1-adamantanol (1-AdOH) as a guest, the fractions of the shorter lifetime forms of **2** and **3** increased, while the fractions of the longer lifetime forms of the hosts decreased (Table 1). These data indicate that the induced-fit conformational change of the chromophore-modified CD occurs in association with the guest accommodation, excluding the chromophore from the inside to the outside of the CD cavity (Fig. 4a). The fact that the fluorescence decay curves of the hosts are still double-exponential indicates that the

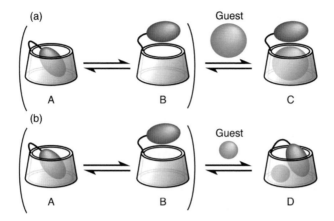

Fig. 4. Conformational equilibria in aqueous solution and guest-induced conformational change of chromophore-modified cyclodextrins; (a) larger guest, (b) smaller guest.

fluorescence lifetimes of species B and C in Fig. 4a are similar to each other and not resolved by our instrument for a lifetime measurement.

On the other hand, the fractions of the longer lifetime components of **4** and **5** increased upon addition of cyclohexanol but decreased upon addition of (-)-borneol (Table 1). These results suggest that the dansyl moiety of **4** and **5** is co-included with cyclohexanol in the cavity (Fig. 4b), whereas it is excluded by (-)-borneol (Fig. 4a). Since the cavity of γ-CD is too large to form a stable intramolecular complex with the dansyl moiety, a small guest such as cyclohexanol may act as a spacer to form stable complexes [11]. However, (-)-borneol is too large to act as a spacer and accommodation of the guest displaces the chromophore from the inside to the outside of the CD cavity.

Although the observed NMR spectra are the averages of the spectra of the species in various conformational states weighted by the fractions of the conformers, the NMR spectra of **2** and **3** provide information about their structures in the self-inclusion state, because the dansyl moieties of **2** and **3** are located in the cavity in the major conformational state. Ikeda *et al.* [12] fully assigned the ^{1}H NMR spectra of **2** and **3** to elucidate their structures by combined use of various 1D and 2D NMR techniques. The structures of **2** and **3** were estimated from the NOE data and the degree of the anisotropic ring current effect from the dansyl moiety in the ^{1}H NMR spectra of the CD protons. The dansyl moiety of **3** is located deeper within the cavity than that of **2**. This deeper inclusion into its cavity makes **3** the more stable self-inclusion complex, and this higher stability is also reflected in their binding abilities.

The difference in the stability of the self-inclusion states produces the difference in the binding abilities, because the guest binding is in competition with binding of the dansyl moiety. In fact, the difference in the stability of the self-inclusion complex between **2** and **3** are estimated to be 1.7-fold from the fluorescence lifetime data and it is almost consistent with the difference in the binding ability (i.e. about half or less) [11].

3. Molecule sensing properties of fluorescent cyclodextrins

We have been using a sensitivity parameter expressed as $\Delta I/I_0$ (where $\Delta I = I - I_0$ with I and I_0 being the fluorescence intensities in the presence and absence of the guest, respectively) to evaluate the sensitivities and molecular recognition abilities of the fluorescent CD [11]. Figure 5 shows that **2** and **3** exhibit similar trends in the guest dependency of the $\Delta I/I_0$ value, but that the $\Delta I/I_0$ value of **2** is larger than that of **3** in any case, because the binding constants of **2** are over two times larger than those of **3** for most of the guests. Four steroidal compounds such as ursodeoxycholic acid (UDCA), chenodeoxycholic acid (CDCA), deoxycholic acid (DCA), and cholic acid (CA) were detected by **2** and **3** with the sensitivity order CA < DCA < CDCA < UDCA, and this order is the same as the order of their binding constants. These compounds have the same steroidal framework, and UDCA and CDCA are isomers with a difference of stereochemistry of the hydroxy group at C-7. Compound DCA is a regioisomer of UDCA and CDCA with one hydroxy group at C-12 in place of the C-7 of UDCA and CDCA. Compound CA has one more hydroxy group than UDCA, CDCA, and CA. These differences were reflected in the different sensitivity values for these steroids. 1-Adamantanecarboxylic acid (1-AdCOOH) and 1-adamantanol (1-AdOH) are also known as good guests for β-CD, and here, they give large $\Delta I/I_0$ values.

4. Cyclodextrin dimers and trimers as chemosensors

As coupling two or three CD offers a larger hydrophobic pocket, CD dimers and trimers would have larger binding affinities for bile acids, thus improving

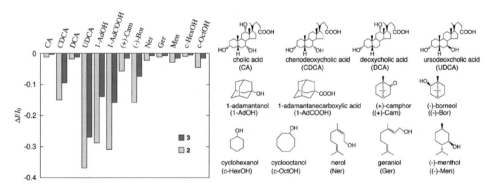

Fig. 5. Sensitivity parameters ($\Delta I/I_0$) of **2** and **3** (2×10^{-6} M) for various guests (1×10^{-5} M).

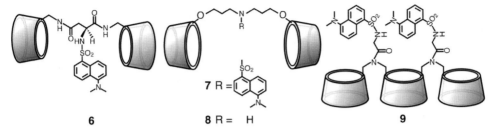

Fig. 6. Dansyl-modified cyclodextrin dimers and trimer.

their sensitivity of molecule recognition abilities. Due to the presence of multiple binding sites on CD dimer and trimer systems, accommodation of a guest in a CD dimer or trimer can occur concurrently with the fluorophore accommodation. However, guest accommodation in the cavity of the fluorophore-modified CD, in most cases, causes displacement of the chromophore from the inside to the outside of the CD cavity as shown in the previous section. It would be expected that this type of host/guest/fluorophore complex formation would influence the sensing selectivity. Nakamura *et al.* [18] prepared a fluorescent CD dimer **6** (Fig. 6). **6** shows markedly different values of $\Delta I/I_0$ for a range of bile acids. $\Delta I/I_0$ is -0.54, -0.21, and $+0.03$ for UDCA, CDCA, and DCA, respectively. No change in fluorescence was observed for CA. The least-squares curve-fitting analysis of the fluorescence titration for the addition of guests suggests that **6** forms 1:1 inclusion complexes with the guests. This indicates that two CD cavities bind the guest cooperatively. Cooperative binding causes exclusion of the fluorophore, decreasing the fluorescence intensity for UDCA and CDCA. On the other hand, the weak responses of **6** for DCA and CA suggest that the guest accommodation occurs along with accommodation of the fluorophore. Additionally, the fluorophore would act as a spacer, aiding the accommodation of DCA.

A fluorescent CD dimer **7** was prepared by de Jong *et al.* [19] in which two CD units are connected through their secondary hydroxy sides (Fig. 6). The guest binding behaviors of **7** and **8** were studied by a calorimetric titration. The binding constants of **8** for CA and DCA are 70- and 700-fold larger, respectively, than those of the native β-CD, while the binding constants of **7** for CA and DCA are only 9- and 50-fold larger, respectively. The lower binging affinities of **7** are mainly due to an unfavorable change in the binding entropy, while the binding enthalpy remains almost unchanged. CA and DCA are cooperatively bound by two CD cavities but the DNS moiety inhibits this cooperation. DCA and UDCA caused a decrease in the fluorescence intensity, while CA and CDCA caused an increase in the fluorescence intensity. During the addition of up to one equivalent of lithocholic acid (LCA), the fluorescence intensity decreases, however on continuing the addition the intensity increases as the equivalents of LCA exceeds one. This behavior suggests a 1:2 host/guest complex. The data for the fluorescence titrations with CA, DCA, CDCA, and UDCA are well fitted to a 1:1 binding model, although the data of calorimetric titration with CDCA and UDCA fit well to a 1:2 binding model. A 1:1 binding model for DCA and CDCA is also confirmed by independent Job's analyses. These results indicate that accommodation of a second guest does not lead to an additional change in fluorescence intensity.

Kikuchi et al. [20] prepared a bis-dansyl-modified CD trimer **9** as a sensory system for bile acids (Fig. 6). The sensitivity parameter ($\Delta I/I_0$) is -83, -192, and -268 for LCA, CDCA, and UDCA, respectively. The fluorescence intensity variations for DCA and CA are almost negligible. The guest-induced variation of the fluorescence intensity was fitted to the equation for a 1:1 complex. The binding constants of **9** for LCA, CDCA, UDCA, DCA, and CA are 153,000, 5300, 21,600, 4450, and 1830 M^{-1}, respectively. The order of binding constants of **9** for the guests does not correlate with the order of the sensitivity parameters. Not every guest displaces the chromophore from the inside to the outside of the CD cavity, because there are three binding sites. Some guests are co-included with the DNS unit and these inclusion complexes result in a weak variation in the fluorescence intensity.

5. Effect of protein environment on molecule sensing of chemosensors

Nakamura et al. [21,22] prepared a monensin-capped dansyl-modified β-CD **10** as a fluorescent chemosensor for molecule recognition (Fig. 7). Monensin is an antibacterial hydrophobic substance that can bind sodium cation to form a ring-like conformation with a sodium ion in its center. If monensin is introduced to the primary hydroxy side of the CD, it is expected that it will act as a hydrophobic cap that is responsive to sodium ion. The guest-dependent fluorescence

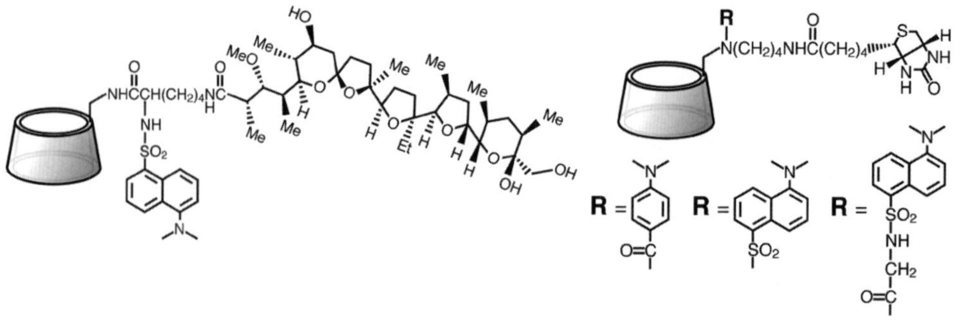

Fig. 7. Monensin-appended and biotin-appended cyclodextrins.

variation of **10** are enhanced by the addition of alkali metal cations in the order of $Na^+ > K^+ > Li^+ > Cs^+$, which is consistent with the expected order for binding of alkali metal cations by monensin itself. The binding constants of **10** for *l*-camphor, nerol, and geraniol increase 2-fold and that for cyclohexanol increase 7-fold on addition of sodium ions.

Wang *et al.* [23] prepared a modified *β*-CD **11**, which has *p-N,N*-dimethylaminobenzoyl and biotin units a fluorophore and a protein-binding site, respectively (Fig. 7). The fluorescence intensity of **11** is 8.4-fold enhanced by the addition of avidin, though this is only normal fluorescence around 375 nm and not TICT emissions. The binding constant of **11** for UDCA increases 8.9-fold by the addition of avidin. The fluorescence intensity of **11** in the presence of avidin increases on addition of guests, while it decreases for the same addition in the absence of avidin. This indicates that in the presence of avidin, accommodation of a guest into the CD cavity displaces the fluorophore unit from the CD cavity to the hydrophobic pocket of avidin instead of the bulk water solution. These results imply that avidin can act as a localized hydrophobic environment that leads to highly responsive sensors.

Ikunaga *et al.* [24] also prepared biotin-appended dansyl-modified CD **12–13** to examine the effects of avidin binding on guest-inclusion phenomena and fluorescence properties (Fig. 7). The fluorescence intensities of these CD were enhanced more than three times by the addition of avidin . The binding constants of **12** and **13** for CDCA increase 1.4-fold and 2.5-fold in the presence of avidin, respectively. The fluorescence decays of the dansyl moiety of **12** both in the presence and in the absence of avidin could be analyzed by simple double-exponential functions. The shorter and longer lifetimes of **12** alone are 7.5 and 14.5 ns, respectively. On the other hand, its shorter and longer lifetimes in the presence of avidin are 12.6 and 22.1 ns, respectively. Even in the presence of hyodeoxycholic acid (HDCA), its shorter and longer lifetimes in the presence of avidin are 12.4 and 20.0 ns, respectively. The ratio of the longer lifetime component of **12** in the presence of avidin increases from 7.3% to 36.2% upon addition of HDCA. These results indicate that the avidin forces the dansyl moiety into the CD cavity in the absence of the guest, and the fluorophore is displaced from the CD cavity to a very hydrophobic pocket of avidin upon addition of the guest. The disappearance of the component at 7 ns in the presence of avidin and the guest suggests that avidin provides a wide hydrophobic environment at the primary hydroxy side of *β*-CD, excluding the bulk water.

6. Cyclodextrin–peptide conjugates as chemosensors

Although it is necessary to introduce multiple functional groups on CD in order to improve sensitivity and selectivity of the CD-based chemosensors, their introduction is difficult to control owing to the many positional isomers for polyfunctionalized CD. If the rigid α-helix scaffold of a peptide was used, two different photoreactive moieties can be easily placed at both the primary hydroxy side and the secondary hydroxy side of CD.

Matsumura *et al.* [25] prepared α-helix peptide systems with the DNS unit and β-CD in the peptide side chain and examined the potential of the CD-peptide hybrids **14–17** as chemosensors (Fig. 8). A 17-residual alanine-based peptide was used due to the α-helix stabilizing property of alanine. In addition, three intramolecular salt bridges of Glu–Lys pairs were introduced into the peptide to stabilize the α-helix. Four kinds of peptides were prepared to examine the effect of the positions and directions of the DNS and β-CD moieties on the α-helix peptide structure and the molecular sensing abilities. The DNS and β-CD units in **EK3** (**14**) and **EK3R** (**15**) are separated by one α-helix turn, while in **EK6** (**16**) and **EK6R** (**17**) they are separated by two α-helix turns. **EK3R** (**15**) and **EK6R** (**17**) have the DNS and β-CD units in the reverse order of **EK3** (**14**) and **EK6** (**16**), respectively. The order of helix formation was **EK3R** (**15**) (75%) > **EK3** (**14**) (61%) > **EK6R** (**17**) (53%) > **EK6** (**16**) (44%). **EK3R** (**15**), **EK3** (**14**), and **EK6R** (**17**) form more helical structures than the peptide **18** (45% helix), which has neither the β-CD nor DNS unit. It appears that the intramolecular inclusion complex of β-CD with the DNS unit promotes higher helical structures. This is supported by the observation that the addition of 1-AdOH reduces their helix contents. The order of the fluorescence intensity is **EK3R** (**15**) > **EK3** (**14**) > **EK6R** (**17**) > **EK6** (**16**) and their maximum wavelengths are 524, 534, 529, and 540 nm, respectively. This order is the same as that of α-helicity. It indicates that the deeper inclusion of the DNS unit into its cavity causes stabilization of the α-helix structure. The time-resolved fluorescence studies suggest that the DNS unit is in an equilibrium between two conformations, located inside and outside the CD cavity similar to DNS-Leu-modified β-CD (**2** or **3**). The order of the binding affinity to guests is **EK6** (**16**) > **EK6R** (**17**) > **EK3** (**14**) > **EK3R** (**15**), which is the reverse of the order for the stability of the inclusion complex between the β-CD and DNS unit. This suggests that the tight intramolecular inclusion complex inhibits guest accommodation. It is noteworthy that the binding constants of **EK3** (**14**) and **EK6** (**16**) are larger than those of **EK3R** (**15**) and **EK6R** (**17**), respectively. This may be related to the orientation of the β-CD unit along the α-helix peptide. The values of the sensitivity parameters ($\Delta I/I_0$) are correlated to the binding affinities and their selectivity for the bile acids as guests is similar to the chromophore-modified CD. This means that the guest selectivity of β-CD on the α-helix is not disturbed by the peptide scaffold and that the CD–peptide conjugate can be used as a chemosensor.

Although it is interesting to know what happens if two chromophores are attached to both primary and secondary hydroxy sites so as to sandwich the CD between two chromophores, it has not been reported because of synthetic difficulties. Toyoda *et al.* [26] synthesized an α-helix peptide system **19** composed of 17 amino acids with γ-CD sandwiched between naphthalene units in the peptide side chain (Fig. 8). The fluorescence spectrum of **19** exhibits the monomer emission around 336 nm and the excimer emission around 390 nm. Because the ratio of the excimer emission intensity to the monomer one was not dependent on the concentration of **19**, the excimer is formed intramolecularly between two naphthalene moieties. Both emissions decrease upon addition of CDCA as a guest. This observation suggests that the two naphthalene units are excluded from the CD cavity upon addition of the guest. The binding constants of **19** for UDCA and LCA are one-third and one-fifth those of the CD–peptide hybrid that have one naphthalene unit, respectively. This result suggests that the two naphthalene units are tightly accommodated in the CD cavity and act as an intramolecular inhibitor for exogenous guests.

Yana *et al.* [27] prepared an α-helix peptide system **20** composed of 17 amino acids with two naphthalene units at the same side of β-CD in the peptide side chain (Fig. 8). This system was designed to signal the detection of a guest by an increase in the excimer emission intensity. **20** by itself exhibits strong monomer emission around 336 nm and relatively weak excimer emission around 390 nm. The excimer is formed between the two intramolecular naphthalene units outside the CD cavity. Upon addition of UDCA, the monomer emission intensity

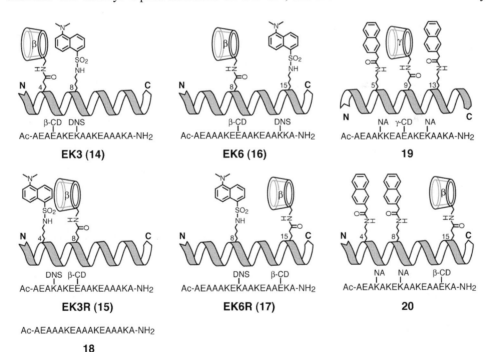

Fig. 8. Cyclodextrin-peptide conjugates (**1**).

decreases and the excimer emission intensity increases. This result suggests that one of naphthalene units was excluded from the CD cavity and two naphthalene units can interact with each other with a face-to-face orientation.

Hossain *et al.* [28,29] prepared CD–peptide hybrids **21–22**, in which pyrene (electron donor) and nitrobenzene (electron acceptor) were placed at both sides of CD on the peptide scaffold (Fig. 9). In an aqueous environment and in the absence of guests, the nitrobenzene moiety is located in the CD cavity, bringing it close to the pyrene moiety. Under these conditions the monomer emission of pyrene is effectively quenched by nitrobenzene. The fluorescence intensity dramatically increases with increasing concentration of a guest (e.g. lithocholic acid). The guest-responsive enhancement in the fluorescence intensity of **21** or **22** can be explained in terms of increased distance between the pyrene and nitrobenzene moieties, which is caused by exclusion of the nitrobenzene moiety from the CD cavity by guest accommodation.

Pyrene (donor) and coumarin (acceptor) were introduced at one side **23** or both the sides **24** of CD on an α-helix peptide scaffold to make a FRET system for a chemosensor (Fig. 9) [30,31]. **23** showed high fluorescence emission of the acceptor coumarin at around 448 nm on irradiating at 340 nm, which corresponds to the absorption wavelength of pyrene, indicating that energy transfer occurs from pyrene to coumarin in **23**. The fluorescence intensity of both the acceptor coumarin and the donor pyrene decreases on addition of a guest, hyodeoxycholic acid (HDCA). This diminishment in the fluorescence intensity is

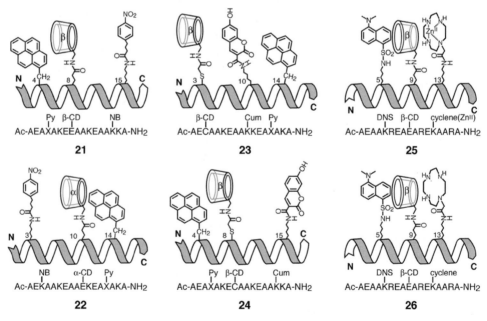

Fig. 9. Cyclodextrin-peptide conjugates (**2**).

thought to be associated with the exclusion of the coumarin unit from inside to outside of the β-CD cavity on accommodating a HDCA molecule. Being excluded from the CD cavity in **23**, coumarin may come into very close contact with the pyrene unit, resulting in the fluorescence quenching.

24 exhibits intramolecular FRET without quenching of two fluorophores in the absence of guests. The fluorescence intensity of the acceptor (coumarin) on irradiating at the absorption wavelength of pyrene (340 nm) is markedly decreased with increasing the concentration of HDCA as a guest, whereas the intensity of the donor (pyrene) emission is increased. Addition of HDCA causes coumarin to be excluded from the CD cavity, and thus the distance between pyrene and coumarin is increased, resulting in less efficient energy transfer.

Furukawa *et al.* [32] prepared a CD–peptide hybrid (**25**) bearing a Zn(II)-cyclen unit as an anion recognition site and the dansyl unit as a fluorescent probe to improve selectivity for organic anionic molecules, such as carboxylates or phosphates (Fig. 9). The binding constant of **25** for UDCA, which has a carboxylate unit, is twice as larger as that of **26**, which has no Zn(II), whereas the binding constant of **25** and **26** for 1-adamantanol (1-AdOH) or 1-adamantanamine (1-AdNH$_2$) are almost the same.

7. A new type of chemosensor based on modified cyclodextrin

Variation in the fluorescence intensity on binding a guest due to chromophore displacement has been shown in the previous sections to be an effective mechanism for chemosensors; however, it does have some defects. (1) Self-inclusion of the chromophore can inhibit accommodation of the guest. (2) Changing the chromophore or spacer unit can alter the selectivity of the chemosensor, but the effect is not large, because the guest selectivity of the chemosensor depends mainly on the selectivity of the CD itself. Both β-CD and γ-CD have greater affinity for bile acid derivatives than for adamantane derivatives. Therefore, adamantanol cannot be selectively detected in the presence of a bile acid using conventional chemosensors. (3) For most conventional chemosensors, the detection of a guest is accompanied by a decrease in the fluorescence intensity, although an increase in the emission intensity in respect to a guest is more effective for chemical sensing systems. Therefore, Ikeda *et al.* [33] have developed a new method to overcome these defects as shown in Fig. 10. If the chromophore is directly connected to the CD, the chromophore cannot be self-included and will remain at the entrance of the CD cavity. In this situation, some water is accommodated in the cavity and the chromophore is surrounded by a hydrophilic environment. If a hydrophobic guest then enters the cavity such that the hydrophobic face of the guest interacts with the chromophore, the chromophore will be located in a more hydrophobic environment. When the

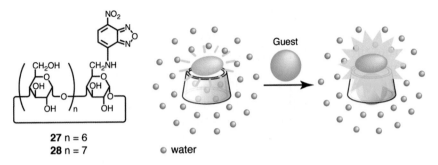

Fig. 10. New type of chemosensors.

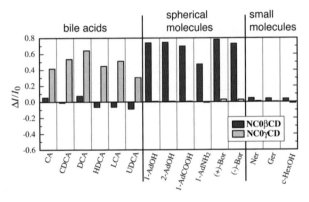

Fig. 11. Sensitivity parameters ($\Delta I/I_0$) of new type of chemosensors **27** and **28** (5×10^{-6} M) for various guests (each at 1×10^{-5} M) in phosphate buffer (200 mM, pH 7.0) at 298.15 K.

chromophore is a fluorophore, its fluorescence intensity will be weak in the absence of guests and stronger in the presence of hydrophobic guests. This mechanism is expected to produce a new selectivity in the chemosensor, because the guest shape that can increase fluorescence intensity is more limited. Furthermore, the fluorophore will act as a hydrophobic cap to increase the affinity of the chemosensor to guests.

The new type of chemosensor **27** is much more sensitive to the adamantane and borneol derivatives (Fig. 11). The shape of these guests is comparatively spherical and their size matches the β-CD cavity more closely. The response of **27** to these guests is a large increase in the fluorescence intensity. It is noteworthy that **27** is not sensitive to bile acids, although bile acids are strongly bound by the native β-CD.

1-AdOH can be detected even in the presence of a bile acid by using **27** as a chemosensor. To the best of our knowledge, this is the first time that adamantanol has been detected in the presence of a bile acid by a CD-based chemosensor. The relative sensitivity parameters (($\Delta I/I_0)_{\text{mix}}/(\Delta I/I_0)_{1-\text{AdOH}}$) of

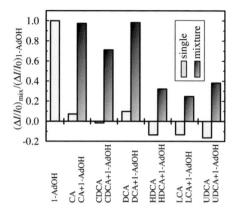

Fig. 12. Relative sensitivity parameters $((\Delta I/I_0)_{mix}/(\Delta I/I_0)_{1-AdOH})$ of **27** $(5 \times 10^{-6}$ M) for mixtures of guests (each at 1×10^{-5} M) in phosphate buffer (200 mM, pH 7.0) at 298.15 K.

27 for a solution containing both a bile acid $(1 \times 10^{-5}$ M) and 1-AdOH $(1 \times 10^{-5}$ M) are positive and are comparable to that for the addition of 1-AdOH alone (Fig. 12).

The sensitivity parameters of **28** for various guests are shown in Fig. 11. **28** is relatively sensitive to each bile acid but has no response to the adamantane derivatives. The responses of **28** to the guests are also an increase in the fluorescence intensity. The guest-responses pattern of **28** differs from that of **27**. Considering the differing guest-response patterns of **27** and **28**, we can now easily discriminate between bile acids and adamantane derivatives at any concentration by the combined use of **27** and **28**.

8. Immobilized fluorescent cyclodextrins on a cellulose membrane

Tanabe *et al.* [34,35] immobilized a dansyl-modified CD on a cellulose membrane for ease of use. Some of glucose rings in the cellulose were opened by oxidation with $NaIO_4$ to produce aldehyde groups . The fluorescent CD, which has an amino group in the spacer, was immobilized by reaction with the aldehyde groups on the membrane. The responses of the immobilized fluorescent CD upon addition of guests are similar to that of CD chemosensors in an aqueous solution.

Conclusion

In conclusion, we have shown that fluorescent CD is useful for detecting molecules via molecular recognition. We can now design chemosensors that show a variety of responses to guests. The application of a series of fluorescent CD in a chemosensor array can be envisaged as one of the many practical uses of the molecule-sensing abilities of these chemosensors.

References

[1] A.P. de Silva, H.Q.N. Gunaratne, T. Gunnlaugsson, A.J.M. Huxley, C.P. McCoy, J.T. Rademacher, T.E. Rice, Chem. Rev. 97 (1997) 1515.
[2] A.W. Czarnik (Ed.), Fluorescent Chemosensors for Ion and Molecule Recognition, American Chemical Society, Washington, DC, 1993.
[3] J.P. Desvergne, A.W. Czarnik (Eds), Chemosensors of Ion and Molecule Recognition, NATOASI Series C492, Kluwer, Dordecht, 1997.
[4] A. Ueno, Supramol. Sci. 3 (1996) 31.
[5] A. Ueno, Adv. Mater. 5 (1993) 132.
[6] A. Ueno, Fluorescent Cyclodextrins for Detecting Organic Compounds with Molecular Recognition, In: A. W. Czarnick (Ed.), Fluorescent chemosensors for ion and molecule recognition ACS, 1993, pp. 74–84.
[7] A. Ueno, H. Ikeda, J. Wang, Signal transduction in chemosensors of modified cyclodextrins, in: J.P. Desvergne, A.W. Czarnik (Eds), Chemosensors of Ion and Molecule Recognition, Kluwer, Dordrecht, 1997, pp. 105–119.
[8] A. Ueno, H. Ikeda, Supramolecular Photochemistry of Cyclodextrin Materials, in: V. Ramamurthy, K.S. Schanze (Eds), Molecular and Supramolecular Photochemistry, Marcel Dekker, New York, 2001, pp. 461–503.
[9] A. Ueno, I. Suzuki, T. Osa, Chem. Lett. 18 (1989) 1059.
[10] A. Ueno, I. Suzuki, T. Osa, J. Am. Chem. Soc. 111 (1989) 6391.
[11] H. Ikeda, M. Nakamura, N. Ise, N. Oguma, A. Nakamura, T. Ikeda, F. Toda, A. Ueno, J. Am. Chem. Soc. 118 (1996) 10980.
[12] H. Ikeda, M. Nakamura, N. Ise, F. Toda, A. Ueno, J. Org. Chem. 62 (1997) 1411.
[13] S. Hashimoto, J.K. Thomas, J. Am. Chem. Soc. 107 (1985) 4655.
[14] G. Nelson, G. Patonay, I.M. Warner, Anal. Chem. 60 (1988) 274.
[15] Y.-H. Li, L.-M. Chan, L. Tyer, R.T. Moody, C.M. Himel, D.M. Hercules, J. Am. Chem. Soc. 97 (1975) 3118.
[16] J. Huang, F.V. Bright, J. Phys. Chem. 94 (1990) 8457.
[17] R.A. Dunbar, F.V. Bright, Supramol. Chem. 3 (1994) 93.
[18] M. Nakamura, T. Ikeda, A. Nakamura, H. Ikeda, A. Ueno, F. Toda, Chem. Lett. 24 (1995) 343.
[19] M.R. de Jong, J.F.J. Engbersen, J. Huskens, D.N. Reinhoudt, Chem. Eur. J. 6 (2000) 4034.
[20] T. Kikuchi, M. Narita, F. Hamada, Tetrahedron 57 (2001) 9317.
[21] M. Nakamura, A. Ikeda, N. Ise, T. Ikeda, H. Ikeda, F. Toda, A. Ueno, J. Chem. Soc. Chem. Commun. (1995) 721.
[22] A. Ueno, A. Ikeda, H. Ikeda, T. Ikeda, F. Toda, J. Org. Chem. 64 (1999) 382.
[23] J. Wang, A. Nakamura, K. Hamasaki, H. Ikeda, T. Ikeda, A. Ueno, Chem. Lett. 25 (1996) 303.
[24] T. Ikunaga, H. Ikeda, A. Ueno, Chem. Eur. J. 5 (1999) 2698.
[25] S. Matsumura, S. Sakamoto, A. Ueno, H. Mihara, Chem. Eur. J. 6 (2000) 1781.
[26] T. Toyoda, S. Matsumura, H. Mihara, A. Ueno, Macromol. Rapid Commun. 21 (2000) 485.
[27] D. Yana, T. Shimizu, K. Hamasaki, H. Mihara, A. Ueno, Macromol. Rapid Commun. 23 (2002) 11.

[28] M.A. Hossain, K. Hamasaki, K. Takahashi, H. Mihara, A. Ueno, J. Am. Chem. Soc. 123 (2001) 7435.
[29] M.A. Hossain, K. Takahashi, H. Mihara, A. Ueno, J. Incl. Phenom. Macro. Chem. 43 (2002) 271.
[30] M.A. Hossain, H. Mihara, A. Ueno, J. Am. Chem. Soc. 125 (2003) 11178.
[31] M.A. Hossain, H. Mihara, A. Ueno, Bioorg. Med. Chem. Lett. 13 (2003) 4305.
[32] S. Furukawa, H. Mihara, A. Ueno, Macromol. Rapid Commun. 24 (2003) 202.
[33] H. Ikeda, T. Murayama, A. Ueno, Org. Biomol. Chem. 3 (2005) 4262.
[34] T. Tanabe, K. Touma, K. Hamasaki, A. Ueno, Anal. Chem. 73 (2001) 1877.
[35] T. Tanabe, K. Touma, K. Hamasaki, A. Ueno, Anal. Chem. 73 (2001) 3126.

Cyclodextrin Materials Photochemistry, Photophysics and Photobiology
Abderrazzak Douhal (Editor)
DOI 10.1016/S1872-1400(06)01013-2

Chapter 13

Cyclodextrin-Based Rotaxanes

He Tian, Qiao-Chun Wang

Laboratory for Advanced Materials and Institute of Fine Chemicals, East China University of Science & Technology, Meilong Road 130, Shanghai 20037, P.R. China

1. Introduction

A rotaxane consists of one or more macrocycles encircling the rod portion of a dumbbell-like component. In the late 1980s and the early 1990s, with the development of supramolecular chemistry and with the better understanding of the processes of the host–guest molecular recognition, it has become possible to take advantages of intermolecular attractive forces, such as hydrophobic effect, hydrogen bonding, metal-ligand coordination, aromatic π–π stacking and [C-H···O] interactions, to develop the so-called "template-directed" synthetic methodology for rotaxanes. From then on, the synthesis of a rotaxane becomes quite routine and a large variety of these intriguing compounds of ever increasing regularity and complexity have been reported. Moreover, rotaxanes are generally with some novel chemical and physical properties, and as a result they have attracted more and more attention.

Cyclodextrins (CDs) are a series of cyclic compounds consisting of glucose units linked by α-1,4 linkages. The most commonly used CDs are known as α-, β- or γ-CD, which have six to eight glucose units, respectively. CDs are water-soluble, nontoxic, commercially available with low price, and their structures are rigid and well defined, most of all, they can bind to a wide variety of guest molecules. As a consequence, CDs continue to be the attractive wheel components in constructing rotaxanes, which have attracted more and more attention because of their challenging constructions and potential applications in fields such as biomedical materials, nanostructured functional materials and molecular devices [1]. Here we summarize the recent works on CDs-based rotaxanes since the year 2000, including their synthesis and the applications.

2. The sorts of cyclodextrin-based rotaxanes

A rotaxane, in narrow sense, is different from a pseudorotaxane because the former has bulky stoppers that lock the macrocycle from breaking away. However, the word "rotaxanes" in this chapter includes rotaxanes and pseudorotaxanes in broad sense because these two species are of a same structural characteristic — that is, one or more rings encircle a rod. These rotaxanes are pseudorotaxane, [2]rotaxane, [1]rotaxane, Janus rotaxane, polypseudorotaxane, polyrotaxane, rotaxane polymer, rotaxane aggregate and poly(polyrotaxane).

2.1. Pseudorotaxane

A pseudorotaxane (Fig. 1) is a type of supramolecular assembly that consists of a cyclic molecule (wheel component) and a rod-like molecule (axle component). The macrocycle may slip off the axle because there are no bulky terminal groups on the axle. The most adopted synthesis strategy for CD-based pseudorotaxanes is the threading of an axle molecule through the cavity of the wheel component with the assistance of noncovalent intermolecular attractive forces. One of the examples was reported by Chen [2] using bisimidazolyl compounds as axle guests. The guest molecules with longer alkylene chain or two benzene rings can thread two β-CDs

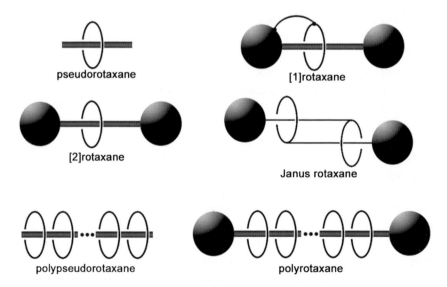

Fig. 1. The cartoon representation of rotaxane species.

to form the [3]pseudorotaxanes, while the shorter chain or one benzene can thread only one *β*-CD ring to become the pseudorotaxanes. These type of pseudo-rotaxanes may be converted to metalo-rotaxanes because of the N atom in terminal imidazole group of the guest molecule can coordinate with transition metal ion.

Cyclodextrins are cone-shaped chiral hosts, and when an asymmetric CD was threaded by a nonsymmetrical dumbbell, it would give two isomers which differ in orientation of CD with respect to the different ends of the dumbbell. As a result, isomeric rotaxanes were always obtained. If a more complicated rotaxane which comprises an asymmetric dumbbell is required, the orientation of the cyclodextrin ring should be controlled during the synthesis to obtain the [2]rotaxane of sole configuration — that is the so-called unidirectional threading strategy. In 2004, Park *et al.* [3] reported the thermodynamic threading of *α*-CD to obtained nonisomeric pseudorotaxane. Owing to an energetically favorable interaction between the carbazole group and secondary surface of *α*-CD, the unidirectional threading of CD with carbazole axle in aqueous solution is obtained. Harada *et al.* [4] also reported a kinetical threading of cyclodextrin ring to realize the unidi-rectional preparation of nonsymmetric pseudorotaxane.

2.2. [2]Rotaxane

A [2]rotaxane is quite similar to a pseudorotaxane with the same structure char-acteristic, except that the wheel component is trapped by two bulky ends so that they cannot detach from one another, as shown in Fig. 1. A [2]rotaxane can be constructed in different template-directed ways including stoppering, clipping and slipping. Stoppering is the addition of bulky ends to a pseudorotaxane; clipping is the formation of a macrocycle around the linear part of the dumbbell-like unit; and slipping is the thermodynamic-controlled self-assembling between a ring compound and a dumbbell-shaped component at relatively high temperature.

One example using the stoppering strategy for synthesizing [2]rotaxane was reported by Anderson [5]. A series of stilbene and tolan cyclodextrin-based symmetrical [2]rotaxanes were obtained in good yield by using aqueous Suzuki coupling. Recently, we developed the synthesis of two nonisomeric cyclodextrin-based [2]rotaxanes [6]. Firstly, two unidirectional [2]pseudorotaxanes were formed thermodynamically and kinetically, respectively. Then Suzuki coup-ling reactions were carried out to introduce stoppers, and the corresponding [2]rotaxanes were thus obtained as one co-conformation.

2.3. [1]Rotaxane

A [2]rotaxane can be transferred into a [1]rotaxane by introducing a covalent bridge between the wheel and the axle of the [2]rotaxane, as shown in Fig. 1. Easton [7] reported the synthesis of two [1]rotaxanes. The succinic anhydride protected amino-α-CD was treated with stilbene guests, then capped by trinitro-benzene to give [2]rotaxanes. Finally, the succinyl group was attached to the axle in each of the [2]rotaxanes, and the corresponding [1]rotaxanes were thus formed. It has been found that incorporation of the methoxyl group and the succinamide restricts the rotational motion of the cyclodextrin around the stil-bene axle, and this system is analogous to a ratchet tooth and pawl to restrict rotational motion.

2.4. Janus rotaxane

A Janus rotaxane comprises two dumbbell-like components, each has a macrocycle and a stopper at the rod portion's ends, and each of the macrocycle of one dumb-bell encircles the rod portion of the other dumbbell, as shown in Fig. 1. The first Janus rotaxane was reported by Kaneda *et al.* [8]. The self-association of the azobenzene-modified α-CD monomer resulted in the formation of a Janus pseu-dorotaxane, then two bulky naphthols stoppers were introduced by bis-azo coupling and Janus rotaxane was thus obtained. Easton et al. [9] also developed another Janus rotaxane. The Janus pseudorotaxane was obtained from the self-assembling of the stilbene-modified α-CD, and then two trinitrobenzene groups were introduced as stoppers and the corresponding rotaxane originated. Harada [10] developed an elegant strategy to construct daisy chain necklace. A cinnamoyl guest was covalently attached to a cyclodextrin host. Such a molecule may form an intermolecular complex to give a trimer in the way of host–guest interaction. The trimer can be stabilized to form a necklace-like tri[2]rotaxane by introducing three bulky trinitrobenzene stoppers.

2.5. Polypseudorotaxane

A polypseudorotaxane comprises of several or more ring components threaded on a linear polymer backbone. There are no bulky stoppers on the polymer ends and the macrocycles may also slip off the linear chain. Yamamoto reported the slipping of cyclodextrin rings to a polythiophene polymer chain to develop a polypseudorotaxane [11]. Stirring the aqueous suspensions of polythiophenes in the presence of β-CD and then centrifugation of the reaction mixtures gave the corresponding water-soluble polypseudorotaxane.

Li *et al.* [12] also reported a polypseudorotaxane based on triblock polymers. The triblock copolymers were added to an excess amount of an aqueous solution of α-CD under ultrasound waterbath. The solutions gradually became turbid, eventually producing the polypseudorotaxanes in the form of crystalline precipitates. Because poly(ethylene oxide) (PEO) forms inclusion compounds with α-CD, while poly(propylene oxide) (PPO) does not, only the middle PEO block in the polypseudorotaxanes is included by α-CD molecules with a channel structure, while the flanking PPO blocks are uncovered and remain amorphous.

2.6. Polyrotaxane

A polyrotaxane is the counterpart of a pseudorotaxane, as shown in Fig. 1. There are macrocycles threaded on a linear polymer chain, and the macrocycles are prevented from slipping off the chain by the two bulky terminals. Zhao and Beckham [13] reported a polyrotaxane based on poly(ethylene glycol) (PEG). A polypseudorotaxane was obtained from the slipping of α-CD rings to a PEG-Ts2 chain in aqueous solution, the consequent endcapping with dimethylbenzene by etherification resulted in the formation of the polyrotaxane.

In general, it becomes more difficult to prepare polyrotaxanes containing β-CD than α-CD, because the cavity of β-CD is large so that it is hard to find suitable stopper groups to prevent their dethreading. Harada *et al.* [14] described an alternative approach for the one-pot preparation of β-CD-based polyrotaxanes by making use of the photocyclodimerization of anthracene.

Liu *et al.* [15] also reported a convenient and efficient method for the termination of CD-based polypseudorotaxanes with two cyclodextrin rings as stoppers to obtained water-soluble polyrotaxanes. The resultant CD-capped polyrotaxanes can be further assembled to larger aggregates by host–guest complexation between these free CD cavities and a wide variety of hydrophobic guests/substrates including C60.

2.7. Rotaxane polymer

Rotaxane polymers can also be deemed as polyrotaxanes or pseudopolyrotaxanes, because they have the common structural characteristics; that is, many macrocycles reside on a polymer chain. However, unlike polyrotaxanes or polypseudorotaxanes, which are synthesized from the slipping of macrocycles into a polymer chain, rotaxane polymers are synthesized from the polymerization of rotaxane (or pseudorotaxane) monomer, and monomer unit on the polymer chain has one macrocycle. Harada [16] reported the preparation of a rotaxane polymer based on the homopolymerization of bithiophene (2T) pseudorotaxane. Bithiophene formed first inclusion compound α-CD-2T with α-CD, and then the polymerizations of α-CD-2T in water using $FeCl_3$ as an oxidative initiator; after centrifugation, the resulting powder was washed with water and THF to remove unreacted monomer and the corresponding rotaxane polymer was found as deep purple powder. Farcas *et al.* [17] developed a novel aromatic polyazomethine polymer with rotaxane architecture by the copolymerization of a diformylcarbazole derivative with a benzenediamine pseudorotaxane in DMF.

2.8. Self-assembling rotaxane aggregates

Harada et al. [18] reported the synthesis of a [2]rotaxane capped by β-CD and a trinitrophenyl group. The β-CD at the end binds the trinitrophenyl group stopper of another [2]rotaxanes gives a rotaxane aggregate, as shown in Fig. 2. Liu et al. [19] also developed a simple strategy to prepare polyrotaxane-like supramolecules. An inclusion compound had been synthesized by the complexation of 4,4'-di-pyridine and β-CD in water; the subsequent coordination of the complex with nickel ions in aqueous solution gave the corresponding pseudopolyrotaxane.

2.9. Poly(polyrotaxane)

A poly(polyrotaxane) can be originated from the polymerization of a poly-rotaxane. Harada et al. [20] reported the preparation of a poly(polyrotaxane) by photodimerization. The polyrotaxane consisting of PEG, β-CD and 9-anth-racene groups was subjected to irradiation by visible light, and the anthracene stoppers underwent photoinduced dimerization and the poly(polyrotaxane) formed. The dimerization is reversible and the polyrotaxane monomer can come back in more than 90% conversion by heating the poly(polyrotaxane) DMSO solution 393 K. Harada [21] also reported the poly(polypseudorotaxane) which is quite similar to the above poly(polyrotaxane), except that the β-CD rings are changed to γ-CD, and the cavity of γ-CD ring is big enough to pass over the anthracene unit.

3. Potential applications of cyclodextrin-based rotaxanes

3.1. Molecular machines — switchable rotaxanes

In a [2]rotaxane, if properly designed, two recognition sites for the macrocycle can be arranged within the dumbbell component. Moreover, the strengths of the noncovalent bonding interactions between two recognition sites and the ring component are quite different, and as a result, the ring component locates preferentially in the site that has stronger interactions with the ring (ground state in Fig. 3). In such a system, when a suitable external stimulus — either chemical or physical — is applied, the binding activity sequence of the two recognition sites altered, as a consequence, the ring component shuttles from the original recognition site to another — now has stronger interactions with the ring component (switched state in Fig. 3). In ideal cases, the properties of the two recognition sites can be shifted reversibly back to the original ones by another stimulus, and the switched state is now again changed to ground state. Such molecule constitutes a switchable rotaxane. Like a macroscopic machine displaying intercomponental relative

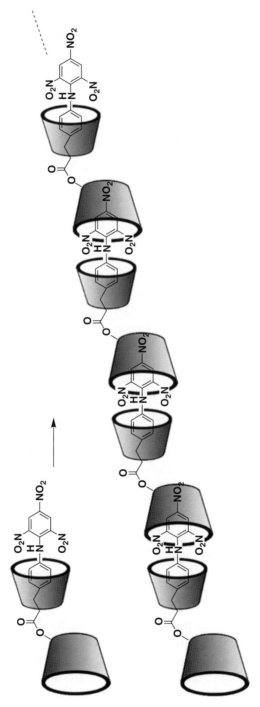

Fig. 2. The formation of rotaxane aggregates via self-assembling.

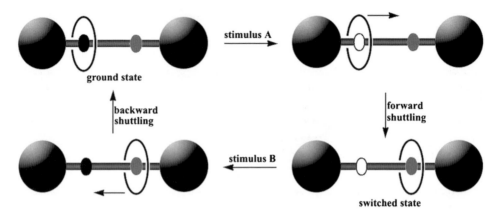

Fig. 3. The stimuli-triggered reversible shuttling motions in a [2]rotaxane between the ground state and switched state.

position changes when applying power, a switchable rotaxane performs shuttling motion in response to external stimuli, from this point of view, a molecular shuttle is also deemed as a molecular machine, and the external stimuli are also the driving forces or energy supplies of the machine system.

In a stilbene-based rotaxane [22], the alternative irradiation at 340 nm or 265 nm results in the E/Z isomerization of the stilbene unit, accompanied by the reversible shuttling motions of the α-cyclodextrin. It should be noted that there are no waste products formed during the photoisomerization-induced reversible shuttling process. We also developed a thermo-driven molecular shuttle [23]. At room temperature, the cyclodextrin ring stays preferentially on the azobenzene station. Upon raising the temperature, more cycledextrin rings move to the biphenyl station. The reversing process can be obtained by lowering the temperature of the system. Recently, Nakashima *et al.* [24] set up a multi-mode-driven molecular shuttle system, which consists of α-CD as a ring, azobenzene as a photoactive moiety and a primary station, viologen as a hydrophilic group and an energy barrier for the slipping of α-CD, and 2,4-dinitrobenzene as a stopper. The molecular shuttle can be driven photochemically and thermally, and the solvent polarity does great influence on the shuttling motion.

3.2. Molecular switches

One of the most important applications of switchable rotaxane is to construct molecular devices that can process and store information. The principle for designing such a molecular device is that, the molecule shuttle switches from a ground state to another metastable state with an output signal in response to an input, and the metastable state can be switched reversibly back to the ground state when another input is applied. From this point of view, a driving force of a molecular shuttle acts as the input and the spectral signals identify the co-conformations and serves as the output.

There are varieties of driving forces for switchable rotaxanes, such as chemical, electrical, heat and light energy, a solvent change as well. A light driving force can lead to a fast response with ease and no waste product is formed and thus has advantages over other inputs.

The UV–Vis absorption, NMR, ICD (induced circular dichroism), fluorescent spectra and cyclic voltammetry are the methods used to monitor the co-conformational changes for switchable rotaxanes. Unlike other spectral signals, a fluorescence signal is highly sensitive so even a single-molecular fluorescence can be detected, and furthermore, a fluorescence output can be easily and remotely detected.

In summarizing the driving forces and the monitoring methods of switchable [2]rotaxanes, it can be found that an optical device, based on a light-driven molecular machine with a fluorescent output, will be characterized by

• fast response;
• low cost because the detection of photons can be easily carried out in a small space; and
• the output signal is highly sensitive and furthermore, which can be remotely detected and thus noncontact readout can be achieved.

However, little attention has been focused on this kind of optical interlocked molecular systems despite that a large number of molecular machine systems have been set up for more than a decade. These facts became our impetus to build up light-driven molecular machines with fluorescence output. A cyclodextrin ring can exert micro-environmental changes because of its rigid hydrophobic cavity, and thus affects the fluorescent properties of a fluorophore coming close. As a consequence, CDs may be served as the useful ring component to construct switchable rotaxanes with fluorescent addresses.

In 2004, we built up a "lockable" light-driven switchable [2]rotaxane [25], which comprises an α-CD ring, a stilbene unit and two stoppers — one 4-amino-1,8-naphthalimide (ANS) fluorophore and one isophthalic acid. Under a basic environment, the cyclodextrin ring resides over the stilbene unit and far away from ANS. Upon irradiation at 335 nm, the photoisomerization of the stilbene unit is induced and the cyclodextrin ring is forced to shuttle to the biphenyl unit and become close to ANS. The rigidity and the displacement of solvent enviroment as well as the shielding from molecular oxygen quenching of the cyclodextrin ring to the methylene and the fluorescent stopper results in an increase in the fluorescence intensity of the fluorophore by 63%. The back isomerization can be reversibly obtained by the irradiation at 280 nm and the fluorescence decrease to the original level.

It should be noted that the switching process could be disabled in the acid form by the hydrogen bonds between the isophthalic acid and the hydroxy groups on the cyclodextrin. It should also be emphasized that this molecular shuttle can serve as a molecular switch and the photons can be explored in this system for both causing the translational changes (input) and monitoring the different co-conformational states (output). It is superior to any other system for the neatness and convenience.

Another optical switchable [2]rotaxane was also developed in our group [26]. The rotaxane consists of an α-CD ring, an azobenzene unit and two fluorophores — ANS and a naphthalimide sulfonic acid (NS). The cyclodextrin ring can be driven from the azobenzene unit to the biphenyl group by the light irradiation at 360 nm, and the fluorescent intensity of ANS at 520 nm decrease as well as the increasing of NS at 395 nm. The backward process can be obtained reversibly by the irradiation at 430 nm and the fluorescence intensities of the two stoppers are recovered.

3.3. Molecular logic gates

When a switchable rotaxane can give outputs (co-conformational changes) in response to inputs (external stimuli), the ability to give outputs in a certain rule on the basis of the inputs raises possibilities for a switchable rotaxane to perform logic calculations.

In 2005, we set up a complicated CD-based molecular shuttle [27]. There are four states — original dynamic state, two photoisomerization states and the static state originates from the excitation with both of the two UV lights. In both the dynamic and the static state, the cyclodextrin ring has little influence on the emitting units, and the fluorescent intensities of the two stoppers are at low level. The photoisomerization of azobenzene unit drives the cyclodextrin ring close to NS stopper, and the fluorescent intensity of NS at 395 nm increases. The photoisomerization of the stilbene unit drives the cyclodextrin ring approaching the ANS unit and the fluorescent intensity at 520 nm increases. All these interconversions are reversible. The relations between the light inputs (313 and 380 nm) and the outputs (absorption and fluorescence changes) indicate that this [2]rotaxane system can be acted as a half-adder logic gate.

Fig. 4 shows the structure of a "Chinese Abacus" and its expression for the digit "15462". It is amazing that this ordinary ancient tool can do complicated arithmetical calculations quickly and precisely, not in the shade of the modern calculator at all, just with the different positions of the abacus beads. The abacus came into being several thousand years ago and now is still used among Chinese accountants. From this point of view, it is possible for a switchable rotaxane to carry out arithmetical calculations because of its abacus-like structure.

Recently, we synthesized a complicated switchable CD-based [3]rotaxane (Fig. 5) to simulate the abacus system [28]. The $E \rightarrow Z$ photoisomerization of azobenzene results in the shuttling of the two macrocycles close to NS stopper accompanied by

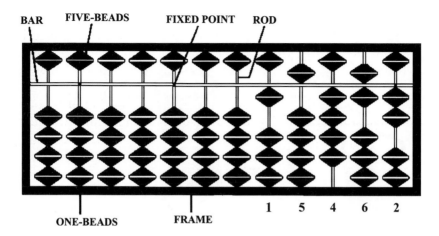

BAR FIVE-BEADS FIXED POINT ROD

1 5 4 6 2

ONE-BEADS FRAME

Fig. 4. The structure of a Chinese abacus and its expression of the digit "15462".

an increase in the fluorescent intensity at 395 nm and the decrease in the fluorescent intensity at 520 nm. The $E \rightarrow Z$ photoisomerization of stilbene shifts the two rings close to ANS stopper, and the fluorescent intensity at 395 nm decreases, at the same time the fluorescent intensity at 520 nm increases. These isomerzations are reversible. It can be seen that the photo-induced shuttling motions of two macrocycles of the [3]rotaxane are analogous to the movements of the beads of a abacus when carrying out " + 2" or "−2" arithmetical calculations. It should be noted that the results of the calculations in this [3]rotaxane can be "read" from the fluorescent changes, as shown in Fig. 5.

3.4. Stability

Anderson *et al.* [29] reported the synthesis of a [2]rotaxane reactive azo dye, and this rotaxane exhibited dramatically enhanced stability toward reductive bleaching, oxidative bleaching and photo-bleaching, which suggested that dyes encapsulated by CDs may have potential applications in areas where chromophore degradation needs to be prevented [30]. Anderson [31] also reported the encapsulation of redox-active cyanine dye with cyclodextrin. The cyclodextrin ring acts as a protective sheath, preventing the oxidized and reduced forms of the cyanine dye from being destroyed, making both redox processes reversible, whereas in the absence of the threaded macrocycle, the free dye is electrochemically bleached rapidly. Similar situations were also found in another series of cyanine dye [2]rotaxanes [32]. The encapsulation of the cyanine dyes by cyclodextrin can result in the increase in the fluorescence and the photostability of dyes under solvent conditions.

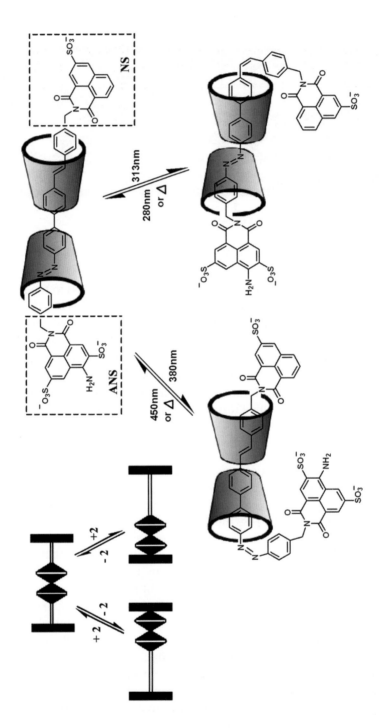

Fig. 5. The photoisomerization-induced shuttling motions in the [3]rotaxane and the corresponding fluorescent intensity changes during the reversible cycles.

3.5. Enhance intervalence charge transfer

Das and co-workers [33] reported the syntheses of [2]rotaxane complexes with Ru(III) and Fe(II) units as the two molecular heads, β-CD as a wheal component and 4,4-bipyridyl derivatives as the bridging ligand. It is found that the inclusion of the bridging ligand into the hydrophobic CD cavity has initiated optical electron transfer from the Fe(II) to the Ru(III) center, which was not observed in the absence of inclusion.

3.6. Insulated molecular wire

Conjugated polymers are often described as molecular wires because of the high charge mobility along individual polymer chain. Insulated molecular wires comprise a conjugated polymer backbone encapsulated at the molecular level by a protective sheath, limiting inter-chain interactions and enhancing the one-dimensional nature of the transport properties. CDs are nonconjugated rings and can serve as an effective insulated and protective outer layer.

Anderson *et al.* [34] reported the synthesis of conjugated polyrotaxanes encapsulated by cyclodextrin rings, as shown in Fig. 6. The hydrophobic diboronic acid monomer was encapsulated inside the cyclodextrin rings and polymerized by Suzuki coupling with a water-soluble diiodide monomer and terminated with bulky naphthalene stoppers to form the corresponding insulated molecular wires.

3.7. Light harvesting

Ueno *et al.* [35] reported the synthesis of a series of polypseudorotaxanes, which were studied as a light-harvesting antenna model, and the naphthalene and anthracene moieties act as energy donor and energy acceptor, respectively. It is found that the antenna effect becomes more marked with increasing number of naphthalene units in the polyrotaxanes, while the energy transfer efficiency from naphthalene to anthracene decreases.

3.8. Construction of nanotube

Liu *et al.* [36] reported the synthesis of a bis-nanotube in the form of a poly-rotaxane. A bis-polyrotaxane was prepared from the assembling of cyclodextrin tetramer with PPG and the subsequent stopping with dinitrobenzoic acid. The cross-linking of the adjacent cyclodextrin rings on the bispolyrotaxane with epichlorohydrin and removal of the polymeric chain through the hydrolysis of the terminal ester groups gave the corresponding bis-nanotube, as shown in Fig. 7.

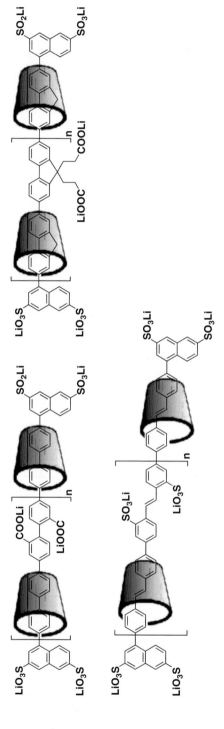

Fig. 6. The chemical structure of the cyclodextrin-based insulated molecular wires.

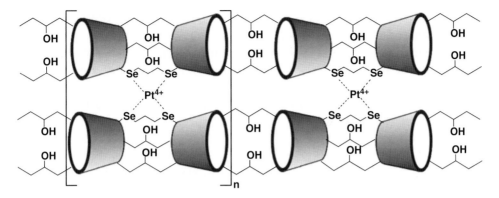

Fig. 7. The cyclodextrin-based bis-nanotube.

3.9. Biological applications

This book contains a contribution by El-Kemary and Douhal on the effect of CD on drugs and a contribution by Ooya on biodegradable polyrotaxanes and hydrogels. Here we would only mention some other applications of CD-based rotaxanes in this area. Liu *et al.* [37] reported the synthesis of the amino-terminated pseudopolyrotaxane, which can be adsorbed on the surface of Au nanoparticles through electrostatic interactions between the terminal amino groups and gold particles. The nanoparticles are fairly soluble in water and are able to capture and enrich [60]fullerene and the resulting fullerene conjugate shows a high DNA cleavage ability under light irradiation.

To mimic saccharide–protein interactions, some PEG-polyrotaxanes were synthesized in Yui's group [38–40]. In these polyrotaxane, many binding ligands were introduced to the CD rings. High association constants were found between these polyrotaxanes and proteins such as concanavalin A and streptavidin.

Yui *et al.* [41] reported a series of polyrotaxanes, in which sulfopropyl-modified CDs are threaded onto PEG–PPG–PEG triblock copolymers and capped with bulky end group. These polyrotaxanes are suggested to be a promising candidate when fabricating blood-compatible medical devices by blending with or coating on clinically used polymers.

4. Conclusion and outlook

A rotaxane, which results from wrapping a molecule with a macrocycle, always generates unexpected chemical and physical properties, and consequently rotaxanes are different from their single molecular counterpart. Great deals of functional materials are thus developed with rotaxane-like structures, and rotaxanes continue to be of great interest for more than a decade because of their beautiful structures and potential applications.

Cyclodextrins are excellent cyclic hosts because of their capability to include a large variety of organic guests into their apolar cavities, and continue to be attractive ring components for the construction of rotaxanes. Many rotaxanes have thus been prepared based on CDs, and cyclodextrin-based rotaxane has become one of the most important rotaxane species. It should be noted that CDs are the oligomers of glucose comprising a lot of hydroxyl groups, and as a result, the threaded CDs can be readily chemically modified and the possibility to introduce distinctive physicochemical properties thus originates. Furthermore, because CDs are water-soluble and nontoxic, they can be used in organisms, especially in the human body. It can be foreseen that, in the near future, more and more functionalized cyclodextrin-based rotaxanes would come forth and find increasing practical applications in scientific areas, and new concepts originated in this area will instill into our life.

References

[1] A. Harada, Acc. Chem. Res. 34 (2001) 456.
[2] X.-J. Shen, H.-L. Chen, F. Yu, Y.-C. Zhang, X.-H. Yang, Y.-Z. Li, Tetrahedron Lett. 45 (2004) 6813.
[3] J.W. Park, H.J. Song, Org. Lett. 6 (2004) 4869.
[4] T. Oshikiri, Y. Takashima, H. Yamaguchi, A. Harada, J. Am. Chem. Soc. 127 (2005) 12186.
[5] C.A. Stanier, M.J. O'Connell, W. Clegg, H.L. Anderson, Chem. Commun. (2001) 493.
[6] Q.-C. Wang, X. Ma, D.-H. Qu, H. Tian, Chem. Eur. J. 12 (2006) 1090. DOI: 10.1002/chem.200500415.
[7] H. Onagi, C.J. Blake, C.J. Easton, S.F. Lincoln, Chem. Eur. J. 9 (2003) 5978.
[8] T. Fujimoto, Y. Sakata, T. Kaneda, Chem. Commun. (2000) 2143.
[9] H. Onagi, C.J. Easton, S.F. Lincoln, Org. Lett. 3 (2001) 1041.
[10] T. Hoshino, M. Miyauchi, Y. Kawaguchi, H. Yamaguchi, A. Harada, J. Am. Chem. Soc. 122 (2000) 9876.
[11] I. Yamaguchi, K. Kashiwagi, T. Yamamoto, Macromol. Rapid Commun. 25 (2004) 1163.
[12] J. Li, X. Ni, K. Leong, Angew. Chem. Int. Ed. 42 (2003) 69.
[13] T. Zhao, H.W. Beckham, Macromolecules 36 (2003) 9859.
[14] M. Okada, A. Harada, Org. Lett. 6 (2004) 361.
[15] Y. Liu, Y.-W. Yang, Y. Chen, H.-X. Zou, Macromolecules 38 (2005) 5838.
[16] Y. Takashima, Y. Oizumi, K. Sakamoto, M. Miyauchi, S. Kamitori, A. Harada, Macromolecules 37 (2004) 3962.
[17] A. Farcas, M. Grigoras, Polym. Int. 52 (2003) 1315.
[18] M. Miyauchi, T. Hoshino, H. Yamaguchi, S. Kamitori, A. Harada, J. Am. Chem. Soc. 127 (2005) 2034.
[19] Y. Liu, Y.-L. Zhao, H.-Y. Zhang, H.-B. Song, Angew. Chem. Int. Ed. 42 (2003) 3260.
[20] M. Okada, A. Harada, Macromolecules 36 (2003) 9701.
[21] M. Okada, Y. Takashima, A. Harada, Macromolecules 37 (2004) 7075.
[22] C.A. Stanier, S.J. Alderman, T.D.W. Claridge, H.L. Anderson, Angew. Chem. Int. Ed. 41 (2002) 1769.

[23] D.-H. Qu, Q.-C. Wang, H. Tian, Mol. Cryst. Liq. Cryst. 59 (2005) 430.

[24] H. Murakami, A. Kawabuchi, R. Matsumoto, T. Ido, N. Nakashima, J. Am. Chem. Soc. 127 (2005) 15891.

[25] Q.-C. Wang, D.-H. Qu, J. Ren, K. Chen, H. Tian, Angew. Chem. Int. Ed. 43 (2004) 2661.

[26] D.-H. Qu, Q.-C. Wang, J. Ren, H. Tian, Org. Lett. 6 (2004) 2085.

[27] D.-H. Qu, Q.-C. Wang, H. Tian, Angew. Chem. Int. Ed. 44 (2005) 5296.

[28] D.-H. Qu, Q.-C. Wang, X. Ma, H. Tian, Chem. Eur. J. 11 (2005) 5923.

[29] M.R. Craig, M.G. Hutchings, T.D.W. Claridge, H.L. Anderson, Angew. Chem. Int. Ed. 40 (2001) 1071.

[30] E. Arunkumar, C.C. Fprbes, B.D. Smith, Eur. J. Org. Chem. (2005) 4051.

[31] J.E.H. Buston, F. Markenb, H.L. Anderson, Chem. Commun. (2001) 1046.

[32] J.E.H. Buston, J.R.Young, H.L. Anderson, Chem. Commun. (2000) 905.

[33] A.D. Shukla, H.C. Bajaj, A. Das, Angew. Chem. Int. Ed. 40 (2001) 446.

[34] J.J. Michels, M.J. O'Connell, P.N. Taylor, J.S. Wilson, F. Cacialli, H.L. Anderson, Chem. Eur. J. 9 (2003) 6167.

[35] M. Tamura, D. Gao, A. Ueno, Chem. Eur. J. 7 (2001) 1390.

[36] Y. Liu, C.-C. You, H.-Y. Zhang, S.-Z. Kang, C.-F. Zhu, C. Wang, Org. Lett. 1 (2001) 613.

[37] Y. Liu, H. Wang, Y. Chen, C.-F. Ke, M. Liu, J. Am. Chem. Soc. 127 (2005) 657.

[38] T. Ooya, H. Utsunomiya, M. Eguchi, N. Yui, Bioconjugate Chem. 16 (2005) 62.

[39] T. Ooya, M. Eguchi, N. Yui, J. Am. Chem. Soc. 125 (2003) 13016.

[40] T. Ooya, N. Yui, J. Control Release 80 (2002) 219.

[41] H.D. Park, W.K. Lee, T. Ooya, K.D. Park, Y.H. Kim, N. Yui, J. Biochem. Mater. Res. 60 (2002) 186.

Cyclodextrin Materials Photochemistry, Photophysics and Photobiology
Abderrazzak Douhal (Editor)
© 2006 Elsevier B.V. All rights reserved
DOI 10.1016/S1872-1400(06)01014-4

303

Chapter 14

Biochemical and Physical Stimuli-Triggered Cyclodextrin Release from Biodegradable Polyrotaxanes and those Hydrogels

Tooru Ooya

School of Materials Science, Japan Advanced Institute of Science and Technology, 1-1 Asahidai, Nomi, Ishikawa 923-1292, Japan
Department of Intelligent System Design Engineering, Toyama Prefectural University, 5180 Kurokawa Imizu, Toyama 939-0398, Japan

1. Introduction

It is known that biological process with large macromolecular assemblies is based on a dynamic consequence of cooperativity. The assembly of large supramolecular complexes produces dynamic features in a construct. Taking biological process into account, the design of supramolecular architecture formed via non-covalent interactions would play an important role in terms of dynamic functions under physiological conditions. Recent progress in supramolecular chemistry [1] has created new supramolecular architecture from synthetic approaches. One of the representative supramolecular-structured polymers to be built up using such intermolecular forces is a polyrotaxane, in which many cyclic compounds are threaded onto a linear polymeric chain capped both terminals with bulky end-groups. The name of rotaxane comes from Latin words indicating wheel and axle, and thus, refer to a molecular assembly of cyclic and linear molecules.

In the last decade, cyclodextrin (CD) has been studied as a host molecule for guest molecule inclusion in aqueous media. Inclusion complexation between cyclic hosts and lower-molecular-weight guests gives us a simple equilibrium equation:

[CD] + [Guest] ⇄ [Complex]

The driving forces to make the inclusion complexes are hydrophobic interactions, van der Waals interaction, hydrogen bondings and electrostatic interactions. When the guest molecules are linear polymers, some or many cyclic molecules can be threaded onto the linear polymers:

m[CD] + [Polymer] ⇄ [(Poly)pseudorotaxane]

where the polypseudorotaxane is an inclusion complex consisting of CD and a linear polymer. The polypseudorataxane formation has been extensively studied by Harada [2,3], Wenz *et al.* [4–6], Baglioni *et al.* [7–10] and Li *et al.* [11]. Generally, CD is threaded on to some linear polymer chains in water. As a result, the polypseudorotaxanes are often obtained as white crystals after precipitation and filtration from water dispersions. As seen in the Harada's report [12], this phenomenon includes

(a) CD threading on a linear polymeric chain.
(b) Sliding on the polymeric chain.
(c) Dethreading from the polymeric chain.
(d) Precipitation of the polypseudorotaxanes via crystallization.

One of driving forces for the processes (a) and (b) is hydrogen bonding between CD molecules [12]. The threading process (a) and the final process (d) cause an increase in turbidity although the other processes are assumed not to increase the absorbance.

The inclusion complexes with bulky blocking groups were called as [*n*]-rotaxane, in which *n*−1 CDs are threaded. Especially, the [*n*]-rotaxane consisting of many CD molecules is called as a polyrotaxane. These polypseudorotaxanes and [*n*]-rotaxanes have been recently recognized as *molecular shuttles* or *muscle* because threading process of cyclic molecules and changing the position of CD in the [*n*]-rotaxanes are of special interest in the sliding motion of CD owing to the non-covalent interactions with the linear polymers, which may lead to the new basis of artificial muscles.

Another interest of [*n*]-rotaxanes is CD release from them. CD molecules can be released from the threaded linear polymers when the blocking group is degraded or cleaved by some external stimuli (Fig. 1). Most of biodegradable

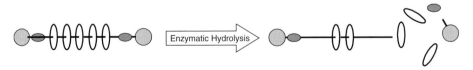

Fig. 1. Image of terminal cleavage-triggered CD release.

polymers such as poly(lactic acid) and polysaccharides must be cleaved at many cleavable bonds to be oligomers. On the other hand, biodegradable poly-rotaxanes easily change the molecular weight from several ten thousands to a thousand via releasing CD only when the terminal linkages are cleaved. This approach provides a novel biomaterials design in the field of drug delivery and tissue engineering. In this chapter, stimuli-triggered CD release from biode-gradable polyrotaxanes and those hydrogels was summarized.

2. Design and chemistry of biodegradable polyrotaxanes and those hydrogels

Polyrotaxanes as biomaterials have focused on preparing biodegradable poly-rotaxanes, in which terminal bulky end-groups are introduced via labile link-ages. So far, the following linkages were introduced to the terminals; peptide bond for protease-catalyzed hydrolysis [13–16], ester bond for simple hydrolysis [17,18], hydrazone bond for acid-catalyzed hydrolysis [19,20] and disulfide bond for reduction with reductive agents or enzymes [20] (Table 1). Biodegradable polyrotaxanes are prepared in two steps: (i) preparing inclusion complexes be-tween α-CD and bi-functional poly(ethylene glycol) (PEG), and (ii) capping both terminals of PEG using amino acid derivatives or oligopeptides via the labile linkages listed in Table 1. Stoichiometry of an inclusion complex between α-CD and PEG has been clarified by Harada *et al.* [12], and two repeating units of ethylene glycol are included within the cavity of one α-CD molecule. Solvent of the polyrotaxanes is limited only to dimethyl sulfoxide (DMSO). However, hydrophilic modifications such as hydroxypropylation [13] and hydrophobic ones such as acetylation [18] lead to change in solubility. Another type of biodegradable polyrotaxane is the combination of α-CD and poly(ε-lysine) that

Table 1
Types of linkages for designing biodegradable polyrotaxanes

Type of linkages	Chemical structure	Biodegradation	Reference
Peptide bond		Protease-catalyzed hydrolysis	[13–16]
Ester bond		Hydrolysis	[17,18]
Hydorazone bond		Acid-catalyzed hydrolysis	[19]
Disulfide bond		Reduction with reductive agents or enzymes	[20]

is degradable by some hydrolytic enzymes [21]. The preparation is simply mixing of α-CD saturated aqueous solution and poly(ε-lysine) dissolved in water. The number of α-CD threaded onto poly(ε-lysine) is controllable by pH due to the ionization of α-amino groups of the poly(ε-lysine) ($pK_a = 8.5$) and hydroxyl groups of α-CDs ($pK_a = 12.2$). More recently, Yui *et al.* have developed polypseudorotaxane formation between α-CD molecules and linear polyethyleneimine (PEI) [22] or triblock copolymer of PEI and PEG (PEI-*b*-PEG-*b*-PEI) [23]. The control of pH affected the polypseudorotaxane formation, and the ionization state of the PEI is the key factor to determine the threading and dethreading process of α-CD onto the polymer chain.

Hydrogels cross-linked by biodegradable polyrotaxanes having ester linkages at the terminal of PEG (Fig. 2) is prepared by activating the hydroxyl groups in the polyrotaxanes using N,N'-carbonyldimidazole and coupling reaction between the activated hydroxyl groups (N-acylimidazole) and amino-terminated PEG [24–27]. Water content of the hydrogels can be controlled by changing molar ratio of the polyrotaxane and the amino-terminated PEG. Recently, a new preparation method has been prepared by radical co-polymerization of methacrylate (MA)-introduced polyrotaxanes and N-isopropylacrylamide (NIPAAm) [28]. The amount of MA attached to the hydroxyl group of α-CD in polyrotaxane could be varied by the feed ratio of glycidyl methacrylate (GMA) and the polyrotaxane. Here, the GMA-introduced polyrotaxane acts as the cross-linking points in hydrogels. Another method for preparing hydrogels is utilization of polypseudorotaxane formation. Inclusion complexation of PEG-grafted polysaccharide (dextran, hyaluronic acid and chitosan) and α-CD can lead to hydrogels [29–31]. The comb-type copolymer is synthesized by coupling reaction using *p*-nitrophenyl choloroformate as mentioned above, and then the obtained copolymer dissolved in water is mixed with α-CD saturated solution. The inclusion complex between α-CD molecules and the PEG graft made physical cross-link points due to the crystalline state of the inclusion complex.

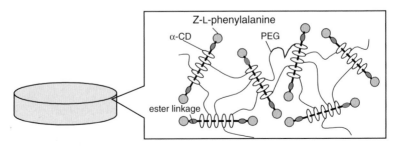

Fig. 2. Illustration of hydrogels cross-linked by biodegradable polyrotaxanes.

3. Principle of cyclodextrin release from polyrotaxanes

As mentioned in the introductive part, important driving force of polyps-eudorotaxane formation is hydrogen bonding between CD molecules [12]. Since the polypseudorotaxane formation is equilibrium reaction, putting polyps-eudorotaxanes in water under diluted conditions leads to dethreading CD from the guest macromolecule. Indeed, α-CD dethreading from PEG chain (the number-average-molecular weight; $M_n = 1000$) in a citrate buffer (2 wt%) was found to be completed within 8–10 s, which was monitored by 1-anilino-8-naph-thalenesulfonic acid (ANS) fluorescence [32,33]. As for polyrotaxanes (i.e. after capping with bulky end-groups), terminal cleavage should be the first step, and then the CDs can diffuse into aqueous medium via the dethreading. However, elimination of hydrogen bonds between CDs is needed because the hydrogen bonds are the driving force to form the polypseudorotaxanes. Chemical mod-ification of hydroxyl groups, on CDs, is one of the ways to eliminate the hy-drogen bonds. For example, hydroxypropylation of biodegradable polyrotaxanes is effective to eliminate the hydrogen bonds. After the hydroxypropylation, sol-ubility of the polyrotaxanes in water was drastically increased up to 83 wt% [13]. Hydroxypropylated (HP)-α-CD was found to be released by the terminal hy-drolysis in the presence of papain as a model enzyme [13,14], suggesting that the terminal cleavage is a dominant step to release CD into the medium (Fig. 3).

4. Hydrolysis-triggered cyclodextrin release

Introducing peptide linkages with appropriate oligopeptide sequences for ter-minal-capping group of polyrotaxanes enable us to design enzyme-specific CD release. When an ester linkage is applied for the terminal labile linkage, ester hydrolysis seems to be regulated by CD complexation with the ester linkage.

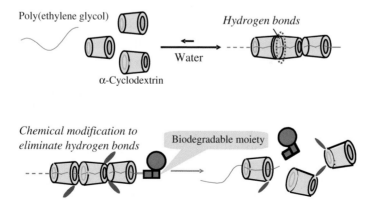

Fig. 3. Role of hydrogen bonds between CDs for polypseudorotaxane formation and CD release.

The regulation of ester cleavage is achieved by cross-linking CDs in the polyrotaxanes.

4.1. Enzyme-mediated hydrolysis of peptide linkages

The most striking feature of polyrotaxanes having peptide linkages at both ends is supramolecular dissociation triggered by terminal hydrolysis: once any of both ends are cleaved via enzymatic hydrolysis, all the CD molecules are to be dethreaded from a PEG chain at once. Interesting feature in these biodegradable polyrotaxanes is their perfect degradation properties [14]. Since Ringsdorf proposed his idea on polymer-drug conjugates [34], many studies on this subject have been reported. However, one of the fates in this kind of polymer designs is that any drug is introduced into a main polymeric backbone via a biodegradable spacer. Hydrophobic nature of both drugs and biodegradable spacers induces inter- and intramolecular association of these conjugates, and this situation makes the accessibility of digestive enzymes toward the spacer groups sterically hindered. Therefore, imperfect degradation has been one of the important issues to be solved. In our approach of biodegradable polyrotaxanes, the biodegradable moieties are only located at the terminals, and intermolecular and intramolecular association of the polyrotaxanes are very limited because the flexibility of PEG chains is sterically restricted by threading α-CD. Therefore, these two unique characters enable the enzymatic degradation perfect, although hydrophobic drugs are incorporated into CD as a side group and oligopeptide sequence is introduced as an enzymatically susceptible linkage at the terminals of inclusion complex with high molecular weight. Indeed, enzymatic degradation of our designed polyrotaxanes has been clarified to proceed perfectly, independent of the molecular weight of the polyrotaxane-drug conjugates [14,35]. Here, papain was used as a model enzyme of protease in lysosome. The terminal hydrolysis in the polyrotaxanes proceeded to over 85%, in contrast to the limited hydrolysis in L-Phe-terminated PEG ($\sim$50%). Since the second virial constant (A_2) value of the polyrotaxane, which is determined by static light scattering measurement, was relatively larger than those of L-Phe-terminated PEG, the complete dissociation of the polyrotaxanes by hydrolysis may be due to the less-association state related to the rod-like structure. On the other hand, the high association state of L-Phe-terminated PEG with strong molecular interaction is considered to decrease the accessibility of papain to the terminals.

Enhanced accessibility of peptide substrates toward membrane-bound metalloexopeptidase (aminopeptidase M) [16] using polyrotaxane structure was achieved. It is clarified that the polyrotaxanes form loosely packed association due to their rod-like structure [14] under physiological conditions, maintains enzymatic accessibility to terminal peptide moieties. A L-phenylalanlylglycylglycine (H-L-PheGlyGly)-terminated polyrotaxane in which many α-CD molecules are threaded onto PEG was synthesized to evaluate the effect of α-CD threading on the degradation of the terminal H-L-PheGlyGly by aminopeptidase M. Michaelis–Menten constant (K_m) value of the polyrotaxane was 1/22 of H-L-PheGlyGly terminated PEG. Presumably, the terminal H-L-PheGlyGly is likely to be exposed to the aqueous environment due to expansion of the linear backbone by the threading of α-CD.

4.2. Hydrolysis of ester linkages and erosion of hydrogels

Yui *et al.* proposed unique degradation and erosion mechanism using PEG-based hydrogels cross-linked by a hydrolyzable polyrotaxane having ester linkages at both ends [24,26]. α-CDs in the polyrotaxane as a cross-linker is covalently bound with another PEG chains (M_n: 600, 2000 and 4000). The obtained hydrogels were characterized by *in vitro* erosion behavior in phosphate buffered saline (pH 7.4). The time to reach complete gel erosion was prolonged with increasing the PEG/α-CD ratio (the number of PEG chains linked with one α-CD molecule) (Fig. 4). These results indicate the enhanced stability of ester hydrolysis in the hydrogels under the condition of highly water-swollen state. Such an anomalous phenomenon is considered to be due to the structural characteristic of the polyrotaxane: ester linkages may be included within the cavity of α-CD, resulting in their enhanced stability. In order to estimate this hypothesis, inclusion conditions of the ester linkages were analyzed by ^{13}C-NMR spectroscopy using a hydroxyethylcarbamoyl polyrotaxane as a model cross-linker of hydrogels [26]. The ester linkages before the chemical modification were exposed to the aqueous environment due to the aggregation of α-CD. On the other hand, the peaks of carbonyl ester groups were hardly detected on the spectrum. These results suggest that the movement of α-CD after the chemical modification from the terminal ester region to another region gives the ester linkages a chance to interact with water (Fig. 5).

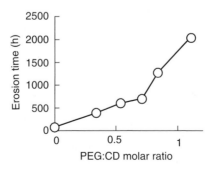

Fig. 4. Relation between PEG/CD ratio and the erosion time in the PEG-PRX gels.

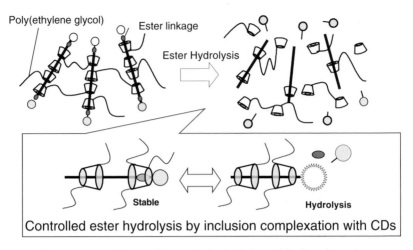

Fig. 5. Image of controlling ester hydrolysis and hydrogel erosion.

5. pH-triggered cyclodextrin release

When hydrazone bond as a labile linkage is introduced into a polyrotaxane, the bond can be cleaved at low pH [36]. From this character, a hydrazone bond-introduced polyrotaxane was synthesized and characterized in terms of pH-triggered release of CDs. A β-cyclodextrin (β-CD)-based biodegradable polyrotaxane was synthesized by capping both terminals of polypseudorotaxane consisting of hydrazode-terminated PEG-*b*-PPG-*b*-PEG (Pluronic) and β-CDs with mono-aldehyde α-CDs [19]. In order to increase the solubility in water, carboxyethyl (CEE) groups were introduced to the hydroxyl groups of β-CDs and/or the terminal α-CDs by reacting with succinic anhydride in dry pyridine. Here, β-CD was used as the cyclic molecule because inclusion complexation between the released CEE–β-CD and 6-(*p*-toluidinyl)naphthalene-2-sulfonate (TNS) can increase the magnitude of TNS fluorescence. This phenomenon can be applied to detect CEE–β-CD release via pH decrease. By the combination of β-CD and PEG-*b*-PPG-*b*-PEG, the inclusion complexation can be occurred in aqueous medium. By using α-CD as a capping molecule, hydrophilic terminals can be maintained without including TNS into α-CD cavity because the cavity size of α-CD is too small to include TNS. From this molecular design, stimuli for the terminal cleavage can be converted to fluorescence intensity change of TNS in solution, which is amplified by exposure of many β-CD cavities to include more than one TNS molecule via the terminal degradation. The pH-responsive exposure of β-CD cavities was evaluated by an increase in magnitude of TNS fluorescent via inclusion complexation with the released CEE–β-CD molecules.

TNS was used as a probe to detect the released CEE–β-CD because inclusion complexation between chemically modified β-CD and TNS increases the

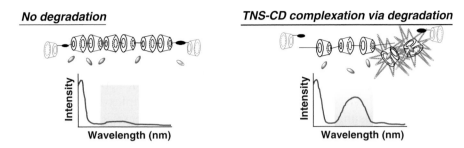

Fig. 6. CD release and following fluorescence amplification via pH decrease.

magnitude of fluorescence [37]. As the reference, CEE–β-CD was used to check the pH-dependent fluorescent intensity (λ_{Ex}: 366 nm, λ_{Em}: 450 nm). The fluorescent intensity of TNS (conc. 5×10^{-5} M) in a phosphate buffer at pH 1, 3, 6 and 8 was 0–1 a.u. In the presence of CEE–β-CD (conc. 1.5×10^{-4} M), the intensity was increased to 190–480 a.u. depending on the pH, indicating inclusion of TNS to the CEE-β-CD cavity. After adding the hydrazode-terminated polyrotaxane (conc. 2.3×10^{-5} M) to the TNS-dissolved buffer (pH 1), the fluorescent intensity of TNS was not changed. Five minutes after the addition, the intensity was slightly increased, and 1 h later, the intensity reached to around 210 a.u. However, increasing the fluorescent intensity was not observed when pH of the buffer was above 5. These results indicate that hydrophobic cavity of the CEE–β-CD exposes in the acidic medium after 1 h, followed by inclusion complexation with TNS to amplify the fluorescence. These observations suggest that cleavage of the terminal hydrozone bonds triggers the dissociation of the polyrotaxane with fluorescence amplification (Fig. 6).

6. Reduction-triggered cyclodextrin release

It is known that disulfide bond can reversibly be cleaved in the presence of reducing agents such as dithiothreitol (DTT) and β-mercaptoethanol. The high redox potential difference between the oxidizing extracellular environment and the reducing intracellular environment makes the intriguing disulfide bond a potential delivery tool. Therefore, a controlled cleavage of the disulfide bond can be applied for intracellular gene delivery by a combination with polycations to form polyion complex with DNA or RNA [38].

Various synthetic and cationic polymers have been focused on complexation with DNA, which is one of the approaches to gene delivery. Although many efforts have been made to develop gene carriers using cationic polymers for gene delivery, transfection activity is not enough for sufficient level. This is due to the poor release of the genes from the polyion complex. Biodegradable polyrotaxanes, in which disulfide bonds are introduced at both the ends of linear polymer, have a potential tool as a gene carrier because hydroxyl groups of CD can be modified to cationic groups to form complexes with DNA. Here, disulfide linkage was introduced at both the terminals of polyrotaxanes. N,N'-dimethylethylenediamine (DMAE)-introduced reducible polyrotaxanes (SS-PRXs) were synthesized and evaluated in terms of complex formation, terminal cleavage of disulfide bond and plasmid DNA release in the presence of dithiothreitol (DTT) [20].

Disulfide bond was introduced to the terminals of α,ω-diaminoPEG ($M_n = 4000$, 35,000) by reacting with N-succinimidyl 3-(2-pyridyldithio) propionate (SPDP), followed by reacting with aminorthanethiol to obtain α,ω-diamino-SS-PEG (SS-PEG-bisamine). The SS-PEG-bisamine dissolved in water was added to saturated solution of α-CDs to obtain a polypseudorotaxane. The terminal amino groups in the polypseudorotaxane were allowed to react with benzyloxycacrbonyl(Z-) tyrosine in the presence of benzotriazol-1-yl-oxytris(dimethylamino)phosphomium hexafluorophosphate (BOP), 1-hydroxybenzotriazole (HOBt) and N, N'-disopropylethylamine (DIEA) in DMF to obtain Z-Tyr-capped polyrotaxanes. Finally, hydroxyl groups of α-CDs in the Z-Tyr-capped polyrotaxanes were activated by N,N'-carbonyldiimidazole (CDI) followed by reacting with DMAE to obtain a tertiary amine-introduced reducible polyrotaxanes (DMAE-SS-PRXs) (Scheme 1). From the ^{1}H-NMR spectra, the number of threading α-CDs and DMAE groups per polyrotaxane was ca. 23 and 40 (M_n of PEG = 4000, DMAE-SS/E4-PRX), and ca. 198 and 865 ($M_n = 35,000$, DMAE-SS/E35-PRX), respectively.

By adding 10 mM DTT to DMAE-SS/E4-PRX or DMAE-SS/E35-PRX solution, DMAE-α-CD was completely released to the medium during 60 min (Fig. 7). This result indicates that the supramolecular structure of DMAE-SS-PRXs was dissociated by cleavage of disulfide bond. When the DMAE-SS-PRXs were complexed with a plasmid DNA, the free DNA band on an agarose gel electrophoretic image was disappeared above the N/P ratio (i.e. the ratio of DMAE cation and phosphate anion) of 0.5. On the other hand, the DMAE-α-CD, the building block of the DMAE-SS-PRXs, did not form the complex at higher N/P ratio. These results indicate that the polycationic nature provides more stable DNA complex than DMAE-α-CD. With increasing the concentration of DTT in the DMAE-SS-PRX/DNA complexes (N/P ratio $= 0.5$), free DNA band in an agarose gel electrophoretic image was observed over 0.15 mM of DTT. Also, free DNA band was observed 3 h after adding DTT. These results suggest that the supramolecular dissociation of the DMAE-SS-PRXs under the reductive condition trigged DNA release. The DNA release was completed within 8 h in the presence of 0.5 mM DTT.

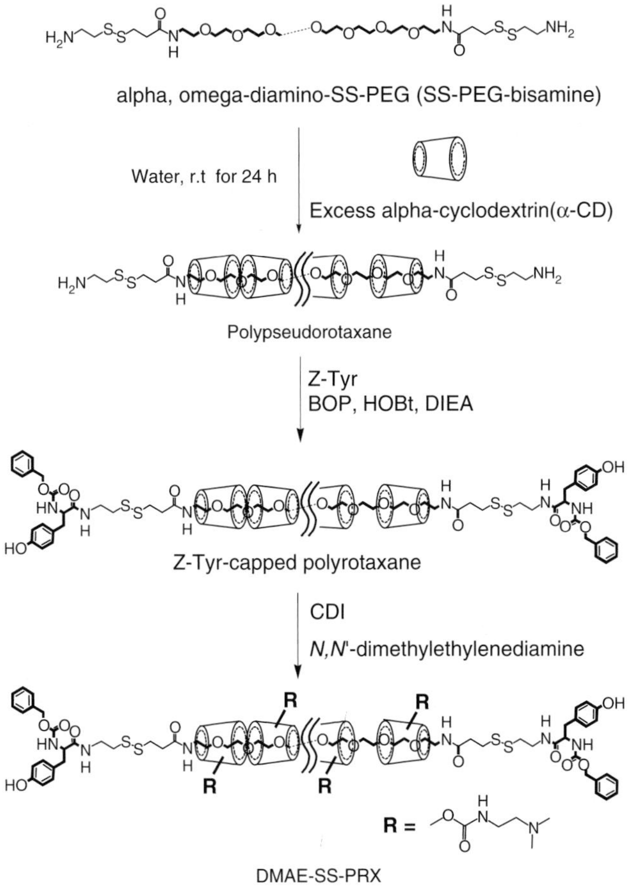

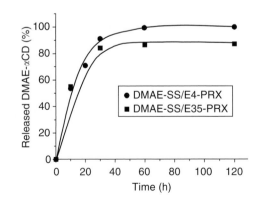

Scheme 1. Synthesis of DMAE-SS-PRXs

Fig. 7. DMAE-alpha-CD release in the presence of DTT.

7. Possibility of photo-triggered cyclodextrin release

Photo-irradiation is another candidate of stimuli to release CD from poly-rotaxanes. Azobenzene (Az) is a well-known photosensitive molecule that transforms from *trans* to *cis* isomer upon UV irradiation. Due to the physico-chemical change of Az accompanied by the isomerization, it maight be possible to regulate CD release from polyrotaxanes having Az at both the ends. Yui *et al.* have demonstrated polypseudorotaxane formation between Az-terminated PEG and α-CD [39]. The Az-terminated PEG was prepared by conjugation between α,ω-diaminoPEG ($M_n = 4000$) and *p*-phenylazobenzoylchloride in the presence of excess amount of magnesium oxide [40]. The polypseudorotaxane formation was analyzed by a change in transmittance at 550 nm in water. The rate of transmittance change was decreased with increasing *cis* isomer content in the Az-terminated PEG. The effect of *cis* isomerization could be quantitatively evaluated, assuming that the Az-terminated PEG having two *cis*-isomerized Az groups does not participate in the polypseudorotaxane formation. The data suggested that α-CD threading onto the PEG chain was hindered by the *cis* isomerization (Fig. 8). This demonstration showed the possibility of photo-triggered CD release from polyrotaxanes.

8. Conclusion

In this chapter, stimuli-responsive cyclodextrin release from polyrotaxanes was described based on the preparation of polyrotaxanes, the principle of cyclo-dextrin release and the recent research development. Introducing labile linkages at both the ends of polymer chain in polyrotaxanes leads to cyclodextrin release

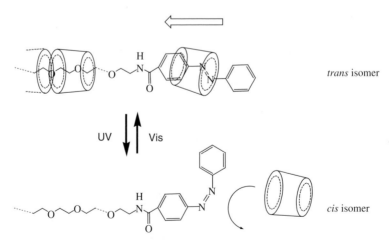

Fig. 8. Schematic illustration of photo-regulated polypseudorotaxane formation [39].

in aqueous medium in response to various stimuli including hydrolysis, pH change, reductive condition and light source. The supramolecular approach to dynamic cyclodextrin diffusion from polyrotaxane molecules to the out medium enable us to design the new biomaterials for drug/gene delivery, tissue scaffolds and biosensors.

Acknowledgments

The author acknowledges Prof. Nobuhiko Yui, JAIST, for advising on all the studies. A part of this work was financially supported by a program for the "Supporting young researchers with fixed-term appointments" in special coordination funds for promoting Science and Technology, and a Grant-in-Aid for young scientists (B) (No. 16700354) from the Ministry of Education, Science, Sports and Culture, Japan.

References

[1] N. Yui (Ed.), Supramolecular Design for Biological Applications, CRC, Boca Raton FL, 2002.
[2] A. Harada, M. Kamachi, Macromolecules 23 (1999) 2821.
[3] A. Harada, Coord. Chem. Rev. 148 (1996) 115.
[4] G. Wenz, B. Keller, Macromol. Symp. 87 (1994) 11.
[5] M. Weickenmeir, G. Wenz, Macromol. Rapid. Commun. 18 (1997) 1109.
[6] W. Herrmann, B. Keller, G. Wenz, Macromolecules 30 (1997) 4966.
[7] M. Ceccato, P.L. Nostro, P. Baglioni, Langmuir 13 (1997) 2436.
[8] P.L. Nostro, J.R. Lopes, C. Cardelli, Langmuir 17 (2001) 4610.
[9] P.L. Nostro, J.R. Lopes, B.W. Ninham, P. Baglioni, J. Phys. Chem. B 106 (2002) 2166.
[10] A. Becheri, P.L. Nostro, B.W. Ninham, P. Baglioni, J. Phys. Chem. B 107 (2003) 3979.
[11] J. Li, X. Ni, Z. Zhou, K.W. Leong, J. Am. Chem. Soc. 125 (2003) 1788.
[12] A. Harada, Kamachi, Macromolecules 23 (1990) 2821.
[13] T. Ooya, N. Yui, J. Biomater. Sci. Polym. Ed. 8 (1997) 437.
[14] T. Ooya, N. Yui, Macromol. Chem. Phys. 199 (1998) 2311.
[15] T. Ooya, K. Arizono, N. Yui, Polym. Adv. Tech. 11 (2000) 642.
[16] T. Ooya, M. Eguchi, N. Yui, Biomacromolecules 2 (2001) 200.
[17] J. Watanabe, T. Ooya, N. Yui, Chem. Lett. 1998 (1998) 1031.
[18] J. Watanabe, T. Ooya, N. Yui, J. Biomater. Sci. Polym. Ed. 10 (1999) 1275.
[19] T. Ooya, A. Ito, N. Yui, Macromol. Biosci. 5 (2005) 379.
[20] T. Ooya, H.S. Choi, A. Yamashita, Y. Sugaya, K. Kano, A. Maruyama, R. Ito, K. Kogure, H. Harashima, J. Am. Chem. Soc. 128 (2006) 3852.
[21] K.M. Huh, T. Ooya, S. Sasaki, N. Yui, Macromolecules 34 (2001) 2402.
[22] H.S. Choi, T. Ooya, A. Sasaki, N. Yui, M. Kurisawa, H. Uyama, S. Kobayashi, ChemPhysChem 5 (2004) 1431.
[23] S.C. Lee, H.S. Choi, T. Ooya, N. Yui, Macromolecules 37 (2004) 7464.
[24] T. Ichi, J. Watanabe, T. Ooya, N. Yui, Biomacromolecules 2 (2001) 2040.
[25] J. Watanabe, T. Ooya, K. Nitta, K.D. Park, Y.H. Kim, N. Yui, Biomaterials 23 (2002) 4041.
[26] T. Ichi, T. Ooya, N. Yui, Macromol. Biosci. 3 (2003) 373.

[27] T. Ichi, W.K. Lee, T. Ooya, N. Yui, J. Biomater. Sci. Polym. Ed. 14 (2003) 567.

[28] T. Ooya, M. Akutsu, Y. Kumashiro, N. Yui, Sci. Tech. Adv. Mater. 6 (2005) 447.

[29] K.M. Huh, T. Ooya, W.K. Lee, S. Sasaki, I.C. Kwon, S.Y. Jeong, N. Yui, Macromolecules 34 (2001) 8657.

[30] T. Nakama, T. Ooya, N. Yui, Macromol. Rapid. Commun. 25 (2004) 739.

[31] K.M. Huh, Y.W. Cho, H. Chung, I.C. Kwon, S.Y. Jeong, T. Ooya, W.K. Lee, S. Sasaki, N. Yui, Macromol. Biosci. 4 (2004) 92.

[32] N. Yui, T. Ooya, in: T. Okano, N. Ogata, J. Feijen, S.W. Kim (Eds), Advances in Biomedical Polymers in Biomedical Engineering and Drug Delivery Systems, Springer, Tokyo, 1996, p. 333.

[33] T. Ooya, H. Mori, M. Terano, N. Yui, Macromol. Rapid. Commun. 16 (1995) 259.

[34] H. Ringsdolf, J Polym. Sci. Symp. 51 (1975) 135.

[35] T. Ooya, N. Yui, J Control. Rel. 58 (1999) 2519.

[36] Y. Bae, S. Fukushima, A. Harada, K. Kataoka, Angew. Chem., Int. Ed. Engl. 42 (2003) 4640.

[37] E. Alvarez-Parrilla, W. Al-Soufi, P.R. Cabrer, M. Novo, J.V. Tato, J. Phys. Chem. B 105 (2001) 5994.

[38] G. Saito, J.A. Swanson, K.-D. Lee, Adv. Drug Del. Rev. 55 (2003) 199.

[39] T. Ikeda, T. Ooya, N. Yui, Polym. J. 31 (1999) 658.

[40] B. Gallot, M. Fafiotte, A. Fissi, O. Pieroni, Macromol. Rapid. Commun. 17 (1996) 493.

Subject Index